Proceedings of the International Conference of Sustainable Production and Use of Cement and Concrete

AF847528

RILEM BOOKSERIES

Volume 22

RILEM, The International Union of Laboratories and Experts in Construction Materials, Systems and Structures, founded in 1947, is a non-governmental scientific association whose goal is to contribute to progress in the construction sciences, techniques and industries, essentially by means of the communication it fosters between research and practice. RILEM's focus is on construction materials and their use in building and civil engineering structures, covering all phases of the building process from manufacture to use and recycling of materials. More information on RILEM and its previous publications can be found on www.RILEM.net. Indexed in SCOPUS, Google Scholar and SpringerLink.

More information about this series at http://www.springer.com/series/8781

Jose Fernando Martirena-Hernandez ·
Adrian Alujas-Díaz · Meylin Amador-Hernandez
Editors

Proceedings of the International Conference of Sustainable Production and Use of Cement and Concrete

ICSPCC 2019

Springer

Editors
Jose Fernando Martirena-Hernandez
Universidad Central de Las Villas
Santa Clara, Cuba

Adrian Alujas-Díaz
Universidad Central de Las Villas
Santa Clara, Cuba

Meylin Amador-Hernandez
Universidad Central de Las Villas
Santa Clara, Cuba

ISSN 2211-0844 ISSN 2211-0852 (electronic)
RILEM Bookseries
ISBN 978-3-030-22036-5 ISBN 978-3-030-22034-1 (eBook)
https://doi.org/10.1007/978-3-030-22034-1

© RILEM 2020
No part of this work may be reproduced, stored in a retrieval system, or transmitted in any form or by any means, electronic, mechanical, photocopying, microfilming, recording or otherwise, without written permission from the Publisher, with the exception of any material supplied specifically for the purpose of being entered and executed on a computer system, for exclusive use by the purchaser of the work.

This Springer imprint is published by the registered company Springer Nature Switzerland AG
The registered company address is: Gewerbestrasse 11, 6330 Cham, Switzerland

Preface

Cement and Concrete are man's most-used materials. Almost anywhere on Earth one can buy a bag of cement, provided there is a store. Cement is a very universal material, with applications in almost everything around us. The volume of resources used for the production of Cement and Concrete is huge, but carbon emissions of concrete—its main application—are much lower compared to any other building material. However, there is still plenty of room for making production and use of these materials more sustainable.

There are many research groups that are working on ways of improving the sustainability of cement and concrete. There are interesting results on new cement formulations and additives to replace clinker and thus make low carbon cement. A great variety of durable Concrete applications are also being developed, with a broad range of applications. Recycling Concrete into new durable products is part of the concepts of the Circular Economy, and it catches the attention of scientists and entrepreneurs.

The proceedings of the International Conference of Sustainable Production and use of Cement and Concrete (ICSPCC) 2019 present a snapshot of the intense work carried out by different research groups in issues related to cement and concrete during the past few years. It brings together written versions of the more than 50 presentation and posters delivered at the conference. Most of the papers have a great significance not only from the scientific viewpoint, but also introduce valuable information for the industry.

Topics covered are: sustainable production of low carbon cements, novelties in development of supplementary cementitious materials, new techniques of microstructural characterization of construction materials, Portland base and Alkaline-activated cementitious systems, development of additives and additions in the sustainable production of Concrete, sustainable production of high-performance concrete, durable concrete produced with recycled aggregates, development of mortars to historical patrimony resaturation, environmental and economic assessment of production and use of cement.

The Editors would like to thank the authors for the outstanding contributions, the sponsors and various other organizations for their contributions and help in making these proceedings and conference a success.

Santa Clara, Cuba

Jose Fernando Martirena-Hernandez
Adrian Alujas-Díaz
Meylin Amador-Hernandez

Contents

RILEM Publications

The following list is presenting the global offer of RILEM Publications, sorted by series. Each publication is available in printed version and/or in online version.

RILEM Proceedings (PRO)

PRO 1: Durability of High Performance Concrete (ISBN: 2-912143-03-9; e-ISBN: 2-351580-12-5; e-ISBN: 2351580125); *Ed. H. Sommer*

PRO 2: Chloride Penetration into Concrete (ISBN: 2-912143-00-04; e-ISBN: 2912143454); *Eds. L.-O. Nilsson and J.-P. Ollivier*

PRO 3: Evaluation and Strengthening of Existing Masonry Structures (ISBN: 2-912143-02-0; e-ISBN: 2351580141); *Eds. L. Binda and C. Modena*

PRO 4: Concrete: From Material to Structure (ISBN: 2-912143-04-7; e-ISBN: 2351580206); *Eds. J.-P. Bournazel and Y. Malier*

PRO 5: The Role of Admixtures in High Performance Concrete (ISBN: 2-912143-05-5; e-ISBN: 2351580214); *Eds. J. G. Cabrera and R. Rivera-Villarreal*

PRO 6: High Performance Fiber Reinforced Cement Composites - HPFRCC 3 (ISBN: 2-912143-06-3; e-ISBN: 2351580222); *Eds. H. W. Reinhardt and A. E. Naaman*

PRO 7: 1st International RILEM Symposium on Self-Compacting Concrete (ISBN: 2-912143-09-8; e-ISBN: 2912143721); *Eds. Å. Skarendahl and Ö. Petersson*

PRO 8: International RILEM Symposium on Timber Engineering (ISBN: 2-912143-10-1; e-ISBN: 2351580230); *Ed. L. Boström*

PRO 9: 2nd International RILEM Symposium on Adhesion between Polymers and Concrete ISAP '99 (ISBN: 2-912143-11-X; e-ISBN: 2351580249); *Eds. Y. Ohama and M. Puterman*

PRO 10: 3rd International RILEM Symposium on Durability of Building and Construction Sealants (ISBN: 2-912143-13-6; e-ISBN: 2351580257); *Eds. A. T. Wolf*

PRO 11: 4th International RILEM Conference on Reflective Cracking in Pavements (ISBN: 2-912143-14-4; e-ISBN: 2351580265); *Eds. A. O. Abd El Halim, D. A. Taylor and El H. H. Mohamed*

PRO 12: International RILEM Workshop on Historic Mortars: Characteristics and Tests (ISBN: 2-912143-15-2; e-ISBN: 2351580273); *Eds. P. Bartos, C. Groot and J. J. Hughes*

PRO 13: 2nd International RILEM Symposium on Hydration and Setting (ISBN: 2-912143-16-0; e-ISBN: 2351580281); *Ed. A. Nonat*

PRO 14: Integrated Life-Cycle Design of Materials and Structures - ILCDES 2000 (ISBN: 951-758-408-3; e-ISBN: 235158029X); (ISSN: 0356-9403); *Ed. S. Sarja*

PRO 15: Fifth RILEM Symposium on Fibre-Reinforced Concretes (FRC) - BEFIB'2000 (ISBN: 2-912143-18-7; e-ISBN: 291214373X); *Eds. P. Rossi and G. Chanvillard*

PRO 16: Life Prediction and Management of Concrete Structures (ISBN: 2-912143-19-5; e-ISBN: 2351580303); *Ed. D. Naus*

PRO 17: Shrinkage of Concrete – Shrinkage 2000 (ISBN: 2-912143-20-9; e-ISBN: 2351580311); *Eds. V. Baroghel-Bouny and P.-C. Aïtcin*

PRO 18: Measurement and Interpretation of the On-Site Corrosion Rate (ISBN: 2-912143-21-7; e-ISBN: 235158032X); *Eds. C. Andrade, C. Alonso, J. Fullea, J. Polimon and J. Rodriguez*

PRO 19: Testing and Modelling the Chloride Ingress into Concrete (ISBN: 2-912143-22-5; e-ISBN: 2351580338); *Eds. C. Andrade and J. Kropp*

PRO 20: 1st International RILEM Workshop on Microbial Impacts on Building Materials (CD 02) (e-ISBN 978-2-35158-013-4); *Ed. M. Ribas Silva*

PRO 21: International RILEM Symposium on Connections between Steel and Concrete (ISBN: 2-912143-25-X; e-ISBN: 2351580346); *Ed. R. Eligehausen*

PRO 22: International RILEM Symposium on Joints in Timber Structures (ISBN: 2-912143-28-4; e-ISBN: 2351580354); *Eds. S. Aicher and H.-W. Reinhardt*

PRO 23: International RILEM Conference on Early Age Cracking in Cementitious Systems (ISBN: 2-912143-29-2; e-ISBN: 2351580362); *Eds. K. Kovler and A. Bentur*

PRO 24: 2nd International RILEM Workshop on Frost Resistance of Concrete (ISBN: 2-912143-30-6; e-ISBN: 2351580370); *Eds. M. J. Setzer, R. Auberg and H.-J. Keck*

PRO 25: International RILEM Workshop on Frost Damage in Concrete (ISBN: 2-912143-31-4; e-ISBN: 2351580389); *Eds. D. J. Janssen, M. J. Setzer and M. B. Snyder*

PRO 26: International RILEM Workshop on On-Site Control and Evaluation of Masonry Structures (ISBN: 2-912143-34-9; e-ISBN: 2351580141); *Eds. L. Binda and R. C. de Vekey*

PRO 27: International RILEM Symposium on Building Joint Sealants (CD03; e-ISBN: 235158015X); *Ed. A. T. Wolf*

PRO 28: 6th International RILEM Symposium on Performance Testing and Evaluation of Bituminous Materials - PTEBM'03 (ISBN: 2-912143-35-7; e-ISBN: 978-2-912143-77-8); *Ed. M. N. Partl*

PRO 29: 2nd International RILEM Workshop on Life Prediction and Ageing Management of Concrete Structures (ISBN: 2-912143-36-5; e-ISBN: 2912143780); *Ed. D. J. Naus*

PRO 30: 4th International RILEM Workshop on High Performance Fiber Reinforced Cement Composites - HPFRCC 4 (ISBN: 2-912143-37-3; e-ISBN: 2912143799); *Eds. A. E. Naaman and H. W. Reinhardt*

PRO 31: International RILEM Workshop on Test and Design Methods for Steel Fibre Reinforced Concrete: Background and Experiences (ISBN: 2-912143-38-1; e-ISBN: 2351580168); *Eds. B. Schnütgen and L. Vandewalle*

PRO 32: International Conference on Advances in Concrete and Structures 2 vol. (ISBN (set): 2-912143-41-1; e-ISBN: 2351580176); *Eds. Ying-shu Yuan, Surendra P. Shah and Heng-lin Lü*

PRO 33: 3rd International Symposium on Self-Compacting Concrete (ISBN: 2-912143-42-X; e-ISBN: 2912143713); *Eds. Ó. Wallevik and I. Níelsson*

PRO 34: International RILEM Conference on Microbial Impact on Building Materials (ISBN: 2-912143-43-8; e-ISBN: 2351580184); *Ed. M. Ribas Silva*

PRO 35: International RILEM TC 186-ISA on Internal Sulfate Attack and Delayed Ettringite Formation (ISBN: 2-912143-44-6; e-ISBN: 2912143802); *Eds. K. Scrivener and J. Skalny*

PRO 36: International RILEM Symposium on Concrete Science and Engineering – A Tribute to Arnon Bentur (ISBN: 2-912143-46-2; e-ISBN: 2912143586); *Eds. K. Kovler, J. Marchand, S. Mindess and J. Weiss*

PRO 37: 5th International RILEM Conference on Cracking in Pavements – Mitigation, Risk Assessment and Prevention (ISBN: 2-912143-47-0; e-ISBN: 2912143764); *Eds. C. Petit, I. Al-Qadi and A. Millien*

PRO 38: 3rd International RILEM Workshop on Testing and Modelling the Chloride Ingress into Concrete (ISBN: 2-912143-48-9; e-ISBN: 2912143578); *Eds. C. Andrade and J. Kropp*

PRO 39: 6th International RILEM Symposium on Fibre-Reinforced Concretes - BEFIB 2004 (ISBN: 2-912143-51-9; e-ISBN: 2912143748); *Eds. M. Di Prisco, R. Felicetti and G. A. Plizzari*

PRO 40: International RILEM Conference on the Use of Recycled Materials in Buildings and Structures (ISBN: 2-912143-52-7; e-ISBN: 2912143756); *Eds. E. Vázquez, Ch. F. Hendriks and G. M. T. Janssen*

PRO 41: RILEM International Symposium on Environment-Conscious Materials and Systems for Sustainable Development (ISBN: 2-912143-55-1; e-ISBN: 2912143640); *Eds. N. Kashino and Y. Ohama*

PRO 42: SCC'2005 - China: 1st International Symposium on Design, Performance and Use of Self-Consolidating Concrete (ISBN: 2-912143-61-6; e-ISBN: 2912143624); *Eds. Zhiwu Yu, Caijun Shi, Kamal Henri Khayat and Youjun Xie*

PRO 43: International RILEM Workshop on Bonded Concrete Overlays (e-ISBN: 2-912143-83-7); *Eds. J. L. Granju and J. Silfwerbrand*

PRO 44: 2nd International RILEM Workshop on Microbial Impacts on Building Materials (CD11) (e-ISBN: 2-912143-84-5); *Ed. M. Ribas Silva*

PRO 45: 2nd International Symposium on Nanotechnology in Construction, Bilbao (ISBN: 2-912143-87-X; e-ISBN: 2912143888); *Eds. Peter J. M. Bartos, Yolanda de Miguel and Antonio Porro*

PRO 46: ConcreteLife'06 - International RILEM-JCI Seminar on Concrete Durability and Service Life Planning: Curing, Crack Control, Performance in Harsh Environments (ISBN: 2-912143-89-6; e-ISBN: 291214390X); *Ed. K. Kovler*

PRO 47: International RILEM Workshop on Performance Based Evaluation and Indicators for Concrete Durability (ISBN: 978-2-912143-95-2; e-ISBN: 9782912143969); *Eds. V. Baroghel-Bouny, C. Andrade, R. Torrent and K. Scrivener*

PRO 48: 1st International RILEM Symposium on Advances in Concrete through Science and Engineering (e-ISBN: 2-912143-92-6); *Eds. J. Weiss, K. Kovler, J. Marchand, and S. Mindess*

PRO 49: International RILEM Workshop on High Performance Fiber Reinforced Cementitious Composites in Structural Applications (ISBN: 2-912143-93-4; e-ISBN: 2912143942); *Eds. G. Fischer and V. C. Li*

PRO 50: 1st International RILEM Symposium on Textile Reinforced Concrete (ISBN: 2-912143-97-7; e-ISBN: 2351580087); *Eds. Josef Hegger, Wolfgang Brameshuber and Norbert Will*

PRO 51: 2nd International Symposium on Advances in Concrete through Science and Engineering (ISBN: 2-35158-003-6; e-ISBN: 2-35158-002-8); *Eds. J. Marchand, B. Bissonnette, R. Gagné, M. Jolin and F. Paradis*

PRO 52: Volume Changes of Hardening Concrete: Testing and Mitigation (ISBN: 2-35158-004-4; e-ISBN: 2-35158-005-2); *Eds. O. M. Jensen, P. Lura and K. Kovler*

PRO 53: High Performance Fiber Reinforced Cement Composites - HPFRCC5 (ISBN: 978-2-35158-046-2; e-ISBN: 978-2-35158-089-9); *Eds. H. W. Reinhardt and A. E. Naaman*

PRO 54: 5th International RILEM Symposium on Self-Compacting Concrete (ISBN: 978-2-35158-047-9; e-ISBN: 978-2-35158-088-2); *Eds. G. De Schutter and V. Boel*

PRO 55: International RILEM Symposium Photocatalysis, Environment and Construction Materials (ISBN: 978-2-35158-056-1; e-ISBN: 978-2-35158-057-8); *Eds. P. Baglioni and L. Cassar*

PRO 56: International RILEM Workshop on Integral Service Life Modelling of Concrete Structures (ISBN 978-2-35158-058-5; e-ISBN: 978-2-35158-090-5); *Eds. R. M. Ferreira, J. Gulikers and C. Andrade*

PRO 57: RILEM Workshop on Performance of cement-based materials in aggressive aqueous environments (e-ISBN: 978-2-35158-059-2); *Ed. N. De Belie*

PRO 58: International RILEM Symposium on Concrete Modelling - CONMOD'08 (ISBN: 978-2-35158-060-8; e-ISBN: 978-2-35158-076-9); *Eds. E. Schlangen and G. De Schutter*

PRO 59: International RILEM Conference on On Site Assessment of Concrete, Masonry and Timber Structures - SACoMaTiS 2008 (ISBN set: 978-2-35158-061-5; e-ISBN: 978-2-35158-075-2); *Eds. L. Binda, M. di Prisco and R. Felicetti*

PRO 60: Seventh RILEM International Symposium on Fibre Reinforced Concrete: Design and Applications - BEFIB 2008 (ISBN: 978-2-35158-064-6; e-ISBN: 978-2-35158-086-8); *Ed. R. Gettu*

PRO 61: 1st International Conference on Microstructure Related Durability of Cementitious Composites 2 vol., (ISBN: 978-2-35158-065-3; e-ISBN: 978-2-35158-084-4); *Eds. W. Sun, K. van Breugel, C. Miao, G. Ye and H. Chen*

PRO 62: NSF/ RILEM Workshop: In-situ Evaluation of Historic Wood and Masonry Structures (e-ISBN: 978-2-35158-068-4); *Eds. B. Kasal, R. Anthony and M. Drdácký*

PRO 63: Concrete in Aggressive Aqueous Environments: Performance, Testing and Modelling, 2 vol., (ISBN: 978-2-35158-071-4; e-ISBN: 978-2-35158-082-0); *Eds. M. G. Alexander and A. Bertron*

PRO 64: Long Term Performance of Cementitious Barriers and Reinforced Concrete in Nuclear Power Plants and Waste Management - NUCPERF 2009 (ISBN: 978-2-35158-072-1; e-ISBN: 978-2-35158-087-5); *Eds. V. L'Hostis, R. Gens, C. Gallé*

PRO 65: Design Performance and Use of Self-consolidating Concrete - SCC'2009 (ISBN: 978-2-35158-073-8; e-ISBN: 978-2-35158-093-6); *Eds. C. Shi, Z. Yu, K. H. Khayat and P. Yan*

PRO 66: 2nd International RILEM Workshop on Concrete Durability and Service Life Planning - ConcreteLife'09 (ISBN: 978-2-35158-074-5; ISBN: 978-2-35158-074-5); *Ed. K. Kovler*

PRO 67: Repairs Mortars for Historic Masonry (e-ISBN: 978-2-35158-083-7); *Ed. C. Groot*

PRO 68: Proceedings of the 3rd International RILEM Symposium on 'Rheology of Cement Suspensions such as Fresh Concrete (ISBN 978-2-35158-091-2; e-ISBN: 978-2-35158-092-9); *Eds. O. H. Wallevik, S. Kubens and S. Oesterheld*

PRO 69: 3rd International PhD Student Workshop on 'Modelling the Durability of Reinforced Concrete (ISBN: 978-2-35158-095-0); *Eds. R. M. Ferreira, J. Gulikers and C. Andrade*

PRO 70: 2nd International Conference on 'Service Life Design for Infrastructure' (ISBN set: 978-2-35158-096-7, e-ISBN: 978-2-35158-097-4); *Ed. K. van Breugel, G. Ye and Y. Yuan*

PRO 71: Advances in Civil Engineering Materials - The 50-year Teaching Anniversary of Prof. Sun Wei' (ISBN: 978-2-35158-098-1; e-ISBN: 978-2-35158-099-8); *Eds. C. Miao, G. Ye, and H. Chen*

PRO 72: First International Conference on 'Advances in Chemically-Activated Materials – CAM'2010' (2010), 264 pp, ISBN: 978-2-35158-101-8; e-ISBN: 978-2-35158-115-5, *Eds. Caijun Shi and Xiaodong Shen*

PRO 73: 2nd International Conference on 'Waste Engineering and Management - ICWEM 2010' (2010), 894 pp, ISBN: 978-2-35158-102-5; e-ISBN: 978-2-35158-103-2, *Eds. J. Zh. Xiao, Y. Zhang, M. S. Cheung and R. Chu*

PRO 74: International RILEM Conference on 'Use of Superabsorsorbent Polymers and Other New Addditives in Concrete' (2010) 374 pp., ISBN: 978-2-35158-104-9; e-ISBN: 978-2-35158-105-6; *Eds. O. M. Jensen, M. T. Hasholt, and S. Laustsen*

PRO 75: International Conference on 'Material Science - 2nd ICTRC - Textile Reinforced Concrete - Theme 1' (2010) 436 pp., ISBN: 978-2-35158-106-3; e-ISBN: 978-2-35158-107-0; *Ed. W. Brameshuber*

PRO 76: International Conference on 'Material Science - HetMat - Modelling of Heterogeneous Materials - Theme 2' (2010) 255 pp., ISBN: 978-2-35158-108-7; e-ISBN: 978-2-35158-109-4; *Ed. W. Brameshuber*

PRO 77: International Conference on 'Material Science - AdIPoC - Additions Improving Properties of Concrete - Theme 3' (2010) 459 pp., ISBN: 978-2-35158-110-0; e-ISBN: 978-2-35158-111-7; *Ed. W. Brameshuber*

PRO 78: 2nd Historic Mortars Conference and RILEM TC 203-RHM Final Workshop – HMC2010 (2010) 1416 pp., e-ISBN: 978-2-35158-112-4; *Eds. J. Válek, C. Groot, and J. J. Hughes*

PRO 79: International RILEM Conference on Advances in Construction Materials Through Science and Engineering (2011) 213 pp., ISBN: 978-2-35158-116-2, e-ISBN: 978-2-35158-117-9; *Eds. Christopher Leung and K.T. Wan*

PRO 80: 2nd International RILEM Conference on Concrete Spalling due to Fire Exposure (2011) 453 pp., ISBN: 978-2-35158-118-6, e-ISBN: 978-2-35158-119-3; *Eds. E. A. B. Koenders and F. Dehn*

PRO 81: 2nd International RILEM Conference on Strain Hardening Cementitious Composites (SHCC2-Rio) (2011) 451 pp., ISBN: 978-2-35158-120-9, e-ISBN: 978-2-35158-121-6; *Eds. R.D. Toledo Filho, F. A. Silva, E. A. B. Koenders and E. M. R. Fairbairn*

PRO 82: 2nd International RILEM Conference on Progress of Recycling in the Built Environment (2011) 507 pp., e-ISBN: 978-2-35158-122-3; *Eds. V. M. John, E. Vazquez, S. C. Angulo and C. Ulsen*

PRO 83: 2nd International Conference on Microstructural-related Durability of Cementitious Composites (2012) 250 pp., ISBN: 978-2-35158-129-2; e-ISBN: 978-2-35158-123-0; *Eds. G. Ye, K. van Breugel, W. Sun and C. Miao*

PRO 84: CONSEC13 - Seventh International Conference on Concrete under Severe Conditions – Environment and Loading (2013) 1930 pp., ISBN: 978-2-35158-124-7; e-ISBN: 978-2- 35158-134-6; *Eds. Z. J. Li, W. Sun, C. W. Miao, K. Sakai, O. E. Gjorv & N. Banthia*

PRO 85: RILEM-JCI International Workshop on Crack Control of Mass Concrete and Related issues concerning Early-Age of Concrete Structures – ConCrack 3 – Control of Cracking in Concrete Structures 3 (2012) 237 pp., ISBN: 978-2-35158-125-4; e-ISBN: 978-2-35158-126-1; *Eds. F. Toutlemonde and J.-M. Torrenti*

PRO 86: International Symposium on Life Cycle Assessment and Construction (2012) 414 pp., ISBN: 978-2-35158-127-8, e-ISBN: 978-2-35158-128-5; *Eds. A. Ventura and C. de la Roche*

PRO 87: UHPFRC 2013 – RILEM-fib-AFGC International Symposium on Ultra-High Performance Fibre-Reinforced Concrete (2013), ISBN: 978-2-35158-130-8, e-ISBN: 978-2-35158-131-5; *Eds. F. Toutlemonde*

PRO 88: 8th RILEM International Symposium on Fibre Reinforced Concrete (2012) 344 pp., ISBN: 978-2-35158-132-2, e-ISBN: 978-2-35158-133-9; *Eds. Joaquim A. O. Barros*

PRO 89: RILEM International workshop on performance-based specification and control of concrete durability (2014) 678 pp, ISBN: 978-2-35158-135-3, e-ISBN: 978-2-35158-136-0; *Eds. D. Bjegović, H. Beushausen and M. Serdar*

PRO 90: 7th RILEM International Conference on Self-Compacting Concrete and of the 1st RILEM International Conference on Rheology and Processing of Construction Materials (2013) 396 pp, ISBN: 978-2-35158-137-7, e-ISBN: 978-2-35158-138-4; *Eds. Nicolas Roussel and Hela Bessaies-Bey*

PRO 91: CONMOD 2014 - RILEM International Symposium on Concrete Modelling (2014), ISBN: 978-2-35158-139-1; e-ISBN: 978-2-35158-140-7; *Eds. Kefei Li, Peiyu Yan and Rongwei Yang*

PRO 92: CAM 2014 - 2nd International Conference on advances in chemically-activated materials (2014) 392 pp., ISBN: 978-2-35158-141-4; e-ISBN: 978-2-35158-142-1; *Eds. Caijun Shi and Xiadong Shen*

PRO 93: SCC 2014 - 3rd International Symposium on Design, Performance and Use of Self-Consolidating Concrete (2014) 438 pp., ISBN: 978-2-35158-143-8; e-ISBN: 978-2-35158-144-5; *Eds. Caijun Shi, Zhihua Ou, Kamal H. Khayat*

PRO 94 (online version): HPFRCC-7 - 7th RILEM conference on High performance fiber reinforced cement composites (2015), e-ISBN: 978-2-35158-146-9; *Eds. H. W. Reinhardt, G. J. Parra-Montesinos, H. Garrecht*

PRO 95: International RILEM Conference on Application of superabsorbent polymers and other new admixtures in concrete construction (2014), ISBN: 978-2-35158-147-6; e-ISBN: 978-2-35158-148-3; *Eds. Viktor Mechtcherine, Christof Schroefl*

PRO 96 (online version): XIII DBMC: XIII International Conference on Durability of Building Materials and Components(2015), e-ISBN: 978-2-35158-149-0; *Eds. M. Quattrone, V. M. John*

PRO 97: SHCC3 – 3rd International RILEM Conference on Strain Hardening Cementitious Composites (2014), ISBN: 978-2-35158-150-6; e-ISBN: 978-2-35158-151-3; *Eds. E. Schlangen, M. G. Sierra Beltran, M. Lukovic, G. Ye*

PRO 98: FERRO-11 – 11th International Symposium on Ferrocement and 3rd ICTRC - International Conference on Textile Reinforced Concrete (2015), ISBN: 978-2-35158-152-0; e-ISBN: 978-2-35158-153-7; *Ed. W. Brameshuber*

PRO 99 (online version): ICBBM 2015 - 1st International Conference on Bio-Based Building Materials (2015), e-ISBN: 978-2-35158-154-4; *Eds. S. Amziane, M. Sonebi*

PRO 100: SCC16 - RILEM Self-Consolidating Concrete Conference (2016), ISBN: 978-2-35158-156-8; e-ISBN: 978-2-35158-157-5; *Ed. Kamal H. Kayat*

PRO 101 (online version): III Progress of Recycling in the Built Environment (2015), e-ISBN: 978-2-35158-158-2; *Eds. I. Martins, C. Ulsen and S. C. Angulo*

PRO 102 (online version): RILEM Conference on Microorganisms-Cementitious Materials Interactions (2016), e-ISBN: 978-2-35158-160-5; *Eds. Alexandra Bertron, Henk Jonkers, Virginie Wiktor*

PRO 103 (online version): ACESC'16 - Advances in Civil Engineering and Sustainable Construction (2016), e-ISBN: 978-2-35158-161-2; *Eds. T.Ch. Madhavi, G. Prabhakar, Santhosh Ram and P. M. Rameshwaran*

PRO 104 (online version): SSCS'2015 - Numerical Modeling - Strategies for Sustainable Concrete Structures (2015), e-ISBN: 978-2-35158-162-9

PRO 105: 1st International Conference on UHPC Materials and Structures (2016), ISBN: 978-2-35158-164-3, e-ISBN: 978-2-35158-165-0

PRO 106: AFGC-ACI-fib-RILEM International Conference on Ultra-High-Performance Fibre-Reinforced Concrete – UHPFRC 2017 (2017), ISBN: 978-2-35158-166-7, e-ISBN: 978-2-35158-167-4; *Eds. François Toutlemonde & Jacques Resplendino*

PRO 107 (online version): XIV DBMC – 14th International Conference on Durability of Building Materials and Components (2017), e-ISBN: 978-2-35158-159-9; *Eds. Geert De Schutter, Nele De Belie, Arnold Janssens, Nathan Van Den Bossche*

PRO 108: MSSCE 2016 - Innovation of Teaching in Materials and Structures (2016), ISBN: 978-2-35158-178-0, e-ISBN: 978-2-35158-179-7; *Ed. Per Goltermann*

PRO 109 (2 volumes): MSSCE 2016 - Service Life of Cement-Based Materials and Structures (2016), ISBN Vol. 1: 978-2-35158-170-4, Vol. 2: 978-2-35158-171-4, Set Vol. 1&2: 978-2-35158-172-8, e-ISBN : 978-2-35158-173-5; *Eds. Miguel Azenha, Ivan Gabrijel, Dirk Schlicke, Terje Kanstad and Ole Mejlhede Jensen*

PRO 110: MSSCE 2016 - Historical Masonry (2016), ISBN: 978-2-35158-178-0, e-ISBN: 978-2-35158-179-7; *Eds. Inge Rörig-Dalgaard and Ioannis Ioannou*

PRO 111: MSSCE 2016 - Electrochemistry in Civil Engineering (2016), ISBN: 978-2-35158-176-6, e-ISBN: 978-2-35158-177-3; *Ed. Lisbeth M. Ottosen*

PRO 112: MSSCE 2016 - Moisture in Materials and Structures (2016), ISBN: 978-2-35158-178-0, e-ISBN: 978-2-35158-179-7; *Eds. Kurt Kielsgaard Hansen, Carsten Rode and Lars-Olof Nilsson*

PRO 113: MSSCE 2016 - Concrete with Supplementary Cementitious Materials (2016), ISBN: 978-2-35158-178-0, e-ISBN: 978-2-35158-179-7; *Eds. Ole Mejlhede Jensen, Konstantin Kovler and Nele De Belie*

PRO 114: MSSCE 2016 - Frost Action in Concrete (2016), ISBN: 978-2-35158-182-7, e-ISBN: 978-2-35158-183-4; *Eds. Marianne Tange Hasholt, Katja Fridh and R. Doug Hooton*

PRO 115: MSSCE 2016 - Fresh Concrete (2016), ISBN: 978-2-35158-184-1, e-ISBN: 978-2-35158-185-8; *Eds. Lars N. Thrane, Claus Pade, Oldrich Svec and Nicolas Roussel*

PRO 116: BEFIB 2016 – 9th RILEM International Symposium on Fiber Reinforced Concrete (2016), ISBN: 978-2-35158-187-2, e-ISBN: 978-2-35158-186-5; *Eds. N. Banthia, M. di Prisco and S. Soleimani-Dashtaki*

PRO 117: 3rd International RILEM Conference on Microstructure Related Durability of Cementitious Composites (2016), ISBN: 978-2-35158-188-9, e-ISBN: 978-2-35158-189-6; *Eds. Changwen Miao, Wei Sun, Jiaping Liu, Huisu Chen, Guang Ye and Klaas van Breugel*

PRO 118 (4 volumes): International Conference on Advances in Construction Materials and Systems (2017), ISBN Set: 978-2-35158-190-2, Vol. 1: 978-2-35158-193-3, Vol. 2: 978-2-35158-194-0, Vol. 3: ISBN:978-2-35158-195-7, Vol. 4: ISBN:978-2-35158-196-4, e-ISBN: 978-2-35158-191-9; *Eds. Manu Santhanam, Ravindra Gettu, Radhakrishna G. Pillai and Sunitha K. Nayar*

PRO 119 (online version): ICBBM 2017 - Second International RILEM Conference on Bio-based Building Materials, (2017), e-ISBN: 978-2-35158-192-6; *Ed. Sofiane Amziane*

PRO 120 (2 volumes): EAC-02 - 2nd International RILEM/COST Conference on Early Age Cracking and Serviceability in Cement-based Materials and Structures, (2017), Vol. 1: 978-2-35158-199-5, Vol. 2: 978-2-35158-200-8, Set: 978-2-35158-197-1, e-ISBN: 978-2-35158-198-8; *Eds. Stéphanie Staquet and Dimitrios Aggelis*

PRO 121 (2 volumes): SynerCrete18: Interdisciplinary Approaches for Cement-based Materials and Structural Concrete: Synergizing Expertise and Bridging Scales of Space and Time, (2018), Set: 978-2-35158-202-2, Vol.1: 978-2-35158-211-4, Vol.2: 978-2-35158-212-1, e-ISBN: 978-2-35158-203-9; *Eds. Miguel Azenha, Dirk Schlicke, Farid Benboudjema, Agnieszka Knoppik*

PRO 122: SCC'2018 China - Fourth International Symposium on Design, Performance and Use of Self-Consolidating Concrete, (2018), ISBN: 978-2-35158-204-6, e-ISBN: 978-2-35158-205-3; *Eds. C. Shi, Z. Zhang, K. H. Khayat*

PRO 123: Final Conference of RILEM TC 253-MCI: Microorganisms-Cementitious Materials Interactions (2018), Set: 978-2-35158-207-7, Vol.1: 978-2-35158-209-1, Vol.2: 978-2-35158-210-7, e-ISBN: 978-2-35158-206-0; *Ed. Alexandra Bertron*

PRO 124 (online version): Fourth International Conference Progress of Recycling in the Built Environment (2018), e-ISBN: 978-2-35158-208-4; *Eds. Isabel M. Martins, Carina Ulsen, Yury Villagran*

PRO 125 (online version): SLD4 - 4th International Conference on Service Life Design for Infrastructures (2018), e-ISBN: 978-2-35158-213-8; *Eds. Guang Ye, Yong Yuan, Claudia Romero Rodriguez, Hongzhi Zhang, Branko Savija*

PRO 126: Workshop on Concrete Modelling and Material Behaviour in honor of Professor Klaas van Breugel (2018), ISBN: 978-2-35158-214-5, e-ISBN: 978-2-35158-215-2; *Ed. Guang Ye*

PRO 127 (online version): CONMOD2018 - Symposium on Concrete Modelling (2018), e-ISBN: 978-2-35158-216-9; *Eds. Erik Schlangen, Geert de Schutter, Branko Savija, Hongzhi Zhang, Claudia Romero Rodriguez*

PRO 128: SMSS2019 - International Conference on Sustainable Materials, Systems and Structures (2019), ISBN: 978-2-35158-217-6, e-ISBN: 978-2-35158-218-3

PRO 129: 2nd International Conference on UHPC Materials and Structures (UHPC2018-China), ISBN: 978-2-35158-219-0, e-ISBN: 978-2-35158-220-6;

PRO 130: 5th Historic Mortars Conference (2019), ISBN: 978-2-35158-221-3, e-ISBN: 978-2-35158-222-0; *Eds. José Ignacio Álvarez, José María Fernández, Íñigo Navarro, Adrián Durán, Rafael Sirera*

PRO 131 (online version): 3rd International Conference on Bio-Based Building Materials (ICBBM2019), e-ISBN: 978-2-35158-229-9; *Eds. Mohammed Sonebi, Sofiane Amziane, Jonathan Page*

RILEM Reports (REP)

Report 19: Considerations for Use in Managing the Aging of Nuclear Power Plant Concrete Structures (ISBN: 2-912143-07-1); *Ed. D. J. Naus*

Report 20: Engineering and Transport Properties of the Interfacial Transition Zone in Cementitious Composites (ISBN: 2-912143-08-X); *Eds. M. G. Alexander, G. Arliguie, G. Ballivy, A. Bentur and J. Marchand*

Report 21: Durability of Building Sealants (ISBN: 2-912143-12-8); *Ed. A. T. Wolf*

Report 22: Sustainable Raw Materials - Construction and Demolition Waste (ISBN: 2-912143-17-9); *Eds. C. F. Hendriks and H. S. Pietersen*

Report 23: Self-Compacting Concrete state-of-the-art report (ISBN: 2-912143-23-3); *Eds. Å. Skarendahl and Ö. Petersson*

Report 24: Workability and Rheology of Fresh Concrete: Compendium of Tests (ISBN: 2-912143-32-2); *Eds. P. J. M. Bartos, M. Sonebi and A. K. Tamimi*

Report 25: Early Age Cracking in Cementitious Systems (ISBN: 2-912143-33-0); *Ed. A. Bentur*

Report 26: Towards Sustainable Roofing (Joint Committee CIB/RILEM) (CD 07) (e-ISBN 978-2-912143-65-5); *Eds. Thomas W. Hutchinson and Keith Roberts*

Report 27: Condition Assessment of Roofs (Joint Committee CIB/RILEM) (CD 08) (e-ISBN 978-2-912143-66-2); *Ed. CIB W 83/RILEM TC166-RMS*

Report 28: Final report of RILEM TC 167-COM 'Characterisation of Old Mortars with Respect to Their Repair (ISBN: 978-2-912143-56-3); *Eds. C. Groot, G. Ashall and J. Hughes*

Report 29: Pavement Performance Prediction and Evaluation (PPPE): Interlaboratory Tests (e-ISBN: 2-912143-68-3); *Eds. M. Partl and H. Piber*

Report 30: Final Report of RILEM TC 198-URM 'Use of Recycled Materials' (ISBN: 2-912143-82-9; e-ISBN: 2-912143-69-1); *Eds. Ch. F. Hendriks, G. M. T. Janssen and E. Vázquez*

Report 31: Final Report of RILEM TC 185-ATC 'Advanced testing of cement-based materials during setting and hardening' (ISBN: 2-912143-81-0; e-ISBN: 2-912143-70-5); *Eds. H. W. Reinhardt and C. U. Grosse*

Report 32: Probabilistic Assessment of Existing Structures. A JCSS publication (ISBN 2-912143-24-1); *Ed. D. Diamantidis*

Report 33: State-of-the-Art Report of RILEM Technical Committee TC 184-IFE 'Industrial Floors' (ISBN 2-35158-006-0); *Ed. P. Seidler*

Report 34: Report of RILEM Technical Committee TC 147-FMB 'Fracture mechanics applications to anchorage and bond' Tension of Reinforced Concrete Prisms – Round Robin Analysis and Tests on Bond (e-ISBN 2-912143-91-8); *Eds. L. Elfgren and K. Noghabai*

Report 35: Final Report of RILEM Technical Committee TC 188-CSC 'Casting of Self Compacting Concrete' (ISBN 2-35158-001-X; e-ISBN: 2-912143-98-5); *Eds. Å. Skarendahl and P. Billberg*

Report 36: State-of-the-Art Report of RILEM Technical Committee TC 201-TRC 'Textile Reinforced Concrete' (ISBN 2-912143-99-3); *Ed. W. Brameshuber*

Report 37: State-of-the-Art Report of RILEM Technical Committee TC 192-ECM 'Environment-conscious construction materials and systems' (ISBN: 978-2-35158-053-0); *Eds. N. Kashino, D. Van Gemert and K. Imamoto*

Report 38: State-of-the-Art Report of RILEM Technical Committee TC 205-DSC 'Durability of Self-Compacting Concrete' (ISBN: 978-2-35158-048-6); *Eds. G. De Schutter and K. Audenaert*

Report 39: Final Report of RILEM Technical Committee TC 187-SOC 'Experimental determination of the stress-crack opening curve for concrete in tension' (ISBN 978-2-35158-049-3); *Ed. J. Planas*

Report 40: State-of-the-Art Report of RILEM Technical Committee TC 189-NEC 'Non-Destructive Evaluation of the Penetrability and Thickness of the Concrete Cover' (ISBN 978-2-35158-054-7); *Eds. R. Torrent and L. Fernández Luco*

Report 41: State-of-the-Art Report of RILEM Technical Committee TC 196-ICC 'Internal Curing of Concrete' (ISBN 978-2-35158-009-7); *Eds. K. Kovler and O. M. Jensen*

Report 42: 'Acoustic Emission and Related Non-destructive Evaluation Techniques for Crack Detection and Damage Evaluation in Concrete' - Final Report of RILEM Technical Committee 212-ACD (e-ISBN: 978-2-35158-100-1); *Ed. M. Ohtsu*

Report 45: Repair Mortars for Historic Masonry - State-of-the-Art Report of RILEM Technical Committee TC 203-RHM (e-ISBN: 978-2-35158-163-6); *Eds. Paul Maurenbrecher and Caspar Groot*

Report 46: Surface delamination of concrete industrial floors and other durability related aspects guide - Report of RILEM Technical Committee TC 268-SIF (e-ISBN: 978-2-35158-201-5); *Ed. Valerie Pollet*

Part I
Cement

Considerations for the Energy Balance and Preliminary Design of an Experimental LC³ Cement Pilot Plant

Ivan Machado-Lopez, F. Bacallao, I. H. Moya and Jose Fernando Martirena-Hernandez

Abstract Associated with cement manufacturing processes, large amounts of carbon dioxide (CO_2) are released into the atmosphere, it is estimated that between 0.65 and 0.90 tons of CO_2 are emitted per ton of cement manufactured. By 2050, the demand for this binder is expected to exceed 5000 million tons, which would contribute to an increase of more than 3% of emissions. However, the emissions can be reduced by using supplementary cementitious materials (SCM), in this sense the calcined clays have a great potential for the reduction of emissions in the manufacture of cement, has been studied in recent years' low carbon cement or LC^3, developed by a joint team of the University of Lausanne and the Central University of Las Villas. The main results are the expansion of production by achieving clinker substitutions of up to 50%. In this sense, the Center for Research and Development of Technologies and Materials (CIDEM) in conjunction with the company *IPIAC-NERY* has committed to the development of a small pilot plant that makes the manufacture of limited quantities of LC^3, which allows the study of the process by interacting different sources of raw materials. In this paper some calculations were for designing a pilot plant, taking into account the balance of mass and energy necessary for its proper functioning, which in turn allows to specify the technology for the scaling of the production in any new industrial plant or adaptation of capacities installed.

Keywords Mass · Energy · Low carbon cement · Clay

1 Introduction

A new technology based on the replacement of clinker by high volumes of supplementary cementitious material, in this case thermally activated clay (*TAC, in Spanish AAT*), has been investigated in recent years by researchers of the international team "Low Carbon Cement". The new proposal is officially recognized by the scientific community as LC^3, whose results and advances are documented in scientific jour-

I. Machado-Lopez (✉) · F. Bacallao · I. H. Moya · J. F. Martirena-Hernandez
Center for Research and Development of Structures and Materials (CIDEM),
Central University Marta Abreu de Las Villas, Santa Clara, Cuba
e-mail: ivanm@uclv.edu.cu

© RILEM 2020

J. F. Martirena-Hernandez et al. (eds.), *Proceedings of the International Conference of Sustainable Production and Use of Cement and Concrete*, RILEM Bookseries 22,
https://doi.org/10.1007/978-3-030-22034-1_1

nals, being in this sense very important the expansion of production to achieve clinker substitutions of up to 50% (Vizcaíno 2015).

The LC^3 is an appropriate option because it considerably reduces CO_2 emissions into the atmosphere by reducing energy consumption, which is economically advantageous [1]. Recently through the management of *IPIAC NERY* the demand for LC^3 commercial production plants has skyrocketed. However, perform the essential industrial tests, although necessary to perform tests in real conditions can be very expensive.

In this sense, this work exposes some elements of the design of a small pilot plant where tests can be carried out in industrial conditions, with the advantage of obtaining reduced yields, however, these small plants replicated in a conscious way can be equivalent to the investment in a large cement plant. The first *IPIAC–CIDEM* plant will be put into operation in facilities of the Central University of Las Villas, its location is shown in Fig. 1.

The balance of mass and energy that is established in the calciner by achieving the thermal modification of the clay is also analyzed. At temperatures between 100 and 400 °C approximately, clay minerals give up their free water (not combined) and adsorbed water, including interlaminar water, at higher temperatures, between 400 and 750 °C, and depending on the types of clay minerals present in the raw material, water combined chemically in the form of hydroxyl groups is released (dehydroxylation).

In the case of kaolinite, the process of mass loss of the material during thermal activation is defined as a function of the activation temperature by the following expressions [2].

$$Al_4\left[(OH)_8Si_4O_{10}\right] \rightarrow 2(Al_2O_3 \cdot 2SiO_2) + 4H_2O \tag{1}$$

$$Al_2O_3 \cdot 2SiO_2(OH)_4 \rightarrow Al_2O_3 + 2SiO_2 + 2H_2O \tag{2}$$

In the dehydration of clays influence factors, such as the type of clay minerals, the nature and quantity of impurities, the size of the particles, the degree of crystallization of the clays, the gaseous atmosphere, and others.

The pilot plant *IPIAC–CIDEM* for the design of experimental tests is conceived to obtain a production flow of TAC of 100 kg/h by dry process, in this work to facilitate the design tasks in terms of mass and energy balances. Data are obtained from the bibliography consulted or from estimates made. The theoretical—methodological approach is exposed to perform the mass and energy balance, within the control limits determined for the kiln system. Figure 2 shows the flow of the process established in the pilot plant for the production of LC^3.

Fig. 1 Location of the pilot plant *IPIAC–CIDEM*. Central University of Las Villas. Faculty of Constructions, Villa Clara, Cuba

2 Analysis and Results

An optimal design of the rotary kiln can considerably reduce the assembly cost of the plant from the initial assumption of the desired production levels. Table 1 shows a summary with the projection in time of the expected production level.

2.1 The Rotary Kiln. Sizing

The rotary kiln consists of a cylindrical tube resting on rolling stations, which has a certain slope with respect to the horizontal and which rotates at speeds of rotation between 1.8 and 3.5 revolutions per minute (rpm).

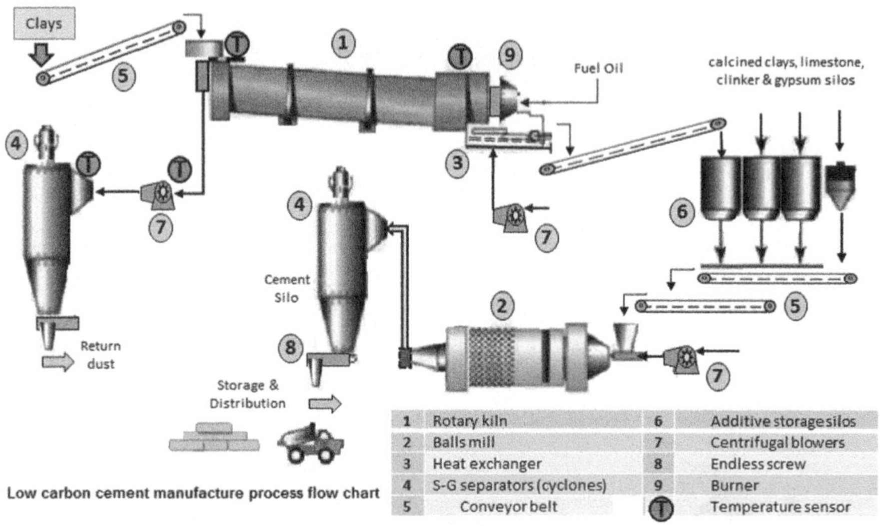

Fig. 2 Flow of the process established in the pilot plant for the production of LC^3. *Source* Esparza, G

Table 1 Estimated TAC production when carrying out the design calculations of the pilot plant *IPIAC–CIDEM*

Productivity calcination—TAC		$CaCO_3$	LC^2	P35	LC^3
100	Kg/h	67	167	167	333
800	Kg/8 h	533	1333	1333	2666.7
2400	Kg/24 h	1600	4000	4000	8000
72,000	Kg/month	48,000	120,000	120,000	240,000
720,000	Kg/year	480,000	1,200,000	1,200,000	2,400,000
720	Tons per year	480	1200	1200	2400

Source Self made

The slope is determined for which there is no rigid rule, its value ranges between 2 and 6%, most furnaces are installed with slopes between 2 and 4% in relation to the horizontal. As a result of the experience of operating rotary kilns, the optimum ratio between slope and filling coefficient (ϕ) has been obtained [3].

For the present design, a slope of 2° is established with a coefficient ϕ of 11%. In addition, the revolutions per minute (r.p.m.) of the rotary kiln will be in the order of 2.5 r.p.m. and a flow rate of 0.045 m^3 of clay/hours.

The calculation of different furnace design parameters is based on the law of transport of material inside an inclined rotary tube given by the US Bureau of Mines, which is expressed by the Eq. 1:

Table 2 Main working parameters and dimensions of the rotary kiln *IPIAC–CIDEM* for obtaining thermally activated clay

Di	L	t	Q	n	ν	ϕ	p	β
0.42	3.745	11	100	2.5	2	11	2	2.751

Source Self made

$$Q = 1.48 * n * Di^3 * \Phi * \rho\alpha * tag(\beta) \tag{3}$$

where: Q: transport capacity per hour (kg / h), n: speed of rotation (r.p.m.)
Di: inner diameter (m), Φ: filling degree, ρα: density of the material
β: movement of the resulting material (degrees °)

Transformed (1) it is possible to obtain an expression for the determination of the inner diameter of the calciner rotary kiln according to 2 [4]:

$$\mathrm{D} = \sqrt{\frac{\mathrm{ws} \cdot 4 \cdot \mathrm{t}}{\mathrm{L} \cdot \mathrm{gr} \cdot \pi \cdot \rho}} \tag{4}$$

where: ws: Solid flow (kg/min). L: Length of the kiln (m).
gr: Furnace filling degree (%). p: kiln slope (in%).
t: Total time of residence of the material inside the kiln (min).

The residence time of the clay material that has to be dehydroxylated is possible to obtain it from Eq. 3, however, there are many other expressions for the calculation of the residence time [3]. The residence time will determine the flow of material inside the oven, so this is one of the basic parameters for the design. When setting this variable, others can be calculated for the sizing of the kiln.

$$\mathrm{t} = \frac{10.62 * \mathrm{L}}{\mathrm{p} * \mathrm{d} * \mathrm{n}} \tag{5}$$

where: t: time of permanence (minutes). L: length of the rotary kiln (m).
d: inner diameter (m). n. rotation speed (r.p.m.).
p: inclination slope of the kiln.

The residence time is inversely proportional to the rotation speed and diameter of the kiln, being directly proportional to the length of the oven. The length and diameter of the furnaces are related by empirical laws such as L/D = 10–13, for long furnaces with different types of support, however prototypes with relation 4–12 can be found [5], in this work we have opted for a relationship in the environment of 5. Table 2 shows the values of the coefficients and factors assumed, as well as the final dimensions adopted.

2.2 Thermal Design

The analysis of the flow processes begins with the choice of the spatial region called "control volume." In the diagram represented in Fig. 3, the border or edge of the analysis performed in this work is shown (Fig. 4).

The thermal balance is defined from the mass balance that occurs in the process, taking into account the elimination of free water and adsorbed water (up to 400 °C), and in the decomposition or loss of the OH- of the clay (up to ≈ 850 °C), given by the equations described above (see 1 and 2). The loss of mass due to the natural humidity of the clay is approximately 6–7% and also the loss of mass due to decomposition or dehydroxylation of clay is 14%

Thermal balance is essentially conducted based on energy and mass conservation laws.

$$E_{in} = E_{out} \tag{6}$$

$$E_{in} = Q_1 = E_{out} = Q_2 + Q_3 + Q_4 \tag{7}$$

where: Q_1, Heat delivered to the kiln; Q_2, Heat needed for the clay transformation process; Q_3, Heat losses with the hot gases through the chimney; Q_4: Radiation heat losses.

$$Q_1 = m_F * LHV/\eta_{comb} \tag{8}$$

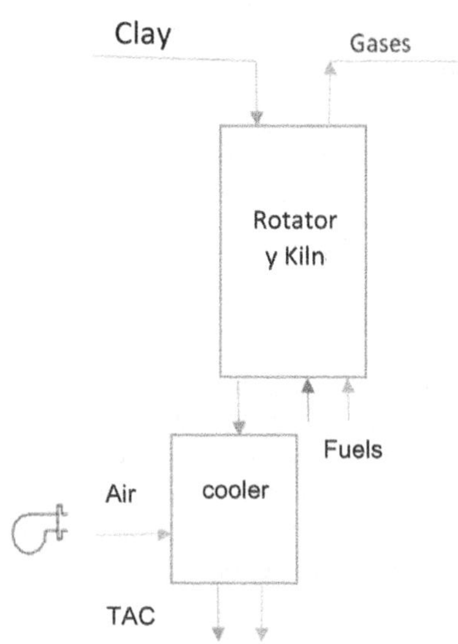

Fig. 3 Calcination of clays in a pilot plant. Analysis and borders. *Source* Self made

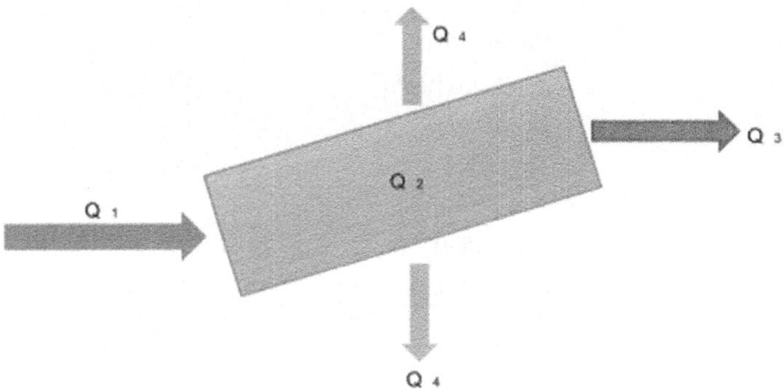

Fig. 4 Energy balance. *Source* Self made

where: m_F, Fuel mass in kg/h; LHV, low heating value (MJ/kg); η_{comb}, combustion efficiency adopted for this case 90%.

$$Q_2 = K_{cap} * Q_e \tag{9}$$

where: K_{cap}, is the kiln capacity in kg/h in this case kiln capacity is 100 kg/h; Q_e is the specific energy needed for clay transformation in the sinterization process, in this case is adopted the value of 0.81 MJ/kg reported by Mayorga (2000).

$$Q_3 = 0.001 * m_g * \overline{Cp} * \Delta t \tag{10}$$

where: m_g is the exit gases mass in kg/h; $\overline{Cp}$ in the average specific heat of the exit gases in the given window temperature in kJ/kgK; Δt is the difference of the exit gases temperature (t_{out}) and the environment temperature ($t_{env.}$).

$$Q_4 = 0.11 * Q_1 \tag{11}$$

The resolution is summarized in Table 3. Calculated the fuel mass consumption can be completed the thermal balance, as shown in Table 4.

3 Conclusions

The design of a rotary kiln for the treatment and thermal modification of clays is advantageous since this technology allows continuous production, by obtaining the optimal sizing according to the desired productivity and characteristics of the material to be obtained, the best behavior of parameters such as fuel consumption and heat losses can be used, extrapolated, for the design and construction of facilities with diverse production capacities.

Table 3 Calculation result and data

Results, Data	Value
m_F (kg/h) [lts/h]	(3.16) [3.76]
Kcap (kg/h)	100
Tg_{out} (°C)	600
T_{env} (°C)	29
Q_e (MJ/kg)	1
LHV (MJ/kg)	42.8
η_{comb}	0.9
Radiation losses	0.11
m_e (kg/kgF)	17
$\overline{Cp}$ (kJ/kgK)	1.10

Table 4 Thermal balance results

Balance	Value	Share (%)
Q_1 (MJ/h)	150.3874	100
Q_2 (MJ/h)	100	66
Q_3 (MJ/h)	33.8448	23
Q_4 (MJ/h)	16.54262	11

References

1. Dairon, J.A.C.: Evaluación económica y ambiental de plantas prototipo de pequeño y mediano tamaño para la producción de cemento LC^3. Facultad de Construcciones. CIDEM. Santa Clara. Cuba, Universidad Central Marta Abreu de Las Villas. TD (2017)
2. Orihuela, E.M.: Análisis energético de las diversas alternativas aplicables a una planta de clinkerizacion y producción de cemento. Escuela Superior de Ingenieros. Sevilla., Universidad de Sevilla. TD, Ingeniería Energética y Mecánica de Fluidos (2007)
3. Duda, W.: Manual tecnológico del cemento. Traducido por Antonio Sarabia González. Editores técnicos asociados, España (1977)
4. Colina, F.G., Caballero, I., López, J.C. (2016). Diseño básico de hornos rotatorios para el tratamiento de minerales. Ingenieria Quimica. Julio/Agosto 2016 Dpto. de Ingeniería Química y Metalurgia. Universitat de Barcelona
5. Aros Oñate, N.: Control predictivo no lineal de un horno rotatorio Ier Congreso Iberoamericano de ingeniería de proyectos. Antofagosta, Chile (2010)
6. Escuela Politécnica Nacional. Trabajo de Diploma
7. Dpto. técnico IPIAC, N.: Estimación del consumo de combustible. Torres Nova, Portugal (2018)

Use of Grinding Aids for Grinding Ternary Blends Portland Cement-Calcined Clay-Limestone

Juan Francisco Garcés Vargas, Marlon Espinosa, Yosvany Diaz Cárdenas, Alina Hereira Diaz and Jose Fernando Martirena-Hernandez

Abstract This article presents the evaluation of three types of grinding intensifier additives which were used for the production of LC3 cement, as well as the addition of LC2 minerals, treated in a laboratory scale ball mill. For its analysis a group of research techniques were applied to study its effect on the rheological and physical—mechanical properties of LC3 cement and LC2 mineral addition. The results of the work indicate that the milling intensifier additives evaluated have a positive impact on the mechanical properties of the samples evaluated, with a slight increase in the water—cement ratio caused by an increase in fineness.

Keywords Calcined clay · Pozzolan · Grinding · Grinding aid · Rheology · Mechanical properties

1 Introduction

Cement will continue to be the key material to satisfy national and international economic development due to its incidence as a raw material in the construction activity [1]. As a result, the cement industry worldwide faces increasing challenges in conserving material and energy resources, as well as reducing its CO_2 emissions.

J. F. G. Vargas
Universidad Estatal Península de Santa Elena, Carretero Libertad- Santa Elena, Cantón La Libertad, Provincia de Santa Elena, Ecuador

M. Espinosa
Hormicreto Paseo Rio Machangara, Bloque 3-18 y Cornelio Vintimilla (Parque Industrial), Cuenca, Ecuador

Y. D. Cárdenas · J. F. Martirena-Hernandez (✉)
Center for Research and Development of Structures and Materials (CIDEM), UCLV, Santa Clara, Cuba
e-mail: martirena@uclv.edu.cu

A. H. Diaz
Universidad Central Marta Abreu de las Villas, CEQA, Carretera de Camajuní Km 5 1/2, Santa Clara, Villa Clara, Cuba

© RILEM 2020

J. F. Martirena-Hernandez et al. (eds.), *Proceedings of the International Conference of Sustainable Production and Use of Cement and Concrete*, RILEM Bookseries 22,
https://doi.org/10.1007/978-3-030-22034-1_2

It is now known that the use of cementitious supplemental materials (MCS) to replace part of the cement clinker has economic advantages and is also an attractive way to reduce CO_2 emissions in cement industries [2]. Calcined clays are one of the possible candidates for MCS which can be used in the production of ternary cements, due to their wide availability as they are considerably scattered throughout the earth's crust, with a relative ease of treatment to be activated by thermal processes that require much less energy than that demanded for the production of clinker in addition to proven pozzolanic properties [3].

Another current trend is the use of technological additives used in the cement grinding process, also known as grinding additives (GA). These are components that improve mill performance by reducing specific energy consumption and increasing cement flow during transport and processing, as well as optimizing the clinker factor [4, 3]. Their actions can be explained fundamentally by a neutralization of the surface charges of cement grains during communition. These grinding additives help decrease the tendency of newly ground particles to re-agglomerate and sticking to the mill walls and media; thereby, increasing grinding efficiency [5].

For more than 10 years, the Centre for Research and Development of Structures and Materials (CIDEM) in collaboration with the Federal Polytechnic School of Lausanne (EPFL) have been working on the use of calcined clays as a partial substitute for clinker, almost doubling the amount of cement produced using the same amount of clinker. This research effort gave rise to a new Low Carbon Cement, also known as LC3. LC3 was originally obtained by intergrinding clinker, gypsum or other calcium sulfate, limestone and calcined clay to achieve beneficial effects that allow maintenance of physical-mechanical properties, such as compressive strength. However, it was noted that the particle size distribution (PSD) of the LC3 was sub-optimal and resulted in an increase in the demand for water. These high water demand was overcome by means of a separate grinding process, where the calcined clay and limestone were first ground to obtain a mineral addition named LC2 and then mixed with Portland cement (PC), to achieve a LC3 with suitable PSD.

Although a cement with more optimal PSD was produced, agglomeration of the LC3 or LC2 were noted. By use of grinding additives, agglomeration was noted to decrease and the time needed for grinding was also noted to decrease, resulting in the production of a final product with a proper particle size.

2 Materials and Methods

This research presents the evaluation of three types of grinding additives in the grinding of LC3 cement and LC2. Both were ground in a laboratory-scale ball mill. The impact of the additives on the rheological and physical-mechanical properties of LC3 cement and LC2 were made by the following methods.

2.1 Materials Used

Clay: The clay used came from a clay deposit with a chemical composition shown in Table 1. This clay is geologically found within the *Biblián* formation. In this mining area, a mixture of clay minerals of type 1:1 (Kaolinite, Dickite, Nacrite) is identified. The clayey mineral content of the kaolinite group is estimated at 55.27%, with a small presence of type 2:1 mineral. Losses in mass were observed between 200 and 300 °C, indicating the presence of natural hydroxides of iron and aluminum, which is commonly found in weathering crust.

From the chemical-mineralogical point of view, the clay deposit 1 provides a clay with excellent reactivity. The presence of accompanying minerals such as calcite and/or pyrite, which could compromise reactivity, is not evident. However, the high content of Fe_2O_3 could be a limiting factor in its use, as it could generate a color problem.

Limestone: The limestone used is of high purity with $CaCO_3$ content of 94%. The chemical composition of the gypsum used is shown in Table 1.

Gypsum: The chemical composition of the gypsum used is shown in Table 1.

Grinding Aids: Three types of grinding aids were supplied by GCP Applied Technologies. According to the manufacturer, these additives were designed to absorb on the newly generated surfaces of cement and other minerals during the grinding process. The adsorbed additives neutralize the surface charges and prevent the particles from re-agglomerating and coating to the walls of the mills and to the grinding media. These benefits provide a more efficient grinding and a reduction in operating costs (Table 2).

Clinker: The Clinker used was manufactured by the cement company Atenas.

2.2 Experimental Procedure

The experimental program focused on the evaluation of three grinding additives on the grinding of LC3 and LC2, which is subsequently mixed with Portland cement (94% clinker, 6% gypsum) to make LC2 2:1 cement. Description of the laboratory tests used to carried out the evaluate of the physical and mechanical properties of the cements are provided below.

2.2.1 Sample Preparation

Sample preparation was carried out on a laboratory scale, starting with the drying of all raw materials to constant weight at 200 °C, then prepared according to the proportions of Table 3, then ground in a ball mill, finally mixed in a 1:1 ratio of Portland cement and LC2.

Table 1 Chemical composition of raw materials used

	CaO	SiO_2	Al_2O_3	SO_3	Fe_2O_3	MgO	NaO_2	K_2O	TiO_2	$CaCO_3$	P.F	Moisture	Al_2O_3/SiO_2
Villa Clara	1.65	43.4	25.78	0.24	12.11	0.64	0.33	0.25	1.07		21.2	16.73	0.59
Gypsum	35.6	2.51	0.67	40.03	0.35	0.25	0.46	0.1	0.01				
Limestone		5.39								94.61			

Table 2 Characteristics of the GA to be used

Designation	TEA-GLCR	GLY	TEA-A
Dosage	0.02–0.06% by weight of cement, or 200 to 600 grams of additive per ton of cement produced	0.02–0.06% by weight of cement, or 200 to 600 grams of additive per ton of cement produced	0.03–0.06% by weight of cement, or 300 to 600 grams of additive per ton of cement produced
Active substance	Mixture of alkanolamine and glycerine	Glycols	Alkanolamine

Table 3 Mixing ratios of CP and LC2

Sample	Total ground (kg)	Proportions					SO_3 (%)
		Clinker (%)	Calcined Clay (%)	Limestone (%)	Gypsum (%)	Total (%)	
LC2 2:1	2.5	0.00	63.00	30.00	7.00	100.00	2.80
Portland cement	10	89.00	0.00	5.00	6.00	100.00	2.40

Table 4 Proportions of LC3-50 and LC2

		Clinker	Calcined Clay	Limestone	Gypsum	Total	SO_3
Portland Cement (%)	50.00	47.00	0.00	0.00	3.00	50.00	1.20
LC2 (%)	50.00	0.00	31.50	15.00	3500	50.00	1.40
LC3-50 2:1 (%)	100.00	47.00	31.50	15.00	6.50	100.00	2.60

On the other hand, LC3-50 2:1 cement was ground by intergrinding (see Table 4). One batch was ground at 15 min and another at 20 min. One batch was prepared for each GA using dosages recommended by the manufacturer. In the same way, the mineral addition LC2 was milled and then mixed with PC to make LC2 2:1. The grinding times and dosage of the grinding additives used were kept the same for the production of cement LC3-50 2:1 cement (see Tables 3, 4 and 5).

2.3 *Mechanical Physical Tests*

The quality of the cement LC3 and LC2 were verified using the procedures established by the ISO, INEN ASTM and NC standards. The tests used are:

- The measurement of the fineness of cement LC3-50 and LC2 2:1 cement was performed according to ISO 13320 using a Laser Granulometer (Dv10%, Dv50%

Table 5 Grinding protocol

#	Sample	Time min	Designation	Dosage (%)
1	Serie LC2 2:1 15	15	TEA-GLCR	0.040
2	Serie LC2 2:1 20	20	TEA-GLCR	0.040
3	Serie LC2 2:1 15	15	GLY	0.040
4	Serie LC2 2:1 20	20	GLY	0.040
5	Serie LC2 2:1 15	15	TEA-A	0.045
6	Serie LC2 2:1 20	20	TEA-A	0.045
5	LC3-50 2:1 inter 15	15	TEA-GLCR	0.040
6	LC3-50 2:1 inter 20	20	TEA-GLCR	0.040
7	LC3-50 2:1 inter 15	15	GLY	0.040
8	LC3-50 2:1 inter 20	20	GLY	0.040
9	LC3-50 2:1 inter 15	15	TEA-A	0.045
10	LC3-50 2:1 inter 20	20	TEA-A	0.045

and Dv90%), the Blaine specific surface according to INEN 196. In addition the compressibility parameter calculated following the procedure described in [5].

- The evaluation of the rheology in pastes and mortars was carried out following the method established in NC 235: 2012 "Cement Pastes-Determination of Plasticity and its Variation in Time by the Minicone Method". The determination of the normal consistency was made according to the standard INEN 157 and the fluidity test in mortars according to the standard INEN 2502.
- The determination of physical-mechanical properties was carried out taking into account the INEN 488 standard at ages 24 h, 3, 7 and 28 days.

3 Discussion of Results

This section presents the analysis of results obtained from the impact of the use of the three types of grinding additives on the fineness, rheology and physical-mechanical properties of LC3, as well as LC2 2:1 cement.

3.1 *Impact of Grinding Aids on Fineness*

As can be seen in Table 6, 12 samples were evaluated with 3 types of grinding additives in times of 15 and 20 min respectively, which were elaborated by means of intergrinding and separate grinding processes. The use of grinding additives were also reported by Prziwara and Toprak, also showing significant differences in the impact imparted by the grinding additives for the grinding processes.

3.1.1 Impact on Particle Size Distribution

Figure 1 shows the impact on PSD of the three types of milling intensifiers used in the elaboration of the different samples of cementitious material, classified in intergrinding (INT) and separate grinding (MS), it indicates the difference that exists between the samples ground at 15 and 20 min.

According to Fig. 1, if one takes into account on the impact of GA in each of the processes (co-grinding and separate grinding), it can be observed that TEA-GLCR type intensifiers show the best performance, followed by TEA-A and GLY type intensifiers.

This result is consistent with the studies of Shakhova, Assaad and Sun [3], who also reported that triethanolamine (TEA)-based active ingredient grinding additives can optimize particle size distribution and unify particle size, resulting in much more efficient grinding.

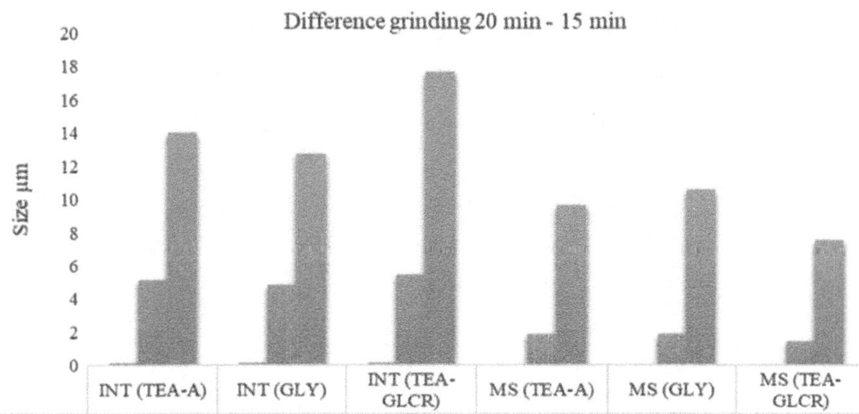

	INT (TEA-A)	INT (GLY)	INT (TEA-GLCR)	MS (TEA-A)	MS (GLY)	MS (TEA-GLCR)
■ Dv 10%	0,14	0,12	0,15	0,07	0,07	0,05
■ Dv 50%	5,15	4,84	5,46	1,88	1,84	1,46
■ Dv 90%	14,02	12,69	17,65	9,65	10,53	7,54

Fig. 1 Impact on DTP

Table 6 Results of the impact of intensifiers on physical properties

Description	TiME (min)	Dv 10%	Dv 50%	Dv 90%	Compressibility (%)	Blaine (cm^2/g)
INT-(TEA-A)-15M 5	15	0.89	13.87	57.86	29.31	5995
INT-(GLY)-15M 4g	15	0.87	13.5	57.24	30.51	6364
INT-(TEA-GLCR)-15M 4g	15	0.88	13.52	57.93	29.58	6193
MS-(TEA-A)-15M 5g	15	0.76	8.44	47.16	31.75	7120
MS-(GLY)-15M 4g	15	0.76	8.34	47.06	34.34	6448
MS-(TEA-GLCR)-15M 4g	15	0.74	8.05	45.19	35.48	6909
INT-(TEA-A)-20M 5g	20	0.75	8.72	43.84	38.91	6385
INT-(GLY)-20M 4g	20	0.75	8.66	44.55	32.61	7213
INT-(TEA-GLCR)-20M 4g	20	0.73	8.06	40.28	32.8	6948
MS-(TEA-A)-20M 5g	20	0.69	6.56	37.51	39.65	7681
MS-(GLY)-20M 4g	20	0.69	6.5	36.53	35.95	6712
MS-(TEA-GLCR)-20M 4g	20	0.69	6.59	37.65	36.41	7681

3.1.2 Impact on the Specific Surface

The results of the PSD tests shown above are consistent with those of the specific surface areas in the case of intergrinding (see Fig. 2). It is observed that the differences that exist between the 15 and 20 min samples were least for the triethanolamine (TEA).

For samples produced by separate milling, the GLY type additive was the one that provided the least difference in terms of the specific surface, leaving the samples containing triethanolamine (TEA-A and TEA-GLCR) to give bigger differences. TEA-A gave the most important impact. This result is also consistent with that reported by Shakhova and Assaad [3] where the authors showed that the use of GA led to a narrower PSD and an increase in the specific surface area of the product.

3.2 Impact of Intensifiers on Rheology (w/c Ratio and Flow)

Figure 3 shows the results of the normal consistency for all the evaluated combinations. The results showed a much lower w/c ratio for the three intensifier types, especially for a grinding time of 15 min, for the interground samples. It is observed that the demand for water was higher with the separately group samples, especially for grinding times of 20 min. This is believed to be a result of the finer particle size which lead to a greater demand for water.

Taking into consideration the fluidity results (see Fig. 3) GLY and TEA-A type intensifiers achieved shorter grinding time and higher flow, while with the TEA-GLCR additive needed a longer grinding time to achieve higher flow. However, all the samples tested at 15 min in the co-grinding process gave having better results, being the samples with GYL in first place followed by the intensifier TEA-A, on the

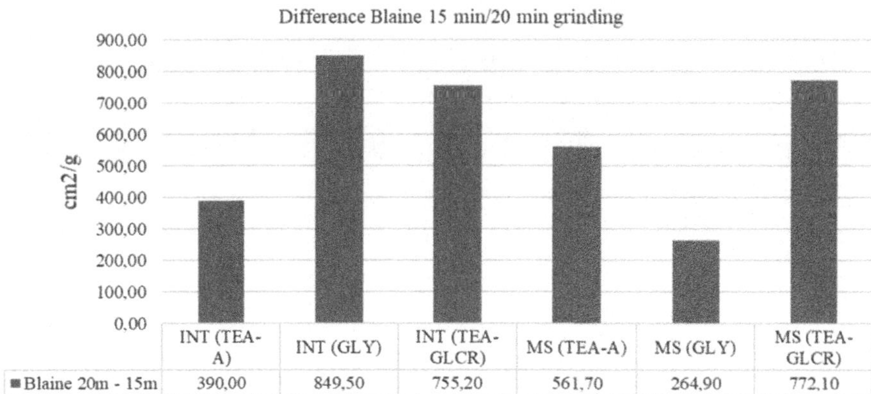

Fig. 2 Difference between specific surface of 15 and 20 min

contrary in the case of separate grinding the intensifier TEA-A and the GYL have less flow at 15 min.

3.3 Impact of Intensifiers on Mechanical Properties

According to Fig. 4, it can be observed that the results obtained with respect to the compressive strength for the different ages, show a superior performance when they are ground to 20 min.

However, if the type of additive is taken into account, the best results are obtained with the TEA-A in the case of intergrind. This is an important factor to consider when selecting the grinding additive. For the separate grinding process, there is no marked difference between the impacts of the three types of additives.

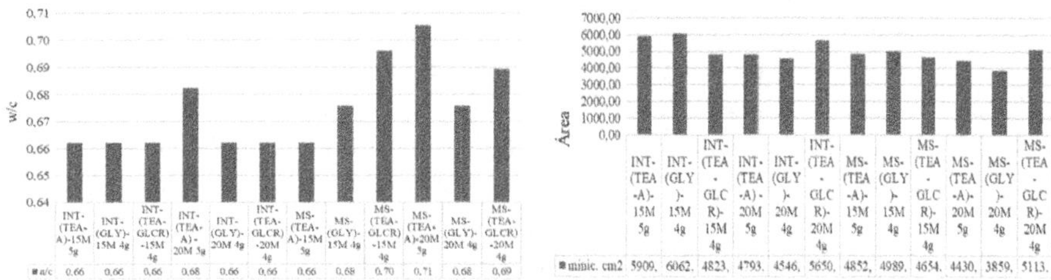

	INT-(TEA-A)-15M 5g	INT-(GLY)-15M 4g	INT-(TEA-GLCR)-15M 4g	INT-(TEA-A)-20M 5g	INT-(GLY)-20M 4g	INT-(TEA-GLCR)-20M 4g	MS-(TEA-A)-15M 5g	MS-(GLY)-15M 4g	MS-(TEA-GLCR)-15M 4g	MS-(TEA-A)-20M 5g	MS-(GLY)-20M 4g	MS-(TEA-GLCR)-20M 4g
a/c	0,66	0,66	0,66	0,68	0,66	0,66	0,66	0,68	0,70	0,71	0,68	0,69

	INT-(TEA-A)-15M 5g	INT-(GLY)-15M 4g	INT-(TEA-GLCR)-15M 4g	INT-(TEA-A)-20M 5g	INT-(GLY)-20M 4g	INT-(TEA-GLCR)-20M 4g	MS-(TEA-A)-15M 5g	MS-(GLY)-15M 4g	MS-(TEA-GLCR)-15M 4g	MS-(TEA-A)-20M 5g	MS-(GLY)-20M 4g	MS-(TEA-GLCR)-20M 4g
minic. cm2	5909.	6062.	4823.	4793.	4546.	5650.	4852.	4989.	4654.	4430.	3859.	5113.

Fig. 3 (Left) water/binder ratio; (Right) minicone slump test

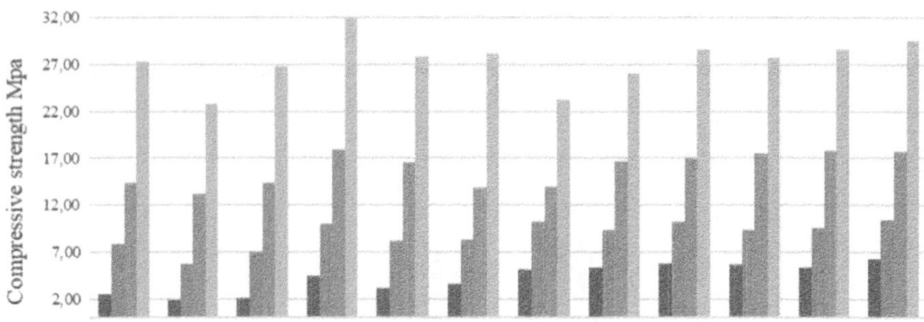

	INT-(TEA-A)-15M 5g	INT-(GLY)-15M 4g	INT-(TEA-GLCR)-15M 4g	INT-(TEA-A)-20M 5g	INT-(GLY)-20M 4g	INT-(TEA-GLCR)-20M 4g	MS-(TEA-A)-15M 5g	MS-(GLY)-15M 4g	MS-(TEA-GLCR)-15M 4g	MS-(TEA-A)-20M 5g	MS-(GLY)-20M 4g	MS-(TEA-GLCR)-20M 4g
24h	2,53	1,83	2,07	4,43	3,19	3,60	5,11	5,25	5,68	5,59	5,32	6,22
3d	7,79	5,67	6,98	10,03	8,16	8,24	10,15	9,35	10,24	9,31	9,57	10,42
7d	14,34	13,16	14,36	17,90	16,52	13,78	13,91	16,62	16,96	17,45	17,73	17,71
28d	27,31	22,82	26,76	31,89	27,86	28,10	23,19	26,01	28,53	27,74	28,62	29,54

Fig. 4 Results of the compressive strength test

4 Conclusions

All evaluated grinding additives show a favorable behavior in the grinding of LC3 cement and LC2 mineral addition. Although if we take into account their impact on each of the processes (co-grinding and separate grinding), it can be stated that triethanolamine based additives (TEA) show the best performance.

Triethanolamine (TEA)-based active ingredient grinding additives can optimize particle size distribution and unify particle size for much more efficient grinding. Further grinding does not compromise flow.

The use of grinding additives help, especially in separate grinding, to improve the mechanical properties of LC3, with a slight increase in the water-cement ratio caused by an increase in fineness.

The increase in grinding time decreases the particle size in the separate grinding and co-grinding processes at the cost of increasing the specific surface, which causes an increase in the demand for water and negatively impacts the compressive strength. In these cases, it is advisable to use superplasticizers.

References

1. Schneider, M., Romer, M., Tschudin, M., Bolio, H.: Sustainable cement production-present and future. Cem. Concr. Res. **41**(7), 642–650 (2011)
2. Suraneni, P., Weiss, J.: Examining the pozzolanicity of supplementary cementitious materials using isothermal calorimetry and thermogravimetric analysis. Cem. Concr. Compos. (2017)
3. Antoni, M., Rossen, J., Martirena, F., Scrivener, K.: Cement substitution by a combination of metakaolin and limestone. Cem. Concr. Res. **42**(12), 1579–1589 (2012)
4. Fernandez, R., Martirena, F., Scrivener, K.L.: The origin of the pozzolanic activity of calcined clay minerals: a comparison between kaolinite, illite and montmorillonite. Cem. Concr. Res. **41**(1), 113–122 (2011)
5. Nair, P.B.R., Paramasivam, R.: Effect of grinding aids on the time-flow characteristics of the ground product from a batch ball mill. Powder Technol. **101**, 31–42 (1999)

Effect of the Addition of Calcined Clay-Limestone-Gypsum in the Hydration of Portland Cement Pastes

María B. Díaz García, Lyannis Aparicio Ruíz and Jose Fernando Martirena-Hernandez

Abstract Currently, mineral additions are widely used as they favor not only the partial replacement of Portland cement, they reduces clinker production, but also provide certain chemical and physical properties that make concrete more durable. The variation of the water-cement ratio, the use of setting retarding additives and the addition of Cementitious Supplementary Materials (MCS) are factors that have a great influence on the hydration of the cement and the formation of the phases of a hydrated cement paste. That is why this research focuses on evaluating the effect of the active mineral addition of calcined clay, limestone and gypsum (LC2), in the cement hydration process. The work consists of several stages: first, a study is made of the variations that are made to the water to define the protocol and the amount of addition to study (30%); In a second stage, calorimetry and chemical shrinkage tests are carried out on cement pastes and a final stage where the influence of the addition on the hydration processes of the pastes is analyzed. This research provides a way of replacing Portland cement through the use of mineral additions, as is the case of LC2, which are in turn within the Construction Industry, to be a sustainable way of production of materials in Cuba.

Keywords Pozzolanas · Concrete · Shrinkage · Cement · Paste

1 Introduction

The replacement of part of the clinker with other mineral constituents has long been investigated for economic and ecological reasons. The amount of clinker that can be replaced by supplementary cementitious materials (MCS) depends on the type of MCS used. MCSs traditionally used are wastes from industrial processes, such

M. B. Díaz García (✉) · J. F. Martirena-Hernandez
Center for Research and Development of Structures and Materials (CIDEM), Santa Clara, Cuba
e-mail: mbetania@uclv.cu

L. A. Ruíz
UCM #4, Matanzas, Cuba

© RILEM 2020

J. F. Martirena-Hernandez et al. (eds.), *Proceedings of the International Conference of Sustainable Production and Use of Cement and Concrete*, RILEM Bookseries 22,
https://doi.org/10.1007/978-3-030-22034-1_3

as fly ash, microsilica and blast furnace slag, as well as natural pozzolans such as zeolitized tuff and volcanic ash [1].

Currently, in Cuba, work is being done on the production of concretes, where cements with active mineral additions are used, with the fundamental objective of improving the properties of the same and reducing the appearance of cracks due to changes in volume. These changes in the volume of the concrete elements, known as shrinkage, cause the appearance of cracks, which not only affect the aesthetics of the buildings, but also represent the malfunctioning of the same. These cracks can be explained as the result of a physical-chemical process that the cement paste undergoes, after being mixed with the aggregates, and which consists mainly of a variation in volume, which may depend on many factors, including the type of cement, the aggregates, the water-cement ratio, and fundamentally the humidity and temperature conditions during curing [2, 3].

In this context, it is of capital importance to identify and develop new sources of pozzolanic materials that allow the production of binders with high volumes of clinker substitution at an affordable cost from an economic, energy and environmental point of view [4].

The Centre for Research and Development of Structures and Materials CIDEM has carried out a study on the evaluation of shrinkage in pastes and concretes of Portland P 35 cement with 30% active mineral additions (LC2), which is no more than a product of calcined clay, limestone and gypsum in suitable proportions that substitute cement in the manufacture of concretes. This is done to know the influence of the hydration process on the volume changes of LC2 as an active mineral addition.

2 Methodology

It is important to define the influence of active mineral additions such as LC2 on the kinetics of hydration in Portland cement pastes from the characterization of materials.

For the microstructure analysis of the cement paste, six samples with cement P 35 were prepared as a reference, varying the water/cement ratio (0.40, 0.45 and 0.50) and the incorporation of Superplasticizer. Then, under the same conditions, the paste samples are elaborated with cement P 35 and the mineral addition LC2 at 30%.

In the study of cement pastes, calorimetry techniques are used to study the hydration of cement with mineral addition LC2 according to NC:525, 2014 Hydraulic Cement (Test methods. Determination of the heat of hydration and chemical shrinkage) to measure volume changes that occur between 0 and 7 days; the standard NC: 504, 2013 (Cement Volume stability) was consulted.

2.1 *Experimental Design*

For the development of the research, a dosage is defined for the tests on cement pastes, where the water/cement ratio is varied (Table 1).

The active mineral addition is formulated in the Faculty of Construction where the thermal activation of the clay is developed by calcination at 800 °C, then proceeds to grinding in a ball mill MB 800, and mixes the clay with limestone in a 2:1 ratio with 7% gypsum total to grind (Tables 2 and 3).

The cement to be used is produced at a cement factory in Cienfuegos, according to NC: 95, 2011 Portland Cement—Specifications, classified as Portland P 35 cement.

3 Discussion of Results

The hydration process of Portland cement with 30% LC2 mineral addition is slow compared to pure Portland cement [5]. Previous studies have shown that these

Table 1 Samples

Samples	Composition
P 40	P 35, Ra/c = 0.40
PA 40	P 35, Ra/c = 0.40, 0.5% SP
P 45	P 35, Ra/c = 0.45
PA 45	P 35, Ra/c = 0.45, 0.5% SP
LC2-40	70%P 35 + 30% LC^2, Ra/c = 0.40
LC2(A)-40	70%P 35 + 30% LC^2, Ra/c = 0.40, 0.5%SP
LC2-45	70%P 35 + 30% LC^2, Ra/c = 0.45
LC2(A)-45	70%P 35 + 30% LC^2, Ra/c = 0.45, 0.5%SP
LC2-50	70%P 35 + 30% LC^2, Ra/c = 0.50
LC2(A)-50	70%P 35 + 30% LC^2, Ra/c = 0.50, 0.5%ASP

Table 2 Raw materials for the production of LC^2

Raw Materials	Origin
Clay	Yaguajay
Limestone	Morejón Quarry, Sancti Spíritus
Gypsum	Punta Alegre

Table 3 Composition of the active mineral addition LC2

Materials	Quantity (Kg)	%
Clay	19.8	62
Limestone	9.6	31
Gypsum	0.6	7

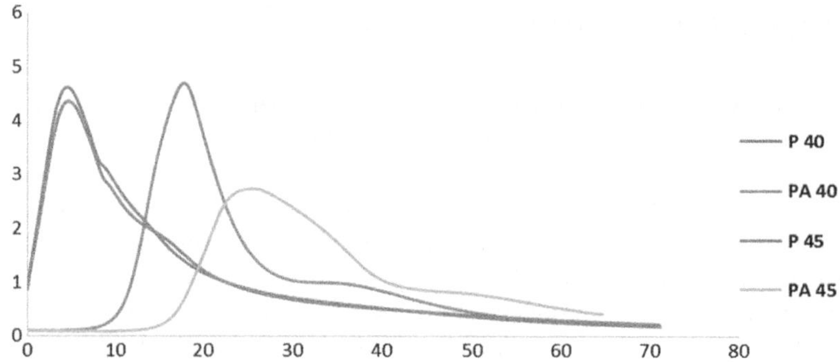

Fig. 1 Heat accumulated during hydration of the standard sample

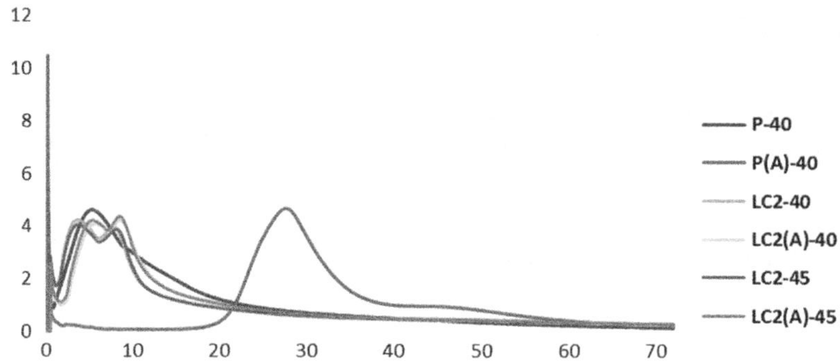

Fig. 2 Heat accumulated during hydration

cements have greater chemical stability and a denser microstructure. The aim of this research is to evaluate the LC2 influence on the hydration of cement pastes with a 30% substitution.

3.1 *Effect in the Hydration of Cement Pastes*

The use of the Superplasticizer favors the delay of the initial setting of the cement, thus prolonging the formation of the phases of a hydrated paste [5]. Therefore, the comparison between the samples with and without additive is carried out in order to analyze the effect of this on the hydration of the cement pastes. the induction period of samples P 40 (Portland Cement, Ra/c 0.40, without additive) and P 45 (Portland Cement, Ra/c 0.45, without additive) which has an approximate duration of two hours. The curves of accumulated heat rise to a maximum point that coincides with the formation of hydrated calcium silicates (C-S-H) that occupy between 50 and 60% of the hydrated cement paste. In samples PA 40 (Portland Cement, Ra/c 0.40, with additive) and PA 45 (Portland Cement, Ra/c 0.45, with additive) a delay in the initial setting stage is observed, due to the effect of the Superplasticizer used. The paste is held in a plastic state for approximately 20 h before moving on to the Acceleration Period, where the hydration rate reaches its maximum value at 36 h, indicating that the initial setting has been completed and hardening of the paste begins (Figs. 1 and 2).

The behavior during hydration of samples with P 35 and mineral addition LC2, with variation of the water-cement ratio. As can be seen, the effect of the additive of delaying the initial setting of the paste is similar to that observed in the standard samples (See Annex I). In samples LC2(A)-4 (30% LC2, Ra/c 0.40, with additive) and LC2(A)-45 (30% LC2, Ra/c 0.45, with additive) the induction period is delayed approximately between five and seven hours; this is due to the fact that, within the composition of the paste, the mineral addition LC2 has a greater specific surface area, a part of the additive is absorbed by it and thus its retarding function is limited.

The samples LC2-40 (30% LC2, Ra/c 0.40, without additive) and LC2(A)-40 (30% LC2, Ra/c 0.40, with additive) in which the formation stage of the hydrated calcium silicate (C-S-H) (first peak of the curve) and the reaction stage of the tricalcium aluminate where ettringite is formed (second peak of the curve) are well defined. In the LC2(A)-40 sample (30% LC2, Ra/c 0.40, with additive) the ettringite formation stage has higher heat release values than LC2-40 (30% LC2, Ra/c 0.40, without additive).

For both samples, a longer induction period is evident with respect to the pattern, because the pozzolanic reaction occurs more slowly. After approximately ten hours, all samples, except PA 40 (Portland Cement, Ra/c 0.4, with additive), which is under the effect of the SikaPlast additive, begin to stabilize. The slopes of the curves decrease and coincide at 72 h. It can be affirmed that, in the hydration of the components of the paste, both the standard and the samples with mineral addition LC2 behave in a similar way for the same water-cement ratio.

3.2 *Effect in the Volume Changes*

In samples LC2(A)-40 (30% LC2, Ra/c 0.40, with additive) and LC2(A)-45 (30% LC2, Ra/c 0.45, with additive) the effect of the additive is evident in the first 10 h. The curves have a behavior similar to the samples that do not have additive, with the difference that in the latter there is an increase in volume variation of 0.005 ml/g. This shows that the additive influences the volume changes that occur in the pastes, which decrease by 40%.

An analysis is made of the variation in volume produced in the samples with mineral addition LC2, with respect to the standard samples, for different water-cement ratios.

The figure shows the influence of the mineral addition LC2 and the variation of the water/cement ratio on the chemical shrinkage of cement pastes. When comparing the pastes with mineral addition LC2, with the standard sample P 40 (Portland Cement, Ra/c 0.40, without additive) an increase of 0.005 ml/g cement is appreciated in

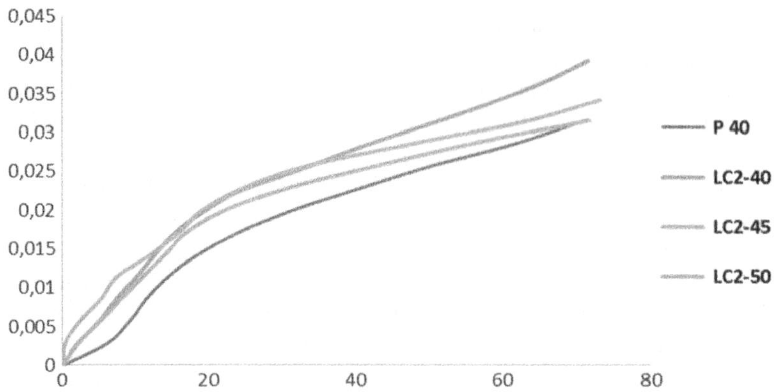

Fig. 3 Effect of variation in water-cement ratio on volume changes

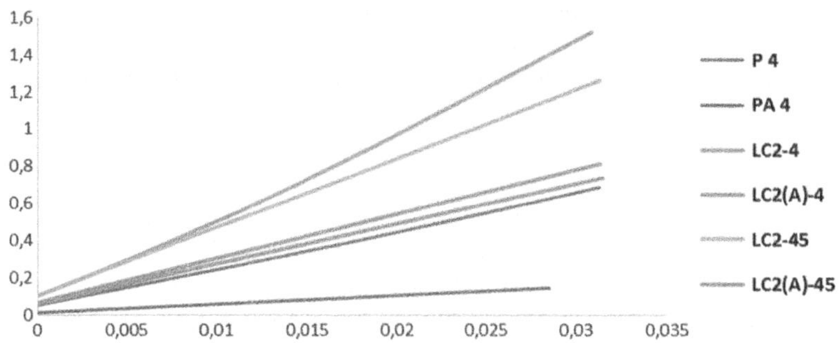

Fig. 4 Comparison of shrinkage values versus calorimetry

the shrinkage during the first 72 h. With respect to the water-cement variation, the samples with LC2 addition with lower water content present a greater variation in volume as a function of time (see Fig. 3).

It is observed that the samples LC2-40 (30% LC2, Ra/c 0.40, without additive) and LC2-50 (30% LC2, Ra/c 0.50, without additive) are the best performing, although the sample LC2-45 (30% LC2, Ra/c 0.45, without additive) in the first 8 h has variations in the deformations with respect to the other samples. After the 30 h the samples are stabilized, where the LC2-50 (30% LC2, Ra/c 0.50, without additive) is the one with the lowest chemical shrinkage values and similar to the sample P 40 (Portland Cement, Ra/c 0.40, without additive) (Fig. 4).

3.3 *Effect the Shrinkage and Calorimetry*

Referring to the comparison between the shrinkage values (ml/g cement) and the calorimetry values (J/g cement) for the samples with P 35 and the samples with 30%

addition of LC2 to Portland cement, it can be seen that there is a linear correlation between the values obtained by both methods, so it is feasible to use them in the analysis of the hydration processes of cement pastes.

It is observed that samples with Superlasticizer, when varying their volume during the three days, give off more heat in the hydration process than samples without additive. In the case of LC2 samples, the one with the best relationship between volume changes and heat release in the component formation process is presented by the LC2(A)-40 sample (30% LC2, Ra/c 0.40, with additive).

4 Conclusions

1. LC2 as an active mineral addition presents less variation in volume and heat release in the hydration process, for a given mass of paste, where the best substitution is with a water-cement ratio of 0.40.
2. The addition of clay + limestone + gypsum (LC2) can be used as a mitigation measure of the cracks produced by shrinkage, since these volume changes occur in the cement paste and with the substitution of 30%, this phenomenon is reduced.
3. The use of additives such as calcined clay and limestone not only favors the partial substitution of cement, reducing the production of clinker, but also provides certain properties that make concrete more durable.

References

1. Baroghel-Bouny, V., Mounanga, P., Khelidj, A., Loukili, A., Rafaï, N.: Autogenous deformations of cement pastes: part II. W/C effects, micro-macro correlations, and threshold values. Cem. Concr. Res. **36**(1), 123–36 (2006)
2. Concrete, C., Concrete, C., Cabinets, M., Rooms, M.: Standard Test Method for Length Change of Hardened Hydraulic-Cement Mortar and Current, vol. 4(C), pp. 1–7 (2010)
3. Darquennes, A., Staquet, S., Delplancke-Ogletree, M.P., Espion, B.: Effect of autogenous deformation on the cracking risk of slag cement concretes. Cem. Concr. Compos. **33**(3), 368–379 (2011). Available from http://dx.doi.org/10.1016/j.cemconcomp.2010.12.003
4. Grassl, P., Wong, H.S., Buenfeld, N.R.: Influence of aggregate size and volume fraction on shrinkage induced micro-cracking of concrete and mortar. Cem. Concr. Res. **40**(1), 85–93 (2010). Available from: http://dx.doi.org/10.1016/j.cemconres.2009.09.012
5. Antoni, M., Rossen, J., Martirena, F., Scrivener, K.: Cement substitution by a combination of metakaolin and limestone. Cem. Concr. Res. **42**(12), 1579–1589 (2012)

Influence of Limestone Content and PSD of Components on Properties of Clinker-Calcined Clay-Limestone Cements Produced by Intergrinding

A. Pérez and Jose Fernando Martirena-Hernandez

Abstract This paper looks at the study of intergrinding for the production of ternary cement based on clinker, calcined clay, limestone and gypsum with 50% of clinker substitution (LC3). The impact of grinding time on clinker, limestone and calcined clay PSD, and how this parameter influences the overall performance of the ternary cement is assessed. Laboratory cement blends were produced by grinding all components in a batch laboratory mill. Industrial cements produced through intergrinding in a continuous ball mill were used for comparison. Three fractions were identified: d < 7 μm, 7 μm < d < 40 μm and d < 40 μm, for each of the cements studied and the amount of each component were assessed. Fresh and hardened state properties of blends were tested. Results indicate that in intergrinding most of clinker remains at the medium fraction, and further grinding cannot improve clinker fineness due to fine calcined clay muffle clinker fineness gaining. PSD of limestone and calcined clay is wider than clinker PSD, with a high amount of each material on fine fraction, having a strong impact on rheology. A change in calcined clay/limestone ratio from 2:1 to 1:1 improves clinker grinding and rheology but has a negative impact on strengths due to the less proportion of calcined clay that impact negatively on the pozzolanic reaction.

Keywords Grinding · Ternary cement · Calcined clay

1 Introduction

The cement industry is responsible for about 5–8% of the global man-made CO_2 emissions. During the production of cement clinker, calcium carbonate from limestone decomposes, liberating CO_2, it represents about 50–60% of the total amount of CO_2 emitted during cement production [1].

A. Pérez (✉)
CIDC, La Habana, Cuba
e-mail: iogiovel@gmail.com

J. F. Martirena-Hernandez
CIDEM, UCLV, Santa Clara, Cuba

© RILEM 2020

J. F. Martirena-Hernandez et al. (eds.), *Proceedings of the International Conference of Sustainable Production and Use of Cement and Concrete*, RILEM Bookseries 22,
https://doi.org/10.1007/978-3-030-22034-1_4

Blended cements are replacing part of the clinker by mineral additions. It causes not only a reduction in fossil fuel consumption and in CO_2 emissions, but can also contribute to better concrete properties in both fresh and hardened state [2]. Medium purity kaolinite clay has proven to be a good alternative to MK in this system; that expands the opportunity to use low grade clays. The synergy of the combined addition of calcined clays and limestone has been set in evidence [1]. Limestone reacts with the alumina in presence of calcium hydroxide and favors the production of carboaluminates reaction products. This allows increasing the clinker substitution until 50%.

Besides components reactivity, PSD of each one plays an important role on final performance of blended cements. PSD of the more grindable component becomes finer and wider and the PSD of the harder to grind one becomes coarser and narrower [2, 3]. Clinker will be the coarser material on the blend due to the presence of relatively soft calcined clay particles that shields harder clinker particles from being ground. Harder and coarser clinker particles also abrade the softer ones increasing its fineness [2]. Due to this, clinker hydration is less effective and early strength most be affected.

Calcined clay had a high specific surface due to its surface structure increasing the water demand. Thus is was of great interest to investigate how intergrinding affects LC3 component PSD and how this parameter affects cement performance on fresh and hardened state in order to adjust final LC3 fineness. The influence of the level of substitution and calcined clay/limestone ratio was also investigated.

2 Materials and Methods

Clinker (CK) was produced at Siguaney cement factory in Cuba. Calcium sulfate (GS) and limestone (LS) originated also from Siguaney. Clay originated from Pontezuela was calcined at 750 °C and the calcined clay (CC) was obtained.

In order to determine the impact of the calcined clay/limestone, four different blended cements were produced for this study. 50% of CK, 29% of CC, 14% of LS, 7% of GS for a CC/LS ratio of 2:1 (50 2:1), 50% of CK, 22% of CC, 22% of LS, 6% of GS for a CC/LS ratio of 1:1 (50 1:1), 65% of CK, 19% of CC, 10% of LS, 6% of GS for a CC/LS ratio of 2:1 (65 2:1) and 65% of CK, 15% of CC, 14% of LS, 6% of GS for a CC/LS ratio of 1:1 (50 1:1). A reference Portland cement was also produced with 5% of GS.

Laboratory grinding was carried out using an ellipsoidal KHD laboratory mill with 50 L volume and 15 L of charge volume.

Using these results, three different blends were grounded to a range of fineness between 91 and 98%.

Industrial grinding was carried out using a continuous semindustrial grinding plant. The mill used has three chambers, 570 L of total volume and 170 L of charge volume with a production regime of 80 L/h. One blend for each calcined clay/limestone ratio and a reference were produced.

3 Results and Discussion

3.1 Fraction Composition

Mass Composition

Figure 1 shows the evolution of the amount of all the LC3 components in the fractions studied. Clinker fineness at fraction d < 7 μm is not increased by further grinding and is similar for all cases. Further grinding on LC3 50 2:1 does not have any major impact on clinker fineness at all fractions. Further grinding on LC3 50 1:1 increases clinker content at 7 μm < d < 33 μm fraction while decreases at d > 33 μm fraction. The higher amount of calcined clay in 2:1 blends increase the shielding process on the mill balls and clinker particles surface produced by calcined clay fine particles. For LC3 65, there is an increment in the medium fraction due to the less substitution level.

Further grinding decreases the amount of limestone at d > 33 μm fraction and increases in the rest of fractions. This increment is remarkable for LC3 50 1:1, For that blend in particular, further grinding increases limestone amount in the finer fractions

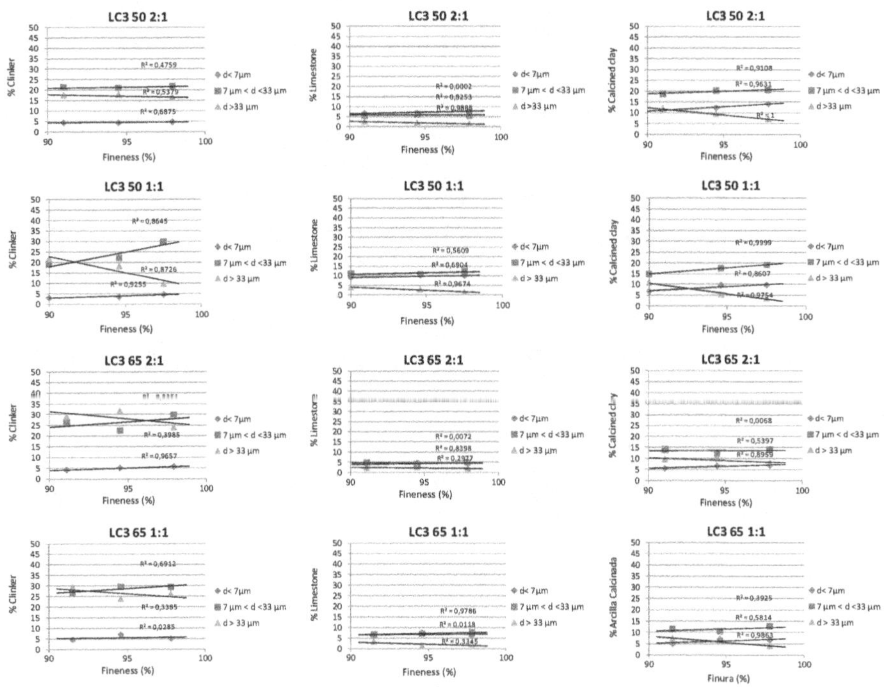

Fig. 1 Evolution of the amount of all the LC3 components in the fractions studied during the grinding process

and is higher than for LC3 2:1 (almost the double). There is a slow increase of fineness at fraction d < 7 μm for LC3 2:1 and no major changes at fraction 7 μm < d < 33 μm.

For all formulations further grinding decreases the amount of coarse calcined clay. The major amount of calcined clay is in the medium fraction with an increment in the finer fraction.

Volumetric Composition

Figure 2 shows the volumetric composition for all fractions. The difference in the amount ratio of the blends component between mass and volumetric composition is evident. The volume of calcined clay is much higher than its contribution in weight. This has a significant effect on rheology due to it depends on the specific surface that is a spacial magnitude and for calcined clays is higher than for clinker. In the finer fractions, for LC3 50, volume of additions is higher than clinker volume. For LC3 50 2:1 the volume of calcined clay is higher than the sum of the other compounds.

In the medium fraction the amount of clinker is higher than the amount of calcined clay; nevertheless, its contribution is higher compared to weight.

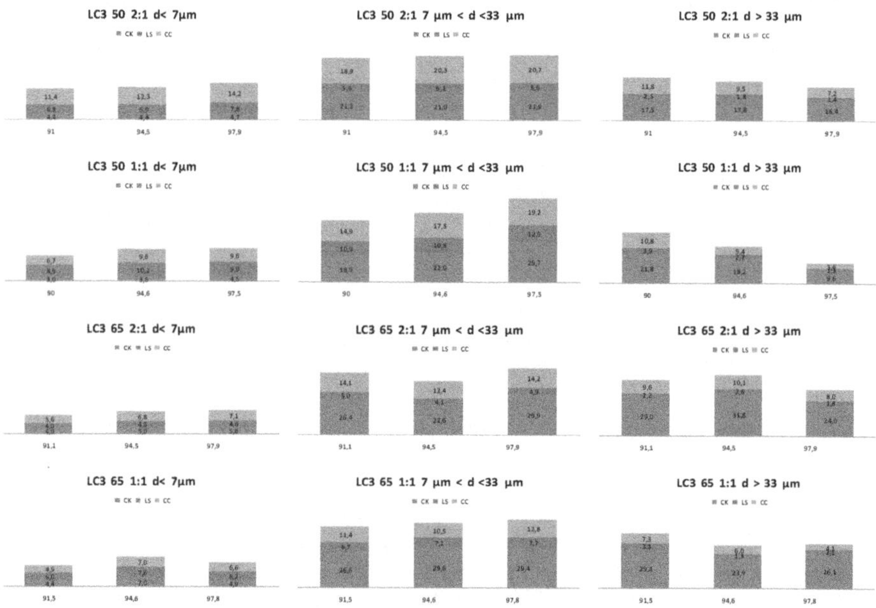

Fig. 2 Volumetric composition in all fractions

3.2 Mechanical Properties

Figure 3a shows the results of compressive strength for all the blends studied and the reference Portland cement; colored lines define the maximum values of compressive strength for each age according to Cuban standards. Increasing fineness there is an increase of strength. For high clinker substitution CC/LS ratio has a strong influence. For LC3 65, clinker fineness is the main factor. For LC3 50 2:1 compressive strengths are higher than for LC3 50 1:1 at all ages. CC/LS ratio is not significant for LC3 65.

For LC3 50 2:1, LC3 65 2:1 and LC3 65 1:1 is possible to reach compressive strength comparable with Portland cement.

Figure 3b shows the increment of compressive strength at all ages for all the blends studied and the reference. For the reference the main increment occurs in the first 2 days, due to the hydration of more amount of clinker. For LC3 65, the high amount of clinker compared to LC3 50, causes higher strength at 2 days. For all blends, the increment on strength at both 7 and 28 days is higher than for the reference due to the contribution of both pozzolanic and synergetic reactions that occurs at late ages. High levels of substitution and a CC/LS ratio 2:1 produce the higher increment of strength at 7 days.

Figure 4 shows the results of specific surface measured by Blaine and rheology measured by minislump. In general terms, increasing the substitution level and fineness specific surface is increased and rheology is reduced. Increasing the amount of limestone rheology is improved. Reducing calcined clay content reduces specific surface, water demand and improves rheology as well as in increasing the amount of limestone

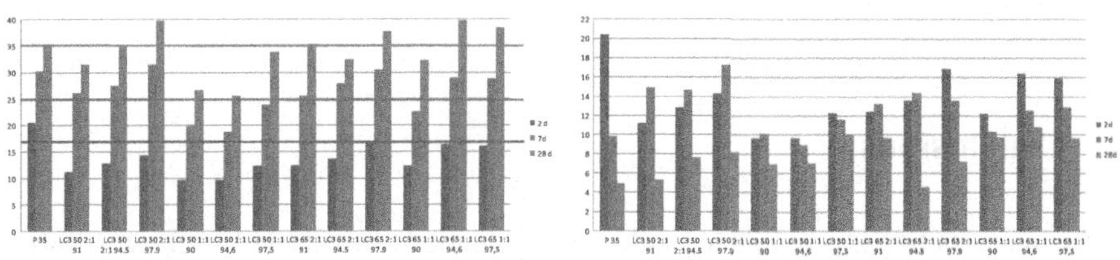

Fig. 3 Mechanical properties, **a** compressive strength for all the cements produced, **b** increment of compressive strength

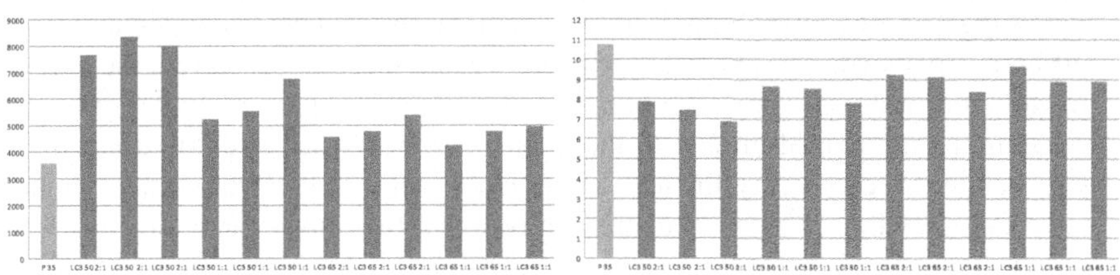

Fig. 4 Rheology, **a** specific surface, **b** minislump diameter

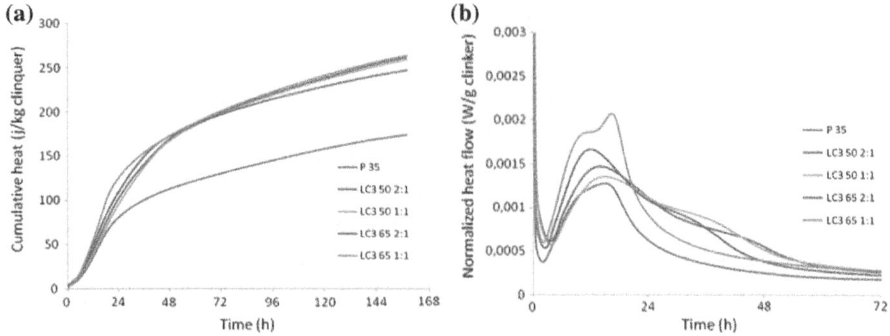

Fig. 5 Isothermal calorimetry curves, **a** cumulative heat, **b** heat flow, normalized

3.3 *Hydration*

Pozzolanic effect of calcined clay is also corroborated by Isothermal calorimetry. Figure 5a shows the normalized cumulative heat/g of clinker for the corresponding cements produced at industrial scale for the first 7days. For all blends the cumulative heat is higher than for the reference cement due to the pozzolanic reaction. Figure 5b shows the normalized heat flow/g of clinker, the peak of silicates hydration is higher for LC3 due to the contribution of fine particles of calcined clay and limestone as nucleation sites for clinker hydration products. For LC3 65 this peak is higher, showing that there is a limit for the substitution level in whish this effect is suitable. For LC3 50 2:1, LC3 50 1:1 and LC3 65 2:1 the aluminates reaction is retarded due to de higher amount of gypsum.

4 Conclusions

- By intergrinding is possible to produce LC3 blended cements with high substitution levels and similar performances than Portland cement.
- For high substitution levels further grinding above 95% of fineness increase energy consumption with no significant changes on performance.
- Increasing grinding time compressive strength is improved but water demand rises. Difference between raw materials grindability and morphology affect the gain of fineness for hard ones. Presence of calcined clay muffles proper clinker grinding, reducing the amount of clinker in the finer fraction, further grinding don't make any change. Calcined clay fineness is the main factor that affects water demand and rheology while compressive strength is driven by clinker fineness.

– Changing calcined clay/limestone ratio, reducing the amount of calcined clay, clinker grinding is improved. By the other hand is possible to find a better compromise between strength and rheology. Increasing limestone amount reduces water demand while is possible to maintain high compressive strength.

References

1. Vizcaino, L.: Effects of particle size distribution and intergrinding in blend of cement, calcined clays and limestone. In: Twin International Conferences 2nd Civil Engineering and 5th Concrete Future, pp. 53–59. Publisher (2013)
2. De Weert, K.: Separate Grinding Versus Intergrindig. Sintef Report, State of the art. In (2007)
3. Tsivilis, S.: Contribution to the problems arising from grinding of multicomponent cements. Cem. Concr. Res. **22**, 95–102 (1992)

Influence of the Limestone Type on the Compression Strength of LC^3 Cements

Felipe Tarquinio Oliveira, Christian Moreira, João Henrique da Silva Rêgo and Valdirene Maria S. Capuzzo

Abstract This article aims to study the influence of the $CaCO_3$ content of the limestone filler on the compressive strength of LC^3 cements. In order to perform this study, the compressive strength test was performed in LC^3 cements containing three types of limestone (A, B and C) with different $CaCO_3$ contents. To characterize the materials, tests were performed as X-ray diffraction, X-ray fluorescence, particle size distribution and BET surface area. Three different types of cements were created, varying the type of limestone filler. All blends contained 50% clinker, 5% gypsum and 30% calcined clay and 15% limestone. Characterization tests found that the filler A presented the highest amount of $CaCO_3$, while the filler B presented medium content and the filler C, low content. The results of the compressive strength showed similar values between the A and B fillets at 28 days and higher results for the filler B at 91 days. Filler C obtained lower scores at all ages. From this, it was possible to conclude that the $CaCO_3$ content of limestone influences the compressive strength of LC^3 cements.

Keywords LC^3 cement · Limestone · Calcined clay

1 Introduction

Portland cement is the most commonly used civil construction material in Brazil. According to data from the National Union of Cement Industry—SNIC, 57 million tons of Portland cement were produced during the year 2016 in Brazil. During its

F. T. Oliveira (✉) · C. Moreira · J. H. da Silva Rêgo · V. M. S. Capuzzo
Universidade de Brasília, Brasilia, Brazil
e-mail: felipetarquinio@gmail.com

C. Moreira
e-mail: concretoleve.eng@gmail.com

J. H. da Silva Rêgo
e-mail: jhenriquerego@unb.br

V. M. S. Capuzzo
e-mail: valdirenecapuzzo@gmail.com

© RILEM 2020

J. F. Martirena-Hernandez et al. (eds.), *Proceedings of the International Conference of Sustainable Production and Use of Cement and Concrete*, RILEM Bookseries 22,
https://doi.org/10.1007/978-3-030-22034-1_5

production large amounts of carbon dioxide are emitted, mainly in the decomposition of calcium carbonate. It is estimated that, in the production of 1 kg of Portland cement, approximately 0.87 kg of CO_2 are emitted into the atmosphere [1]. According to [2], the cement industry is responsible for the emission of 5–8% of CO_2 in the world.

Replacing clinker with supplementary cementitious materials (SCMs) is a promising alternative to reduce these environmental impacts. Among these materials, we can highlight the limestone, blast furnace slags, the fly ash and the calcined clay. Blast furnace slag and fly ash calcined clay are SCM with pozzolanic activity that give the cement extra C-S-H, as well as being environmentally friendly because it is an industrial waste.

The addition of limestone and calcined clay, blast furnace slag and fly ash are widely used by the cement industry, resulting in a large decrease in released CO_2. However, the blast furnace slag and fly ash are not produced in the same quantity of Portland cement [3]. Limestone, in turn, as supposedly inert material, has its content of limited addition, not achieving large levels of substitution.

Synergistic reactions are noted between the limestone and the calcined clay forming the hydrated monocarboaluminate. The synergistic reaction was described by [4], showing that the monocarboaluminate comes from the calcium carbonate from the limestone with the alumina from the clays.

$$A\ from\ clay + \mathrm{Cc} + 3\mathrm{CH} \rightarrow \mathrm{C_3A.C_{c.}H_{11}} \tag{1}$$

where:

A is Alumina;
Cc is calcium carbonate;
CH is Calcium hydroxide; and
$C_3A.Cc.H_{11}$ is hydrated monocarboaluminate.

Calcareous filler has chemical effects when used in cement. Due to this characteristic, its chemical composition may influence LC^3 cement performance. A study carried out by Lins [5] established the relationship between the $CaCO_3$ content of the filler and the mechanical strength of the LC^3 cement. According to the author, higher levels of $CaCO_3$ in the film lead to higher mechanical resistances.

2 Methodology

In this study were used Cement, Brazilian normal sand, calcined clay, limestone A, B and C, Gypsite and superplasticizer additive.

To make this cement, clinker and gypsum must be ground together. These materials were then milled in a ball mill with 25 kg of steel and capacity of 7 kg of material. The grinding time was set to obtain a standard interval that would result in a cement thin enough to comply with the limits established by NBR 5732 [6]. This fineness was verified through its surface area obtained with the Blaine method. After attempts, 180 min were defined as the best milling time, using 7 kg of material.

The fineness of the limestone influences the compressive strength of the LC^3 cement [7]. In order to avoid that this characteristic interferes in the analysis of the influence of the limestone content, the 3 types of limestone should have similar fineness. The limestone A, ground for 60 min in the ball mill, was chosen as the standard. And then for the limestone B and C to reach the same fineness, the time required in the ball mill was defined by means of attempts, so that the films B and C acquired granulometry similar to the film A. The defined time was of 68 min for the film B and C.

The characterization tests of the chosen materials were X-ray diffraction, X-ray Fluorescence, Laser Granulometry and BET surface area. These tests have as main function to obtain the chemical composition and the size of the particles of each material.

Figure 1 shows that the peak of calcite grows from filler A to C. This indicates that there is a larger amount of that compound in filler A, and a lower amount in filler C.

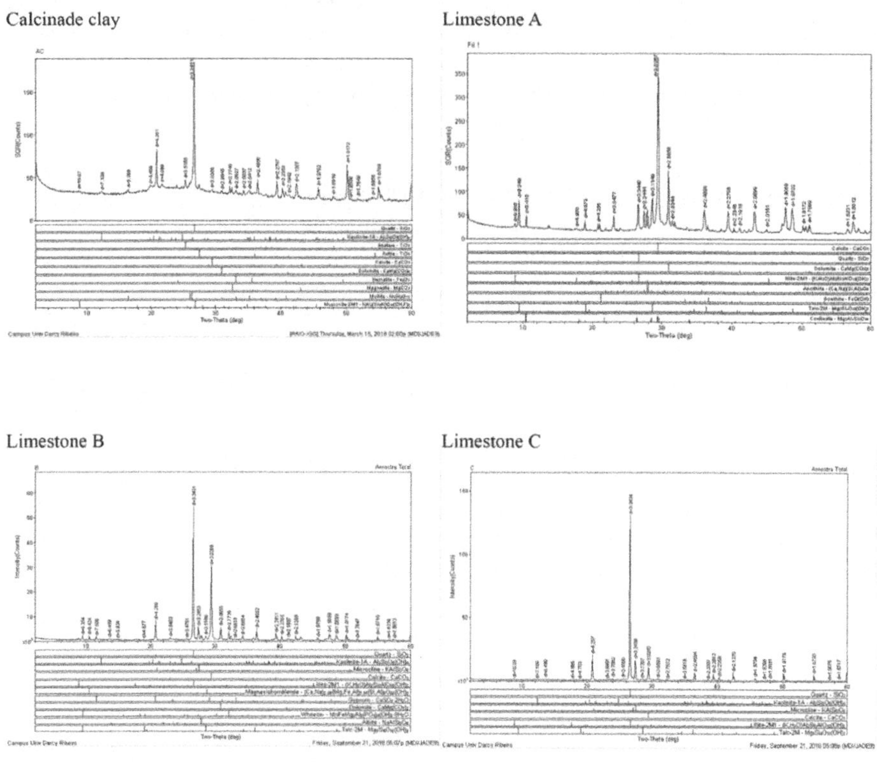

Fig. 1 Results of X-Ray diffraction test

The BET test showed that the Limestone A has a lower surface area than the Limestone B and C, and these showed similar results among themselves. According to [7], this lower surface area may negatively influence the mechanical performance of Limestone A (Table 1).

The laser granulometry reported a similarity in the particle size distribution of the Limestones, but the Limestone A is slightly thinner with smaller diameters in the percentages of 10, 50 and 90% and also smaller average diameter. Limestones B and C presented very close average diameter, showing great similarity in the granulometric distribution of these materials. These results differed from the surface area values found in the BET test, in which the Limestone A presented the lowest surface area and was expected to obtain the largest mean diameter different from that observed in the granulometry. It can be suggested, therefore, that the B and C must have larger roughness, increasing their surface area even though they have larger diameters.

The results of the X-ray diffraction and X-ray fluorescence show that Limestone A has the highest $CaCO_3$ content, that of Limestone B has a medium $CaCO_3$ content and that of Limestone C has a low CaCO3 content. Thus, it will be possible to analyze the influence of this compound on the compressive strength of LC^3 cements, the main objective of this work.

In order to analyze the compressive strength, four mortars were executed, one of reference and the other three using limestone filters with different levels of Calcium. Table 1 shows the composition of the elements. Note that the calcium content increases from filler A to filler C. The mortars were shaped and ruptured according to the Brazilian standard NBR 7215 - 1996 [8]. For each mortar were made 4 speci-

Table 1 Fluorescence-ray results and size granulometry results

	Calcined clay	Limestone A	Limestone B	Limestone C
BET surface area (m^2/g)	13.98	3.77	5.04	5.43
Average particle diameter (μm)	25.14	29.16	33.76	32.76
SiO_2 (%)	61.13	4.01	49.01	71.07
Al_2O_3 (%)	27.9	0.35	2.97	3.97
MgO (%)	0.4	3.9	1.97	0.91
Fe_2O_3 (%)	2.2	0.13	1.15	1.25
CaO (%)	3.45	48.73	25.82	11.64
Na_2O (%)	0.04	0.009	0.24	0.36
K_2O (%)	0.51	0.012	1.82	2.66
TiO^2 (%)	1.9	0.015	0.48	0.63
P_2O_5 (%)	0.04	0.009	0.04	0.04
MnO (%)	0	0.002	0.02	0.03
L.O.I. (%)	2.2	42.37	16.2	6.76
Total (%)	99.77	99.54	99.72	99.32

Table 2 Composition of LC3 cement

Name	Calcined clay	Metakaulim	Limestone A	Limestone B	Limestone C	Cement	
						Gypsum (%)	Clinker (%)
OPC	–	–	–	–	–	5	95
CCA	30%	–	15%	–	–	5	50
CCB	30%	–	–	15%	–	5	50
CCC	30%	–	–	–	15%	5	50

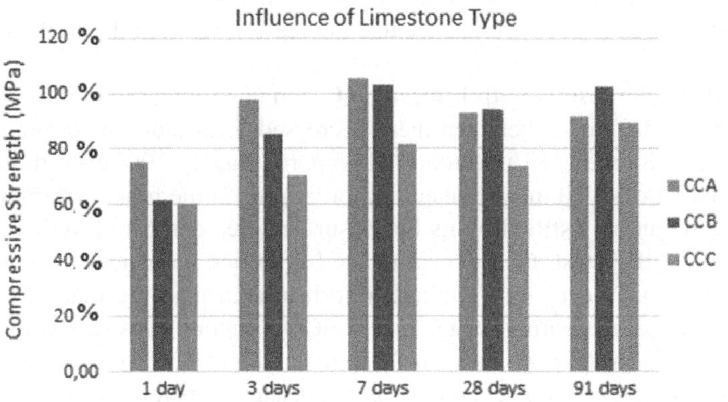

Fig. 2 Results of compressive strength of LC3 cements in percentage relation with the reference cement

mens, ruptured at the ages of 1, 3, 7, 28 and 91 days. The means of resistance at each age were calculated, excluding the results that presented a standard deviation greater than 6%. The compositions of the different cements tested are shown in Table 2.

3 Results and Discussion

3.1 Compressive Strength (MPa)

Figure 2 shows the results of compressive strength of LC^3 cements, in percentage relation with the reference cement.

It is observed that, on the first day, Limestone A presented greater performance indices. In the laser granulometry test, Limestone A presented the smallest mean diameter, which may justify its better performance in the early ages as it intensified the physical effects generated by the limestone.

At 7 days, the CCA and CCB cements presented similar resistance indices, both larger than the CCC. According to [9, 10], the formation of monocarboaluminates and hemicarboaluminates, from the $CaCO_3$ of the limestone and the aluminates of the clays, from the 7 days, can be more easily noticed, justifying the greater resistance indices of the binders with the limestone A and B, taking since they have larger amounts of $CaCO_3$.

At 28 days, the CCA and CCB cements presented resistance indexes also higher than the CCC, with differences of around 20%, respectively. This difference, similarly, to the analysis carried out for 7 days, is probably due to the higher amount of $CaCO_3$ present in the A and B fillers, as shown by X-ray fluorescence. However, Limestone B has a smaller amount of carbonate of calcium compared to Limestone A, presenting 50% less CaO, a compound bound to the CaCO3 content. Still, the Limestone B presented results of resistance to compression very similar to the Limestone A at 28 days.

At 91 days, the results with Limestone C remained lower than the other Limestones, but the difference between them decreased. Limestone B in turn surpassed the performance index of Limestone A by approximately 10% even though it had a lower content of calcium carbonate. According to [7] the best performance of the Limestone B can be justified by its larger surface area compared to the Limestone A found in the BET test, the fineness of the Limestone influences the compressive strength of the LC^3 cements. The larger surface area possibly made limestone B more reactive, compensating for its lower $CaCO_3$ content. However, it is important to note that the limestone C has a surface area and a mean diameter similar to that of the limestone B, and this fact did not compensate for its low $CaCO_3$ content. It is suggested that the influence of the lower limestone content can be reduced or even surpassed by the finer, however, there must be a lower limit in which the low $CaCO_3$ content becomes predominant.

4 Conclusions

This article sought to contribute to the studies about LC^3 cement, a new type of cement that, when exploring the synergy between the calcined clay and the limestone, seeks to reduce the amount of clinker and consequently the environmental impact of Portland cement.

Through characterization tests it was shown that the limestones were similar in their sizing compositions and their main difference was the amount of calcium carbonate in their compositions. Limestone A had the highest amount of $CaCO_3$, the lowest limestone C and limestone B was between the other two. Thus, the differences found in the values of compressive strength were attributed to this aspect. It was observed that the limestone C presented a lower result than the others, the others (A and B) like 28 days.

References

1. Damtoft, J.S., et al.: Sustainable development and climate change initiatives. Cem. Concr. Res. **38**(2), 115–127 (2008)
2. Gartner, E., Hirao, H.: A Review of Alternative Approaches to the Reduction of CO_2 Emissions Associated with the Manufacture of the Binder Phase in Concrete Cement and Concrete Research (2015)
3. Avet, F., Scrivener, K.: Investigation of the calcined kaolinite content on the hydration of Limestone Calcined Clay Cement (LC3). Cem. Concr. Res. **107**, 124–135 (2018)
4. Scrivener, K.L.: 202SPECIAL ISSUE—Future cements options for the future of cement. Indian Concr. J. **88**(7), 11–21 (2014)
5. Lins, D.N.: Análise da Resistência à Compressão do cimento LC^3 e a influência das adições minerais em seu desempenho. Dissertação de mestrado, Departamento de Engenharia Civil e Ambiental, Universidade de Brasília, DF, 141 p
6. ASSOCIAÇÃO BRASILEIRA DE NORMAS TÉCNICAS. NBR 16697: Cimento Portland - Requisitos Rio de Janeiro: ABNT (2018)
7. Scrivener, K.L., et al.: Effect of fineness in clinker-calcined clays-limestone cements. Adv. Cem. Res. **27**(9), 546–556 (2015)
8. ASSOCIAÇÃO BRASILEIRA DE NORMAS TÉCNICAS. NBR 7215: Cimento Portland - Determinação da resistência à compressão. Rio de Janeiro (1996)
9. Lothenbach, B., et al.: Influence of limestone on the hydration of Portland cements. Cem. Concr. Res. **38**(6), 848–860 (2008)
10. Antoni, M., et al.: Cement substitution by a combination of metakaolin and limestone. Cem. Concr. Res. **42**(12), 1579–1589 (2012)

Pozzolanic Reactivity of the Calcination Products Obtained from Yaguajay Clay Deposit

Karina Duverger Escobar, Adrián Alujas Díaz and Luis Alberto Pérez García

Abstract The influence of calcination temperature and mineralogical composition on the pozzolanic reactivity of calcination products obtained from Yaguajay clay deposit is assessed in this research. The samples were characterized by ATG, DRX and petrography. The pozzolanic reactivity of calcination products was evaluated by compressive strength (NC95:2011, NC506:2013, NC-TS527:2013, NC-TS528:2013) and its influence in the hydrate kinetics by isothermal calorimetry in pastes with 30% of cement substitution for calcined clay. All identified technological types present quartz, calcite, goethite and kaolinite like main mineral phases. The kaolin contents oscillate among 40% (Types A and B) and 50% (Type C). The calcite contents are lower than 5%; except in the layer B, with 20%. The pozzolanic reactivity of the calcination products of all the technological types is strongly influenced by the activation temperature, being reached the best reactivity at 800 °C. Other factors that could affect negatively the pozzolanic reactivity are high calcite contents (Type B) and high contact degree between the calcite and the kaolinite (Type C). For a sustainable exploitation of the deposit, it is recommended that the composite dedicated to the industrial production has an approximate composition of 37.5% Type A, 25% Type B and 37.5% Type C.

Keywords Calcined clay · Pozzolanic reactivity · Isothermal calorimetry · Mixture criteria

K. D. Escobar (✉) · A. A. Díaz
Center for Research and Development of Structures and Materials (CIDEM), UCLV, Santa Clara, Cuba
e-mail: kduverger@uclv.cu

L. A. P. García
Mining-Metallurgical Higher Institute of Moa (ISMMM), Moa, Cuba

© RILEM 2020

J. F. Martirena-Hernandez et al. (eds.), *Proceedings of the International Conference of Sustainable Production and Use of Cement and Concrete*, RILEM Bookseries 22,
https://doi.org/10.1007/978-3-030-22034-1_6

1 Introduction

Portland cement (PC) is among the most used materials and with the highest volume of production worldwide. However, its production is associated with high energy consumption and large volumes of carbon dioxide (CO_2) emissions on a global scale, mainly during the production of clinker, factors that negatively influence its costs and environmental sustainability of PC [1]. One of the most extended solution to this problem is the use of Supplementary Cementitious Materials (SCM), which allows to maintain or even increase the mechanical properties and durability of the final products [2]. At the same time, there are environmental benefits through the reduction of greenhouse gases emissions into the atmosphere [1]. Among the wide variety of MCS that can be used, there is growing interest in the use of thermally activated clays [2], especially the clay minerals of the kaolinite group, due to its high pozzolanic reactivity and relative abundance. Using a mixture of calcined kaolinitic clay and limestone in a 2:1 ratio allows raising up to 50% the degree of substitution of clinker. The cement obtained has a mechanical behavior similar to cement P-35 and higher durability [3]. In 2013, an industrial production test was carried out and a ternary cement was produced based on clinker–calcined clay–limestone. It took place took place in Siguaney Cement Factory and satisfactory results were achieved [3]. Several deposits of kaolinitic clays were studied later, but the Yaguajay deposit was the one that met the parameters required for exploitation on an industrial scale under technically and economically viable conditions. However, its structure is not homogeneous due to the processes that gave rise to it and it is divided into layers interspersed, grouped into three technological types, which have variable contents of kaolinite, iron oxides and calcite. For these reasons, it is necessary to know how each technological type responds to the calcination conditions, in order to work with optimal calcination conditions and obtain the maximum pozzolanic reactivity of the calcination products.

2 Materials

The clay deposit Yaguajay has documented presence of clay minerals of the kaolinite group in contents never lower than 40% and reserves that guarantee its exploitation on an industrial scale for not less than 20 years. Based on its coloration and chemical composition (content of CaO and Al_2O_3), three technological types can be differentiated [4]:

- Type A: they are very plastic and yellowish clays; with pellets of iron (Fe) and manganese (Mn) of size up to 5 mm, occasionally abundant fragments of white limestones, dispersed nests of blue-white kaolin of 3–5 cm.
- Type B: they are clays from ocher to brown to reddish, due to the abundant content of iron oxide. They have iron oxide and manganese pellets, whose diameters are up to 1 cm. Nests and lenses of blue-white kaolin from 5 to 10 cm and isolated

fragments of silicite and limestones are present in all their mass. These clays are plastic and easily dispersed in the presence of water.

- Type C: they are clays of yellowish color, with very fine structure, with incipient stratification and sometimes well stratified. They have scattered pellets of iron oxide, Mn and fragments of $CaCO_3$, nests and lenses of kaolin up to 15 cm and throughout its mass gypsum crystallites of 0.5–1.5 cm confirming its marine origin. They are very plastic and easily dissolve in water.

3 Experimental Procedure

The mineralogical characterization was carried out by X-ray diffraction (XRD), using a diffractometer Panalytical Xpert Pro MPD, with radiation $Cu_{K\alpha}$ and a divergence grid of 0.5°. The samples were scanned between 4 and 65° (2θ), at an angular step of 0.017° and a time per step of 80 s. The diffractograms were processed using X Pert High Score Plus software (2011). For the analysis of thermal behavior by Thermogravimetric Analysis (ATG) a Mettler–Toledo instrument was used TGA/SDTA 851, at a heating rate of 10 °C/min from 30 to 1000 °C and the mass losses were calculated using the peak area integration method.

The pozzolanic reactivity of the calcination products was evaluated by compression strength tests in PC—Pozzolan mortars. The mortars we prepared with 30% substitution of PC for calcined clays and Sika Plast, a plasticizer in an experimental phase, was added, to achieve the required fluidity. A series prepared with 100% PC was used as reference. The compression resistance was evaluated at the ages of 3, 7 and 28 days, using the Cuban standards [5–8]. To evaluate the influence of calcined clays in the hydration kinetics of a mixed system, isothermal calorimetry tests were carried out, for which 70% PC and 30% calcined clay with a water/solid ratio = 0.45 pastes were made and 100% PC (P-35) pastes were prepared as a reference. The pastes were mixed at 1600 rpm for two minutes. A TAM AIR 3238 calorimeter was used. Readings were performed for 72 h.

4 Experimental Results and Discussion

4.1 Mineralogical Characterization of Raw Materials

In Table 1, the mineralogical composition derived from the preliminary semiquantitative analysis of the diffractograms of each technological type sample is presented. In all cases, kaolinite predominates as clay mineral and quartz, calcite and goethite are the main non-clay minerals, all of which corresponds to geological and mineralogical descriptions of clay deposit made in previous studies. In addition, the relatively frequent presence of vermiculite, anatase, birnesite and gypsum is reported.

Table 1 Semiquantitative mineralogical composition of the different technological types

Semiquantitative (%)	CA_01	CA_02	CB_01	CB_02	CC_01	CC_02
Kaolinite	31	37	39	47	64	58
Calcite	9	2	26	11	4	–
Quartz	41	42	18	22	22	18
Goethite	11	12	11	14	–	14
Vermiculite	3	2	1	1	–	–
Anatase	4	4	5	5	7	7
Anorthite sodian	1	–	–	–	–	–
Birnessite	–	–	–	–	3	–
Gypsum	–	–	–	–	–	2

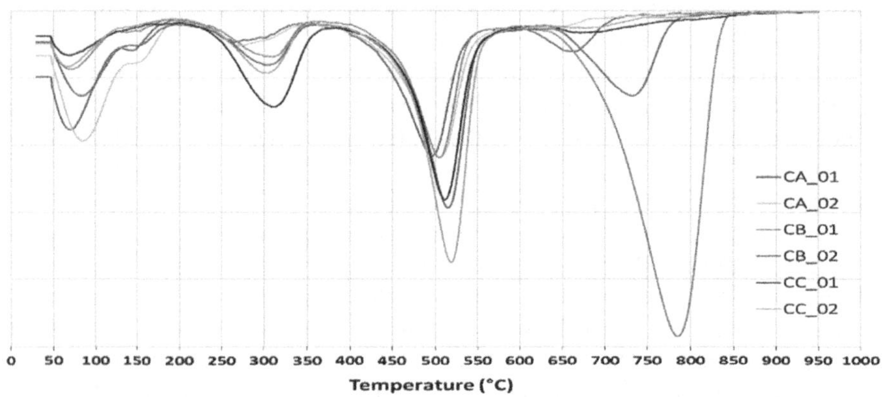

Fig. 1 ATG curves for the samples of the different technological types in their natural state

In Fig. 1, the main changes occurring in the samples of the different technological types due to the increase in temperature can be observed. For all the samples, a significant weight loss is observed at temperatures below 200 °C, mainly due to the loss of water associated with cations present in the interlaminar regions, probably divalent cations, because this thermal effect clearly occurs in two stages. This behavior corresponds to the detected presence of vermiculite in the samples, clay mineral where Mg^{2+} occurs predominantly as a cation in the interlaminar region. The loss of mass that occurs between 200 and 370 °C is attributed to the decomposition of the iron hydroxides present, mainly goethite. The magnitude of this effect increases progressively from the samples corresponding to the technological type A (less oxidized layer, of lighter coloring) to the samples corresponding to the technological type C (more oxidized layer, of darker coloration).

In all the samples analyzed, the loss of structural OH^- of the clay minerals of the kaolinite group manifested between 370 and 600 °C. This thermogravimetric effect can be taken as a basis to quantify the content of the clay minerals of the kaolinite group, as reported in Table 2, where it is observed that technological types A and B present average contents of clay minerals of the group of the kaolinite close to 40%, while the technological type C has an average contents close to 50%. Between 600 and 850 °C, an effect mainly associated with the decomposition of calcite is observed, although a slight contribution of clay minerals type 2:1 at temperatures higher than 700 °C is not ruled out. In technological types A and C, the calcite content varies between 2 and 4%. However, technological type B has relatively high calcite contents, with great variability between one sample and another. This is directly associated to the relatively high presence of limestone fragments interspersed in this layer.

The presence of calcite can negatively influence the pozzolanic reactivity of the calcination products [9]. The causes of this anomalous behavior are related to the interaction of calcite and kaolinite during calcination and the probable melting effect of calcite at high temperatures. The association of these factors leads to the formation, at temperatures above 800 °C, of vitreous phases associated with calcium and aluminum silicates, with a compact structure and low specific surface, which decreases the availability of reactive phases rich in Si and Al to participate in the pozzolanic reaction [10]. Although calcite normally decomposes at temperatures above 800 °C, but when it is found in contents of less than 20% or has low crystallinity, the decomposition temperature can decrease significantly [11], a phenomenon that can be observed in the ATGs of all technological types.

4.2 *Evaluation of the Pozzolanic Reactivity of the Calcination Products*

The compression test in mortars is used as a reference for the evaluation of the general quality of a construction material. Also, this essay is found in the body of Cuban norms [5–8] to evaluate the pozzolanic reactivity. In Table 3, the evolution of the compressive strength in mortars with partial substitutions of 30% clay calcined at 750, 800 and 850 °C is shown. The compression resistance values of the PC series, established by the standard, are also included [5–7] for type P 35 cements. In all cases, the standard deviation values were less than 5%.

After 3 days, the mortars of practically all the series present values of compressive strength higher than those of the PC reference series, complying with the normed specifications. However, technological types B and C calcined at 850 °C do not meet this criterion. The results for this age do not allow to offer criteria on the pozzolanic reactivity in the systems, because at early ages the effects produced are also associated with purely physical transformations such as the filler effect (nucleation and early growth of hydration products) [12]. However, the effect filler itself is strongly influenced by the surface area of the calcination products, what follows that this parameter was significantly lower for the products of calcination at 850 °C of technological types B and C. This behavior can be partly explained by the formation

Table 2 Mineralogical composition of the different technological types determined by ATG

	CA_01	CA_02	Layer A	CA_COMP	CB_01	Layer B	CA_02	CC_01	CC_02	Layer C
% Humidity	4.26	5.12	4.49	4.08	2.19	2.81	3.43	1.86	2.00	1.93
% OH-associated with iron hydroxides	1.59	1.73	1.57	1.39	1.98	2.17	2.37	3.80	2.37	3.09
% Kaolinite	40.23	41.44	40.62	40.18	35.67	42.17	48.67	45.94	56.92	51.43
% Calcite	3.52	2.00	3.72	5.54	32.61	20.62	8.64	3.62	2.73	3.17

Table 3 Compression strength values (MPa) of the different technological types

	PC	CA750	CA800	CA850	CB750	CB800	CB850	CC750	CC800	CC850
3 days	17.60	20.44	30.25	25.97	20.03	24.66	15.97[a]	24.59	26.91	15.59[a]
7 days	28.70	29.34	41.69	31.31	23.28[a]	34.81	28.38	28.59	36.25	23.03[a]
28 days	35.20	37.53	56.83	35.84	27.66[a]	44.63	30.72	35.16	45.62	25.03[a]

[a]Less than the standard values established for a PC P-35

at high temperatures of reaction products of kaolinite and calcite, typically vitreous phases rich in Ca, Si and Al and of very low specific surface area. In the case of technological type B, it is the one with the highest average calcite content. Technological types A and C differ significantly in terms of their kaolinite content and have similar average calcite contents. However, in the case of layer C, both mineral phases are intimately mixed, but not in technological type A, where kaolinite occurs in the form of relatively homogeneous layers. Therefore, the solid state reactions of the decomposition products of kaolinite and calcite are favored in technological types B (high calcite content) and especially in technological type C (high degree of contact between calcite and kaolinite).

In general, the observed trends are maintained at 7 and 28 days, with the lowest values of compressive strength compared to the PC series for the CC850, CB850 and CB750 series. For all technological types, the values of compressive strength were strongly influenced by the calcination temperatures, with the highest values achieved for calcined clays at 800 °C. In all cases, the highest resistance was obtained for the calcination products of layer A. The anomalous behavior of technological type C in relation to its content of clay minerals of the kaolinite group is attributed to the high degree of contact that exists in this layer between the calcite and the clay minerals of the kaolinite group.

4.3 *Evaluation of the Effect of Clays on the Hydration Kinetics of Portland Cement*

The results of heat evolution curves normalized to the PC content in the pastes, representative of the hydration kinetics, are presented in Fig. 2a. When comparing the behavior of the PC-calcined clay mixtures with the reference series, we observed that there is no appreciable change in the duration of the induction period, although in the slope of the acceleration period, which corresponds to a higher speed in the formation of hydration products, specifically CSH, since this first stage is dominated by the hydration of the PC alite to produce hydrated calcium silicates. During both, the induction period and the main reaction of the calcium silicates (alite), it is observed that the heat released normalized to the PC content in the paste is higher in the systems with calcined clay incorporation than in the reference series. This behavior is the result of the filler effect of the calcined clays, which can be divided into two components: the increase in the degree of hydration of the clinker at an early age due to the dilution effect (higher water/cement ratio effective in the mixed systems) and the contribution of the additional specific surface associated with the calcined clays that facilitates heterogeneous nucleation mechanisms for the formation of hydration products [2, 9].

The second and intense maximum in the hydration heat curve, which extends approximately from 5 to 15 h, corresponds to the reaction of the aluminates and the secondary formation of ettringite, product of the desorption of the sulphates adsorbed on the surface of hydration products. This effect is strongly influenced by the specific surface area of the reactants and by the availability of reactive alumina. The start of this reaction and the total accumulated heat associated with it were calculated from the calorimetric curves and are presented in Table 4, where it is evident that the accumulated total heat associated with the reaction of the aluminates follows the following order: CA < CB < CC. The greater intensity of this effect for the calcining products of Layer C is based on its relatively high content of kaolinite compared to other technological types, which could be a potentially negative factor in the development of mechanical strength if not accompanied of an adequate adjustment of the sulfates. Due to the high amount of reactive alumina provided by the calcined clay, the formation of type AFt and AFm phases interferes with the normal hydration of the PC, which could partially explain the relatively low mechanical strength of this series compared to its kaolinite content. For each technological type, as the calcination temperature of the sample increases, the start of the reaction of the aluminates is delayed and a slight decrease in the amount of heat released during said reaction occurs. This is related to the decrease of the specific surface area that occurs with the increase in the calcination temperature, product of the morphological changes that occur in the calcination [2, 9].

For times between 18 and 72 h (Fig. 2b), it is observed that the heat released normalized to the PC content is higher in systems with calcined clay incorporation with respect to the reference series. This area of the calorimetric curve is associated with the formation of phases AFm type, which are favored in the presence of calcined clays by the additional contribution of alumina reactive to the system [2].

4.4 Mixture Selection Criteria

From the results of previous research [2, 13], it is known that to obtain a material with high pozzolanic reactivity, the mineral content of the kaolinite group should not be less than 40% by mass. Based on the above, it is possible to estimate in what proportions the three technological types of clay must be mixed in such a way that a material with potentials for the production of MCS of high reactivity is obtained.

(a)

(b)

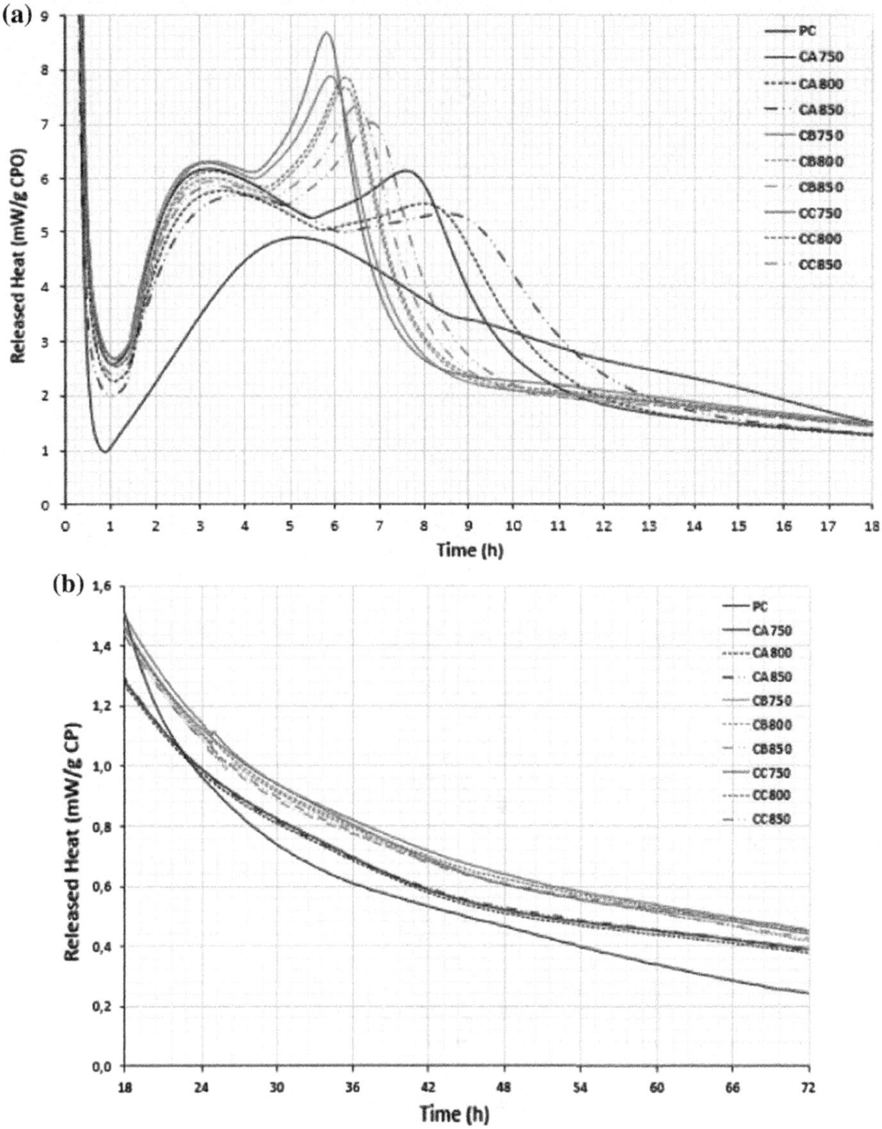

Fig. 2 **a** Released heat curves for the calcination products during the first 18 h. **b** Released heat curves for the calcination products for times between 18 and 72 h

Table 4 Total accumulative heat and apparent start time of the second thermal effect

	CA750	CA800	CA850	CB750	CB800	CB850	CC750	CC800	CC850
Total accumulative heat (mW/g PC)	129.03	126.07	127.68	145.48	143.38	140.6	151.25	145.26	139.38
Apparent start time	5.5	5.77	6.2	4.22	4.4	4.6	4.15	4.47	4.82

According to its content of kaolinite, the three technological types that make up the clay deposit can be mixed in any proportion; in all cases, a kaolinite content of more than 40% is obtained. To maximize the content of kaolinite, the limit criterion can be extended up to 45%. In this case, the maximum contents of the technological types A and B in the mixture should not be higher than 55 and 60% respectively, while the technological type C, richer in minerals of the kaolinite group, can be present in any proportion. The main limitation in the exploitation of the deposit is to minimize the calcite content in the mixture, while maximizing the use of each of the technological types and generating a minimum of rejection during the exploitation of the deposit, in correspondence with the criteria of exploitation of mineral resources and non-selective exploitation that govern mining activity in our country. In a recent study [10], established that the calcite content in the sample should not exceed 8%. Considering this criterion as a reference, it is proposed to limit the content of technological type B in the mixture to 25%, with the remaining 75% contributed in equal parts by technological types A (37.5%) and C (37.5%). In this way, a mixture of the three technological types with a mineral content of the group of kaolinite not less than 45% and calcite not higher than 8% is obtained.

5 Conclusions

1. The Yaguajay clay deposit is formed by technological types A, B and C, each with different lithology. In all, kaolinite predominates as clay mineral and quartz, calcite and goethite as the main non-clay minerals. The mineral content of the kaolinite group ranges from 40% (A and B) to 50% (C). The contents of calcite are generally less than 5%, except in technological type B, with 20%. The contents of iron-rich phases increase progressively, according to the sequence A < B < C.
2. In all cases, the pozzolanic reactivity of the calcination products is strongly influenced by the calcination temperature, reaching the best reactivity at 800 °C and the lowest at 850 °C. Other factors that can negatively affect pozzolanic reactivity are the high calcite content (technological type B) and the high degree of contact between calcite and kaolinite (technological type C). The highest reactivity is reached for the calcination products of technological type A.
3. For a sustainable exploitation of the deposit, it is recommended that the composite sample destined to industrial production have an approximate composition of 37.5% Type A, 25% Type B and 37.5% Type C.
4. The calcination products of the different technological types affect the hydration kinetics of the PC, acting predominantly as filler at early ages and significantly increasing the intensity of the hydration reactions of the aluminates.

References

1. Martirena, J.: Una alternativa ambientalmente compatible para disminuir el consumo de aglomerantes de clínker de emento Pórtland: el aglomerante cal-puzolana como adición mineral activa (2009)
2. Alujas, A.: Obtención de un material puzolánico de alta reactividad a partir de la activación térmica de una fracción arcillosa multicomponente. Ingeniería Civil Universidad Central "Marta Abreu" de Las Villas Santa Clara (2010)
3. Vizcaíno, L., Sánchez, S., Pérez, A., Damas, S., Scrivener, K., Martirena, F.: Industrial trial to produce low clinker, low carbon cement. Materiales de Construcción **65** (2015)
4. C. d. Autores: Instructivo técnico. Prueba industrial de producción de cemento de bajo carbono LC^3. C. D. I. Y. D. D. E. Y. M. (CIDEM), p. 17 (2018)
5. NC95:2011: CEMENTO PORTLAND—ESPECIFICACIONES (2011)
6. NC506:2013: CEMENTO HIDRÁULICO—MÉTODO DE ENSAYO—DETERMINACIÓN DE LA RESISTENCIA MECÁNICA (2013)
7. NC-TS527:2013: CEMENTO HIDRÁULICO—MÉTODOS DE ENSAYO—EVALUACIÓN DE LAS PUZOLANAS (2013)
8. NC-TS528:2013: CEMENTO HIDRÁULICO—PUZOLANAS—ESPECIFICACIONES (2013)
9. Almenares, R.S.: Potencialidades de arcillas caoliníticas cubanas para ser empleadas como fuente de materia prima en la obtención de materiales cementicios suplementarios. Ingeniería Civil Universidad Central "Marta Abreu" de Las Villas Santa Clara (2017)
10. Zunino, F., Scrivener, K.: Assessing the effect of calcite impurities in clay on optimal dehydroxylation parameters for enhanced reactivity. In: Calcined Clays for Sustainable Concrete, RILEM Bookseries, vol. 16, p. 7 (2018)
11. Todor, D.N.: Thermal Analysis of Minerals. Abacus Press, Kent, Inglaterra (1976)
12. Cyr, M., Lawrence, P., Ringot, E.: Efficiency of mineral admixtures in mortars: quantification of the physical and chemical effects of fine admixtures in relation with compressive strength. Cem. Concr. Res. **36**, 264–277 (2006)
13. Avet, F., Snellings, R., Alujas, A., Ben, M., Scrivener, K.: Development of a new rapid, relevant and reliable (R3) test method to evaluate the pozzolanic reactivity of calcined kaolinitic clays. Cem. Concr. Res. **85**, 1–11 (2016)

Mechanical Strength Analysis of Ternary Cement Pastes Containing Nanosilica and Metakaolin

Matheus Ian Castro Sousa and João Henrique da Silva Rêgo

Abstract Supplementary cementitious materials' (SCM) usage in cement based composites is increasing in construction, aiming to improve performance, sustainability and cost of those materials. Among those SCM's, Metakaolin (MK) and Nanosilica (NS) have been studied by many authors with interesting results, usually an increase in mechanical strength and microstructure refinement. However, the combined use of those two in a ternary mixture with Portland cement was studied by few authors. The present work aims to evaluate this type of mixture through an analysis of compressive strength of 6 different mix proportions. The mix proportions used were a reference paste with 100% Portland cement, two pastes with 1 and 2.5% replacement of clinker by Colloidal Nanosilica and three pastes with 15% replacement of clinker for complementary contents of NS and MK (0 + 15%, 1 + 14% and 2.5 + 12.5%). The compressive strength tests were performed at 1, 3, 7 and 28 days of hydration. The results were analyzed by statistical methods, such as ANOVA test and Duncan Test. At early ages, the use of nanosilica increased the compressive strength. The ternary mixtures presented the highest strength values at 7 and 28 days, and a synergistic effect between these SCM's was observed at 7 and 28 days, for the 12.5MK2.5NS paste.

Keywords Metakaolin · Nanosilica · "Ternary mixtures"

1 Introduction

Sustainability is a growing concern in cement production, since the industry is responsible for about 8% of total CO_2 emissions worldwide, with 5% produced as a direct result of the decarbonation process, and 3% from the fuels used [1]. For that reason, one strategy to reduce CO_2 emissions is to use Supplementary cementitious materials

M. I. C. Sousa (✉) · J. H. da Silva Rêgo
Postgraduate Program in Structural Engineering and Construction, University of Brasília, Campus Universitário Darcy Ribeiro, 70, 910-900 Brasília, DF, Brazil
e-mail: matheusiancs@gmail.com

© RILEM 2020
J. F. Martirena-Hernandez et al. (eds.), *Proceedings of the International Conference of Sustainable Production and Use of Cement and Concrete*, RILEM Bookseries 22,
https://doi.org/10.1007/978-3-030-22034-1_7

(SCM's) as a replacement for Portland Cement Clinker. The use of these materials in concrete can also significantly increase its technical characteristics [2].

Among the SCM's, metakaolin and silica fume are widely used in high-performance concretes, improving concrete's mechanical strength and durability. These effects occur because both materials are highly reactive, due to their high pozzolanic activity [3]. Furthermore, with the development of nanomaterials for construction, the use of nanosilica as an SCM has increased. When used in cementitious composites, these SiO_2 nanoparticles contribute to the formation of a denser, more compact microstructure, when compared with other SCM's, such as Silica Fume [4].

Metakaolin's influence on cement hydration occurs due to two distinct effects. The first is the filler effect, which increases hydration speed, and the second is the chemical effect, consequence of the pozzolanic reaction. The filler effect is mainly a consequence of the nucleation effect, since Metakaolin is a finer powder than cement [5]. The pozzolanic reaction occurs between SiO_2 from the SCM and the CH from the cement paste to form C-S-H. However, when the SCM is rich in aluminum, as Metakaolin, there is an incorporation of aluminum in the C-S-H, forming the C-A-S-H [3]. These changes in the hydration affect the properties of the cementitious material, including compressive strength. Many authors [6–11] have studied the compressive strength of cements incorporating different metakaolin content. The results showed a consistent increase in strength at later ages (28–91days), while at early ages, the results were dependent on the metakaolin content. For higher Metakaolin content (10–20%), the strength values decreased. This occurred because at early ages, the pozzolanic reaction of metakaolin isn't significant yet, and the filler effect can't compensate the decrease in clinker content that is hydrating [8, 11].

Nanosilica has some of the same effects in cement hydration; however, as it is composed of much smaller particles and has a higher surface area, those effects are increased. The hydration reaction is faster because of the filler effect and pozzolanic reaction is observed at earlier ages [12, 13]. Many studies [14–16] have observed the compressive strength of cementitious materials containing nanosilica. At early ages (1–3 days), the strength was significantly higher, but at later ages, results diverged, some showing an increase in strength, others showing values similar to the reference.

The combined use of nanosilica and metakaolin, forming a ternary mixture with cement, has been topic of recent studies [11, 17], with promising results. Strength values of the ternary mixture were higher than reference paste at all ages, while at later ages (28–91 days), strength values were also higher than the binary mixtures, containing either nanosilica or metakaolin. This increase suggests a synergistic effect between the two SCM [11], and the present work intends to complement the study of this ternary mixture.

2 Materials and Methods

2.1 Materials

The following materials were used in this work:

- Cement Type I (CPI), according to the requirements of Brazilian Standard [18], produced by Votorantim Cimentos;
- Metakaolin (MK), produced by Metacaulim do Brasil;
- Colloidal Nanosilica (NS) with 30% of solid content, produced by Akzonobel;
- Superplasticizer (SP) Visconcrete 6900, based on a polycarboxylate solution with 46% solid content, produced by Sika;
- Potable water.

2.2 Cement Paste Composition and Preparation

In order to analyze the influence of clinker replacement by MK and NS, six different pastes were produced. Their composition is shown in Table 1. The Metakaolin content of 15% was chosen because it yielded the best compressive strength results in a study using the same material [11]. The ternary mixtures were composed with the same clinker replacement content (15%), so MK and NS contents were complementary, adding up to that value. This composition was chosen to compare pastes with the same clinker content, allowing for a better analysis of the synergistic effect in the Ternary mixtures. Water/binder ratio was fixed at 0.40, and the superplasticizer was dosed to maintain the same consistency (94 $\pm$ 4 mm), measured by the mini-slump test. This value was found in the literature as ideal for molding the pastes [19]. Water content in the mixture was adjusted due to water content in the Colloidal Nanosilica and the superplasticizer.

It is worth remarking that the superplasticizer content needed to maintain the same consistency increased with the use of SCM's. This increase was more pronounced when NS was used. This is explained by the higher surface area of these materials. In all pastes, the maximum SP content prescribed by the manufacturer was respected.

The pastes were produced in a planetary mortar mixer. The mixing procedure was to add first the water/NS/SP, and then the CPI/MK mix. This mix was homogenized by 60 s with the mixer's slow rotation, and 90 s with the fast rotation. The paste produced was used to mold cylindrical test specimens, with 50 mm diameter and 100 mm height. The specimens were cured in a humid chamber for 24 h, demolded, and kept immersed in lime saturated water solution until the testing age. The curing process followed the Brazilian Standard [20].

Table 1 Composition of cement pastes

Pastes	Abbreviation	CPI (g)	MK (g)	NS (g)	SP (%)	Water (g)
100%CPI	REF	2400	–	–	0.17	957.83
99%CPI + 1%NS	1NS	2376	–	80	0.60	896.34
97%CPI + 2.5%	2.5NS	2340	–	200	1.30	803.15
85%CPI + 15%MK	15MK	2040	360	–	0.69	951.20
85%CPI + 14%MK + 1%NS	14MK1NS	2040	336	80	1.10	889.94
85%CPI + 12.5%MK + 2.5%NS	12.5MK2.5NS	2040	300	200	1.77	797.06

2.3 *Test Methods*

Compressive Strength and Statistical Analysis

The compressive strength tests were performed with three specimens at the ages of 1, 3, 7 and 28 days of hydration. A universal equipment for mechanical testing was used, following the procedures of the Brazilian Standard [20].

The results were than analyzed by the ANOVA test, performed with the software Statistica 10, in order to confirm the significance of the paste composition in compressive strength. The significance can be concluded with the p-value. For a given confidence level (0.05 was chosen for the present work), the difference between two group means is significant if the p-value is equal or lower than the confidence level.

Furthermore, the Duncan Test was performed to determine the homogenous groups, considering the significance level of 0.05. This was used to classify the different pastes, at each age. The statistical analysis was performed separately for each age.

3 Results and Discussion

3.1 *Compressive Strength*

The compressive strength results, for all pastes and ages, are presented in Table 2 and the strength development over time is shown in Fig. 1.

At the early ages (1 and 3 days), the highest strength values were observed in pastes containing NS (1NS and 2.5NS), which is consistent with the literature. For the 2.5NS paste, there was an increase of 42 and 66% in strength at 1 and 3 days, respectively, when compared to the REF paste. These pastes had compressive strengths

Table 2 Mean compressive strength of pastes at 1, 3, 7 and 28 days of hydration

Paste	1 day (MPa)	3 days (MPa)	7 days (MPa)	28 days (MPa)
REF	23.0	35.8	46.0	59.1
1NS	30.4	54.1	58.1	60.0
2.5NS	32.7	59.4	64.0	65.4
15MK	16.1	36.8	42.9	61.1
14MK 1NS	22.2	41.5	49.9	67.3
12.5MK 2.5NS	27.7	41.7	68.0	79.2

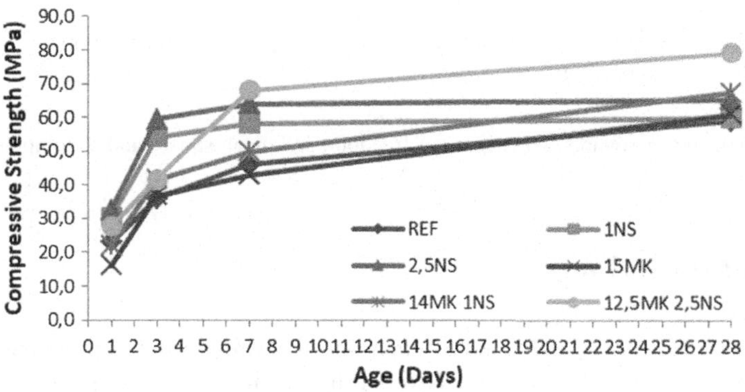

Fig. 1 Strength development of cement pastes over time

considerably higher than the REF paste at 7 days, by 26 and 39% (1NS and 2.5NS respectively), but had results closer to the reference at 28 days.

The MK15 paste had the lowest strength at 1 day, 30% lower than the reference, and at later ages, was higher than the REF only at 3 and 28 days, by a small margin of 3%. According to the literature, the result at early ages was compatible, but at 28 days, higher strength values were expected, when compared to ordinary Portland cement pastes [6–11].

In the ternary pastes, it was observed a different behavior, when comparing the 12.5MK2.5NS paste with the 14MK1NS. The first had higher strength values than the REF paste at all ages, and was the highest overall at 7 and 28 days (48 and 34% higher than REF at those ages, respectively). The second, however, had lower strength at 7 days, than both pastes with only NS.

These results may indicate that there is, in fact, a synergistic effect between the two SCM's, but this effect was only perceptible when there was a higher NS/MK ratio, with the same clinker content, and after 7 days. The 12.5MK2.5NS paste has a 0.2 NS/MK ratio, while the 14MK1NS has a ratio of 0.07. For the first, the increase in strength far surpassed those of the sum of the binary mixtures with the same replacement content, both at 7 and 28 days.

Table 3 Statistical evaluation of compressive strength of pastes

Source of variation	Parameters	SS	DF	MS	F	*p*-value	Significance
Compressive strength 1 day	Paste composition	556.11	5	111.22	113.97	0.00000	Yes
Compressive strength 3 days	Paste composition	1392.97	5	278.59	23.66	0.00001	Yes
Compressive strength 7 days	Paste composition	1537.90	5	307.58	24.66	0.00001	Yes
Compressive strength 28 days	Paste composition	842.40	5	168.48	11.86	0.00026	Yes

SS sequential sum of squares, *DF* degree of freedom, *MS* mean square and *F* computed value of F-distribution

3.2 *Statistical Analysis*

The ANOVA test was performed at each age to evaluate the significance that the paste compositions had on compressive strength. The results are shown in Table 3. From the results, since all *p*-values were lower than the confidence level, it can be concluded that the paste composition was significant at all ages.

The Duncan test was used to evaluate the different homogeneous groups between cement pastes at each age. The results are presented in Table 4.

At 1 day of hydration, most pastes were statistically different, with the 2.5NS paste isolated as the highest strength group. At 3 days, there are only two homogenous groups. The highest strength group was composed of the binary NS pastes. At 7 days, there were 4 groups, and the highest strength one had the pastes with 2.5 NS, both binary and ternary, but the binary paste was also classified in the second highest group, differing from the ternary. Finally, at 28 days, there were three groups and the only paste in the highest strength group was the ternary 12.5MK2.5NS.

This test confirms that the binary pastes with nanosilica had the highest strength at 1 and 3 days with statistical difference, and that the ternary paste 12.5MK2.5NS had the highest strength at 7 and 28 days.

4 Conclusions

The following conclusions may be drawn from the test results presented in this work:

- Superplasticizer content needed to maintain consistency was higher when SCM's were used;

Table 4 Homogeneous groups by Duncan Test for the cement pastes

Age	Paste	Mean compressive strength (MPa)	Standard deviation (MPa)	G1	G2	G3	G4	G5
1 day	REF	23.0	0.4876		X			
	1NS	30.4	0.9186				X	
	2.5NS	32.7	1.2927					X
	15MK	16.1	0.4235	X				
	14MK + 1NS	22.2	0.2795		X			
	12.5MK + 2.5NS	27.7	1.6867			X		
3 days	REF	35.8	1.7764	X				
	1NS	54.1	4.5446		X			
	2.5NS	59.4	4.8440		X			
	15MK	36.8	0.2430	X				
	14MK + 1NS	41.5	3.8347	X				
	12.5MK + 2.5NS	41.7	2.9346	X				
7 days	REF	46.0	1.1272	X	X			
	1NS	58.1	1.7725			X		
	2.5NS	64.0	4.4836			X	X	
	15MK	42.9	4.0870	X				
	14MK + 1NS	49.9	4.0804		X			
	12.5MK + 2.5NS	68.0	4.1190				X	
28 days	REF	59.1	4.2810	X				
	1NS	60.0	3.2358	X				
	2.5NS	65.4	3.5933	X	X			
	15MK	61.1	5.1058	X	X			
	14MK + 1NS	67.3	0.9816		X			
	12.5MK + 2.5NS	79.2	4.0627			X		

The groups are different at each age

- The use of nanosilica in binary mixtures significantly increased compressive strength at early ages and yielded equal (1NS) or higher (2.5NS) strengths at 28 days of hydration, when compared with the reference paste;
- Binary mixtures with metakaolin had lower performance than the reference at 1 day, and nearly similar performance at later ages, regarding compressive strength;

- Ternary mixtures had the highest strength at 28 days. The results also indicated that the synergistic effect between these SCM's occurred, but only in the paste with higher NS/MK ratio and after 7 days.

References

1. Andrew, R.M.: Global CO_2 emissions from cement production, 1928-2017. Earth Syst. Sci. Data **10**, 2213–2239 (2018)
2. Hou, P., Kawashima, S., Kong, D., et al.: Modification effects of colloidal nanoSiO2 on cement hydration and its gel property. Compos. Part B Eng. **45**, 440–448 (2012)
3. Juenger, M.C.G., Siddique, R.: Recent advances in understanding the role of supplementary cementitious materials in concrete. Cem. Concr. Res. **78**, 71–80 (2015)
4. Singh, L.P., Karade, S.R., Bhattacharyya, S.K., et al.: Beneficial role of nanosilica in cement based materials —a review. Constr. Build. Mater. **47**, 1069–1077 (2013)
5. Lothenbach, B., Scrivener, K., Hooton, R.D.: Supplementary cementitious materials. Cem. Concr. Res. **41**, 1244–1256 (2011)
6. Poon, C.S., Kou, S.C., Lam, L.: Compressive strength, chloride diffusivity and pore structure of high performance metakaolin and silica fume concrete. Constr. Build. Mater. **20**, 858–865 (2006)
7. Güneyisi, E., Gesoğlu, M., Karaoğlu, S., Mermerdaş, K.: Strength, permeability and shrinkage cracking of silica fume and metakaolin concretes. Constr. Build. Mater. **34**, 120–130 (2012)
8. Mermerdaş, K., Gesoğlu, M., Güneyisi, E., Özturan, T.: Strength development of concretes incorporated with metakaolin and different types of calcined kaolins. Constr. Build. Mater. **37**, 766–774 (2012)
9. Borosnyói, A.: Long term durability performance and mechanical properties of high performance concretes with combined use of supplementary cementing materials. Constr. Build. Mater. **112**, 307–324 (2016)
10. Dadsetan, S., Bai, J.: Mechanical and microstructural properties of self-compacting concrete blended with metakaolin, ground granulated blast-furnace slag and fly ash. Constr. Build. Mater. **146**, 658–667 (2017)
11. Andrade, D.S., Rêgo, J.H.S., Morais, P.C., Rojas, M.F.: Chemical and mechanical characterization of ternary cement pastes containing metakaolin and nanosilica. Constr. Build. Mater. **159**, 18–26 (2018)
12. Shah, S.P., Bhattacharyya, S.K., Mishra, G., et al.: Studies on early stage hydration of tricalcium silicate incorporating silica nanoparticles: Part II. Constr. Build. Mater. **102**, 943–949 (2015)
13. Ji, T.: Preliminary study on the water permeability and microstructure of concrete incorporating nano-SiO_2. Cem. Concr. Res. **35**, 1943–1947 (2005)
14. Shih, J.Y., Chang, T.P., Hsiao, T.C.: Effect of nanosilica on characterization of Portland cement composite. Mater. Sci. Eng., A **424**, 266–274 (2006)
15. Tobón, J.I., Payá, J.J., Borrachero, M.V., Restrepo, O.J.: Mineralogical evolution of Portland cement blended with silica nanoparticles and its effect on mechanical strength. Constr. Build. Mater. **36**, 736–742 (2012)
16. Gaitero, J.J., Campillo, I., Guerrero, A.: Reduction of the calcium leaching rate of cement paste by addition of silica nanoparticles. Cem. Concr. Res. **38**, 1112–1118 (2008)
17. Jamsheer, A.F., Kupwade-Patil, K., Büyüköztürk, O., Bumajdad, A.: Analysis of engineered cement paste using silica nanoparticles and metakaolin using 29Si NMR, water adsorption and synchrotron X-ray Diffraction. Constr. Build. Mater. **180**, 698–709 (2018)

18. ASSOCIAÇÃO BRASILEIRA DE NORMAS TÉCNICAS, NBR 5732: Cimento Portland comum [Portland Cement—Specification], Rio de Janeiro (1991)
19. Kantro, D. L.: Influence of water-reducing admixtures on properties of cement paste—a miniature slump test. Cement Concrete Aggregate **2**(2), 95–102 (1980)
20. ASSOCIAÇÃO BRASILEIRA DE NORMAS TÉCNICAS, NBR 7215: Cimento Portland - Determinação da resistência à compressão [Portland Cement—Determination of Compressive Strength], Rio de Janeiro (1997)

Hydration of Cement Pastes Using the Cement LC³

Myrelle Y. F. Câmara, Yuri S. B. Fraga and Valdirene M. S. Capuzzo

Abstract Nowadays, cement is still one of the most consumed products in the world, and has grown more significantly. Thus, the use of supplementary cementitious materials (SCMs) to some extent reduces the carbon dioxide (CO_2) emissions in the environment, contributing to the formation of a sustainable cement. In this context, Limestone Calcined Clay Cement (LC^3) has a mechanical performance similar to the ordinary Portland cement (OPC). This work aims to analyze the evolution of the hydration of the LC^3 cement in comparison with the reference cement, according to the existing materials in Brazil. For this, the characterization of Portland cement of high initial resistance (CP V-ARI), metakaolin, and limestone filler was carried out. Pastes with a 45% substitution of Portland cement by these SCMs were prepared in a ratio of 2:1, with 30% of the metakaolin and 15% of the limestone filler. Pastes were prepared according to ABNT NBR 16606 [8] in water/binder ratio of 0.40. To follow the hydration, the X-ray diffraction (XRD) assays were performed, at the ages of 1, 3 and 7 days. The results indicated a change in the formation of the hydration products of the LC^3 cement compared to the reference paste over the ages.

Keywords LC^3 cement · Portland cement · Hydration of cement

1 Introduction

In the current scenario, Portland cement still presents itself as one of the most consumed products on the planet. Thus, this material has been the subject of several researches in order to reduce the emissions of carbon dioxide (CO_2) in the environment and the consumption of natural resources.

The most viable option for this problem would be the use of supplementary cementitious materials (SCMs). Among the materials already studied, calcined clays in combination with limestone (LC^3) are promising compared to blast furnace slag and

M. Y. F. Câmara (✉) · Y. S. B. Fraga · V. M. S. Capuzzo
Post-Graduation Program in Structures and Civil Construction,
University of Brasília, Brasília, Brazil
e-mail: myrelleyasmine@hotmail.com

© RILEM 2020

J. F. Martirena-Hernandez et al. (eds.), *Proceedings of the International Conference of Sustainable Production and Use of Cement and Concrete*, RILEM Bookseries 22,
https://doi.org/10.1007/978-3-030-22034-1_8

fly ash, which are not present in sufficient quantity to meet the growth of cement production in the coming years [14].

In view of the above, this article aims to study the evolution of hydration of this cement by means of X-ray diffraction to understand the increase of strength even with the low content of clinker.

2 Methodology

2.1 *Materials*

The following materials were used for the preparation of the cement pastes: initial high strength Portland cement (CP V-ARI), metakaolin, limestone filler, and water.

2.2 *Methods*

2.2.1 Characterization of Materials

Portland Cement

Characterization of this material was performed by the manufacturer. The physical, chemical and mechanical analyzes are presented in Tables 1, 2 and 3.

Supplementary Cementitious Materials

The results of characterization of the supplementary cementitious materials were presented in Tables 4 and 5. A similarity was observed in the chemical analyzes between the materials, but with differences in the chemical analyzes.

Mineralogical Analysis by X-Ray Diffraction

The X-ray diffraction analysis was used to verify the evolution of the hydration of the cementitious pastes, allowing the identification of crystalline compounds.

The equipment used was the RIGAKU–ULTIMA IV diffractometer, by the non-oriented powder method. The tube used was copper (CuKα) and nickel filter, under voltage of 30 kV and amperage of 15 mA. A monochromator with scanning from 2° to 60°, with step 2θ of 0.05° per minute was used.

Table 1 Physical characterization of Portland cement

Test	Methodology	Unit	Standard deviation	Result	Regulatory requirements
Blaine	ABNT NBR 16372	cm^2/g	108	4.684	Not applicable
Specific mass	ABNT NBR 16605	g/cm^3	–	3.12	Not applicable
Fineness #75 μm	ABNT NBR 11579	%	0.0	0.1	≤6.0
Fineness #45 μm	ABNT NBR 9202	%	0.5	0.4	Not applicable
Water of paste consistency	ABNT NBR 16606	%	0.3	30.1	Not applicable
Initial setting time	ABNT NBR 16607	min	21	154	≥60
Final setting time		min	21	214	≤600
Expandability to hot	ABNT NBR 11582	min	–	4	≤5

Table 2 Chemical characterization of Portland cement

Test	Methodology	Unit	Standard deviation	Result	Regulatory requirements
Insoluble residue—IR	ABNT NBR NM 15	%	0.4	1.9	≤3.5
Loss of ignition—LOI	ABNT NBR NM 18	%	0.4	4.0	≤6.5
Magnesium oxide—MgO	ABNT NBR 14656	%	0.2	1.5	≤6.5
Sulfur trioxide—SO_3		%	0.2	3.2	≤4.5
Sodium oxide—Na_2O		%	0.0	0.07	Not applicable
Potassium oxide—K_2O		%	0.04	0.99	Not applicable
Alkaline equivalent—Na_2O_{eq}	$Na_2O + 0.658*K_2O$	%	0.02	0.72	Not applicable

Table 3 Mechanical characterization of Portland cement

Test	Methodology	Unit	Standard deviation	Result	Regulatory requirements
Compressive strength—1 day	ABNT NBR 7215	MPa	0.6	29.4	≥14.0
Compressive strength—3 days		MPa	1.5	41.7	≥24.0
Compressive strength—7 days		MPa	1.8	46.6	≥34.0
Compressive strength—28 days		MPa	2.1	55.7	Not applicable

Table 4 Physical characterization of SCMs

Property	Test method	Limestone filler	Metakaolin
Specific Mass (g/cm^3)	ABNT NBR NM 23 [3]	2.7	2.6
D10 (μm)	Laser granulometry	1.88	1.77
Mean Diameter (μm)		19.94	18.27
D90 (μm)		43.86	39.36

Table 5 Chemical characterization of SCMs

Chemical compost	% in mass	
	Limestone filler	Metakaolin
SiO_2	50.54	6.63
CaO	0.00	83.10
Al_2O_3	42.28	2.12
Fe_2O_3	3.23	2.16
TiO_2	1.63	0.26
K_2O	1.07	0.94
MgO	1.02	3.61
SO_3	0.09	0.23
ZrO_2	0.03	0.00
Cr_2O_3	0.03	0.00
SrO	0.00	0.96

3 Results and Discussion

3.1 Mineralogical Analysis by X-Ray Diffraction

Figures 1 and 2 show the diffractograms obtained from the reference paste and the LC^3 paste, respectively. In the diffractograms we have identified the corresponding peaks and the main phases found, which are: calcite, quartz, ettringite, portlandite, larnite, calcium silicate, calcium aluminate, calcium hydrogen silicate, katoite, glaucofan, gordonite and nacrite.

In the reference samples, it can be seen that there was not much variation between the peaks in relation to their location. However, a change in the intensity of the peaks was observed, which was already expected, due to the advance of the hydration of the pastes.

The peaks of the portlandite, represented by the interplanar distance of 4915 Å, decreased with increasing ages. This decrease indicates that it is still in the formation phase of the hydration products. It also realizes that although in these pastes there is no substitution for some additional cementitious material, the peaks of calcite have increased. This calcite can come from the composite limestone filer in the cement. The alita and belite peaks, which are characterized by having structure of the larnite mineral, remained at similar intensities, throughout the ages.

In the diffractogram of the samples of the LC^3 cement, we noticed that the portlandite peak was decreasing along the ages, indicating the frequent pozzolanic reac-

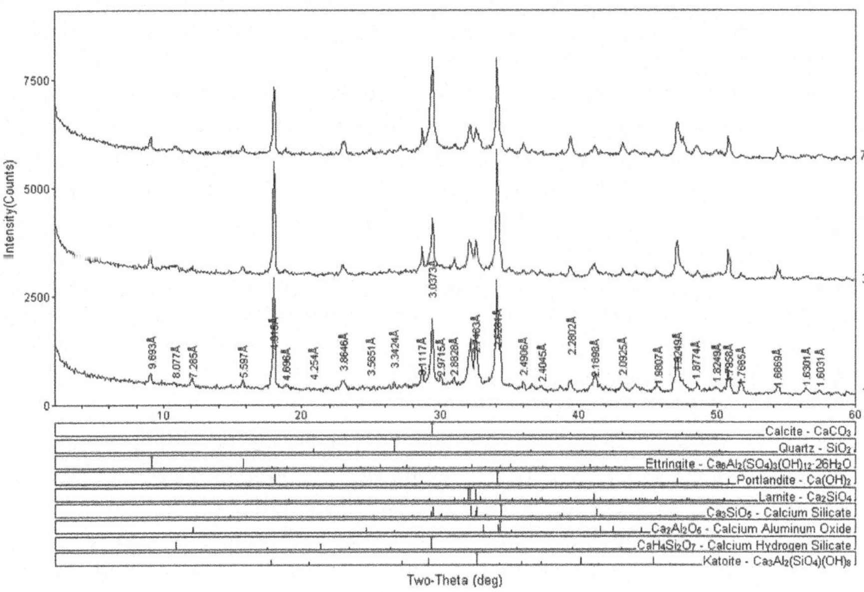

Fig. 1 Diffratogram of refrence paste at 1, 3 and 7 days

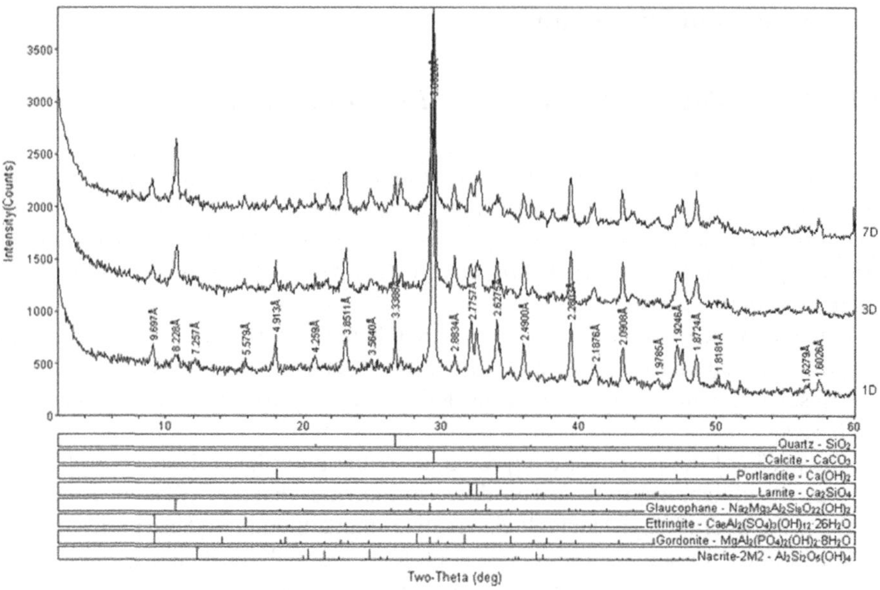

Fig. 2 Diffratogram of 30MK15FC paste at 1, 3 and 7 days

tion in advanced ages, since the hydration was paralyzed at 7 days. The peak of the calcite, in turn, presented the highest intensity, due to the additions have limestone in its composition. The alite and belite peaks decreased on a smaller scale than the others, indicating that the additions influence the rate of reaction of the silicates, especially the belite that reacts at advanced ages.

In the samples with SCMSs, it was observed that the quartz peak had a higher intensity, this being from the addition of metakaolin to the mixture. He also noticed the tendency of ettringite in all the samples. This hydration product may persist indefinitely or may be replaced by calcium monosulfate (AFm phase), which was not identified in the samples. The AFm phase may not have been identified in samples containing additions, because the latter appears at later ages.

In addition to forming the hydration products in the pastes, metakaolin forms the hydrated calcium monosulfuruminate (3CaO Al2O_3 $CaSO_4$ 12H_2O) and the hydrated calcium hemicarboaluminate (3CaO Al_2O_3 0.5$Ca(OH)_2$ 0.5$CaCO_3$ 11.5H_2O). These compounds have not been identified in the pastes, however, it is understood that they are the minor peaks.

According to tests performed by Antoni et al. [1], the period of higher formation of carboaluminates comprises between 28 and 90 days, with the most intense peaks at the advanced ages. According to Durdzínski et al. [15] in this range the amount of AFt generated is higher in the LC^3 cement compared to the reference sample because the limestone reacts with the additional alumina and avoids the dissolution of the ettringite to give the amorphous monosulfate phase.

4 Conclusions

According to the study, it is concluded that the use of SCMs is interesting for the development of new civil construction products. Regarding the behavior of the reference pastes and with substitution, it was evident that the calcium hydroxide was consumed with the advancement of the ages. This behavior is characteristic of the pozzolanic reaction, to form hydrated calcium silicate (C-S-H). In the presented results, there was a reduction in the calcium hydroxide content, and is a great problem in the use of the cement LC^3. This effect, may facilitate carbonation and compromising the use in structures of reinforced concrete.

References

1. Antoni, M., et al.: Cement substitution by a combination of metakaolin and limestone. Cem. Concr. Res. **42**(12), 1579–1589 (2012)
2. ABNT, NBR 16372: Cimento Portland e outros materiais em pó - Determinação da finura pelo método de permeabilidade ao ar (método Blaine), Rio de Janeiro (2015)
3. ABNT, NBR NM 23: Cimento Portland e outros materiais em pó - Determinação da massa específica, Rio de Janeiro (2001)
4. ABNT, NBR 11579: Cimento Portland – Determinação do índice de finura por meio da peneira 75 m (n° 200). Rio de Janeiro ABNT (2013)
5. ABNT, NBR 11582: Cimento Portland - Determinação da expansibilidade de Le Chatelier, Rio de Janeiro (2016)
6. ABNT, NBR 14656: Cimento Portland e matérias-primas - Análise química por espectrometria de raios-X - Método de ensaio. Rio de Janeiro (2001)
7. ABNT, NBR 16605: Cimento Portland e outros materiais em pó - Determinação da massa específica. 1 ed. Rio de Janeiro: ABNT (2017)
8. ABNT, NBR 16606: Cimento Portland - Determinação da pasta de consistência normal. Rio de Janeiro (2017)
9. ABNT, NBR 16607: Cimento Portland – Determinação do tempo de pega, utilizando o aparelho utilizando o aparelho de Vicat. Rio de Janeiro (2017)
10. ABNT, NBR 7215: Cimento Portland – Determinação da Resistência à Compressão, Rio de Janeiro (1997)
11. ABNT, NBR NM 15: Cimento Portland - Análise química - Determinação do resíduo insolúvel, Rio de Janeiro (2004)
12. ABNT, NBR NM 18: Cimento Portland - Análise química - Determinação de perda ao fogo. Rio de Janeiro (2004)
13. ABNT, NBR 9202: cimento portland e outros materiais em pó: determinação da finura por meio da peneira 0,044 mm (nº 325) - Método de ensaio. Rio de Janeiro (1985)
14. Avet, F., Scrivener, K.: Hydration study of limestone calcined clay cement (LC^3) using various grades of calcined kaolinitic clays. In: Rilem Bookseries, pp. 35–40 (2017)
15. Durdzínski, P.T., et al.: Phase assemblage of composite cements. Cem. Concr. Res. **99**, 172–182 (2017)

Potentialities of Different Materials from the Eastern Region of Cuba as Partial Cement Substitutes: A Comparison of Their Performance

Roger S. Almenares Reyes, Adrián Alujas Díaz, Carlos A. Leyva Rodríguez, Lisandra Poll Legrá, Yosbel Guerra González, Luis A. Pérez García, Sergio Betancourt Rodríguez and Jose Fernando Martirena-Hernandez

Abstract The prospects for the use of supplementary cement materials depend to a large extent on their potential and their availability. In this work, the potentialities of three materials (calcined clays, zeolitized tuffs, vitreous tuffs) as sources of supplementary cementitious materials are presented. Each material was characterized from the chemical and mineralogical point of view. The reactivity was determined in standardized mortars and their behavior was assessed in mixed cements and hollow concrete blocks. Based on the results obtained, a comparison between the performances was made. The best results were shown by the calcined clay, followed by the zeolitized tuff and vitreous tuff, which corresponds to the reactivity of each material.

Keywords Calcined clay · Natural pozzolans · Volcanic tuff

R. S. Almenares Reyes (✉) · L. Poll Legrá · Y. Guerra González
Departamento de Metalurgia – Química, Instituto Superior Minero Metalúrgico de Moa, 83330 Moa, Cuba
e-mail: ralmenares@ismm.edu.cu

A. Alujas Díaz
Centro de Estudios de Química Aplicada, Universidad Central de Las Villas, 54830 Santa Clara, Cuba

C. A. Leyva Rodríguez · L. A. Pérez García
Departamento de Geología, Instituto Superior Minero Metalúrgico de Moa, 83330 Moa, Cuba

S. Betancourt Rodríguez
Facultad de Construcciones, Universidad Central de Las Villas, 54830 Santa Clara, Cuba

J. F. Martirena-Hernandez
Centro de Investigación y Desarrollo de Estructuras y Materiales, Universidad Central de Las Villas, 54830 Santa Clara, Cuba

© RILEM 2020

J. F. Martirena-Hernandez et al. (eds.), *Proceedings of the International Conference of Sustainable Production and Use of Cement and Concrete*, RILEM Bookseries 22,
https://doi.org/10.1007/978-3-030-22034-1_9

1 Introduction

The partial replacement of clinker by Supplementary Cementitious Materials (SCMs) has become one of the most effective ways to reduce CO_2 emissions to the environment and increase cement production [1]. In some regions, the demand for SCMs outstrips supply, particularly in those regions with projected development of their infrastructure.

The main known additions are blast-furnace slag, fly ash, silica fume, calcined clays, natural pozzolans and limestone [2]. Clays and volcanic tuffs are widely available in the eastern region of Cuba [3]. These materials are considered an encouraging alternative to implement as local production. However, not all clays and volcanic rocks are suitable as SCMs.

Given the diversity of SCMs in the eastern region and the growing interest in their use. The aim of this paper is to provide, under a simple approach the potentialities of three supplementary cementitious materials through characterization and comparison of their performance in cements and hollow concrete blocks.

2 Materials and Methods

2.1 Materials Selection and Sample Preparation

Clay material, zeolitized tuff and vitreous tuff were selected. The three materials of greater perspectives for the production of cementitious materials supplementary by its availability, easy access, little competition in others industrial uses. The clay sample was calcined at 850 °C for one hour and cooled rapidly. The samples of vitreous and zeolitized tuffs were crushed in a jaw machine. Each material was ground in a ball mill separately until it reached 90% passed in the sieve 90 μm approximately.

2.2 Characterization of Materials

The chemical composition of the whole material sample was determined by X-ray Fluorescence (XRF), using a Bruker AXS S4 spectrophotometer. The mineralogical composition of clays was analyzed by X-ray Diffraction (XRD) using a diffractometer Simens-D5000 operating with Cu radiation at 30 kV and 30 mA with a fixed divergence slit size of 1°. The samples were scanned between 5° and 80° 2θ. The step size and time per step were set to 0.05° 2θ and 1 s, respectively. A combination of technique were used to determine the active phases. Clay minerals content was determined by Thermogravimetric Analysis [4], zeolite content and volcanic glass phase was determined by XRD and Optic Microscopy. The specific surface of the ground calcined clay and volcanic tuffs was estimated from the particle size distribution, determined in a HORIBA LA-910 analyzer, assuming a spherical symmetry.

2.3 Pozzolanic Activity Test

The pozzolanic activity of SCMs was measured by Strength Activity Index in Portland cement–Pozzolan systems, reported in the Cuban specification NC 527 [5]. The cement system was composed by 30% w/w of SCM and 70% w/w of Ordinary Portland cement (OPC). Control mortars were prepared using OPC.

2.4 Assessment of Performance in Ternary Blended Cements and Hollow Concrete Blocks

Three ternary cements were formulated. A Portland Pozzolanic cement (PP-35) was used as reference. The designation and constitution are presented Table 1. The samples of limestone, clinker and gypsum were ground, each material separately until achieving values of fineness of approximately 90% passed through the sieve of 90 μm, with different time intervals according to the hardness of the material. The compressive strength was assessed on standard mortars according to NC-506 [6]. The formulated binders was used in the manufacture of hollow blocks having size 400 × 150 × 200 mm, produced at a prefabricated plant in Moa. Mix design was made under standard manufacturing conditions.

Table 1 Formulation of blended systems

Blended designation	Zeolitized tuff (%)	Vitreous tuff (%)	Calcined Clay (%)	Limestone (%)	Clinker (%)	Gypsum (%)
LC3	–	–	30	15	50	5
PZ	30	–	–	15	50	5
PV	–	30	–	15	50	5
PP-35	–	–	–	6	89	5

3 Results and Discussion

3.1 Chemical and Mineralogical Composition

Tables 2 and 3 presents the chemical and mineralogical composition of the three materials analyzed. Clay is characterized by the high aluminum content as part of the kaolinite clays mineral group and gibbsite as associated phase. The zeolitized tuff composition is related mainly with the minerals of the zeolite group, quartz and clays phases. On the other hand, the vitreous tuff present a composition related with the feldspar, clay mineral of the smectite group. In the three materials is identified the hematite as iron main phase.

3.2 Pozzolanic Reactivity in Portland Cement–Pozzolan System

Strength activity index of mortars prepared using 30% w/w of SCMs are shown in Table 4. The order of reactivity of the three SCMs analyzed was as follows: calcined clay, zeolitized tuff and vitreous tuff. Assessing from the results, the difference can be attributed to the nature and quality of the active phase since there does not seem to be a directly proportional relationship between the content of active material and the strength activity index (Fig. 1). Although, in this case, the pozzolanic reactivity seems to be directly related with the specific surface of the analyzed SCMs (Fig. 2).

Table 2 Chemical composition of the raw samples

Materials	SiO_2	Al_2O_3	Fe_2O_3	CaO	MgO	SO_3	Na_2O	K_2O	TiO_2	P_2O_5	PPI
Clay	40.55	29.23	12.68	0.01	0.75	0.13	0.13	0.17	0.50	0.03	14.18
Zeolitized tuff	61.62	13.32	4.69	5.25	2.23	–	1.63	2.02	0.40	–	8.70
Vitreous tuff	60.51	14.03	6.24	4.12	2.38	–	2.23	1.24	0.41	–	8.80

Table 3 Mineralogical composition of raw materials

Materials	Active phase content[a] (%)	Main phases
Clay	81	Kaolinite, gibbsite, goethite, hematite
Zeolitized tuff	85	Heulandite, clinoptilolite, quartz, hematite
Vitreous tuff	75	K-Feldpars, Plagioclase, montmorillonite, quartz, hematite

[a]Clay: Equivalent kaolinite, Vitreous tuff: volcanic glass, Zeolitized tuff: zeolite phases

Table 4 Strength activity index of the mortars with 30% w/w of SCMs

	Calcined Clay	Zeolitized tuff	Vitreous tuff
28 days	102	84	80

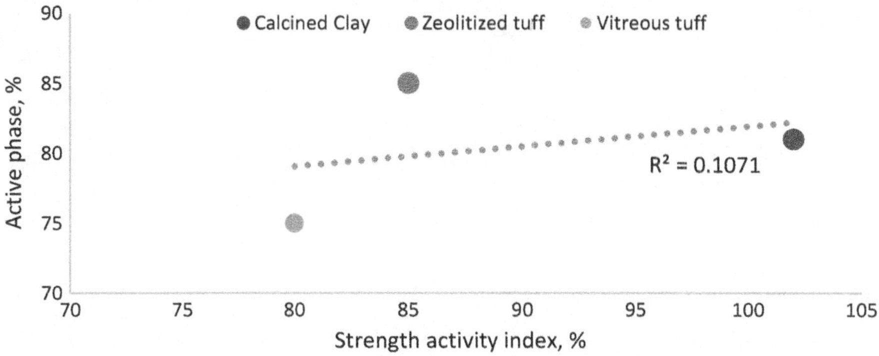

Fig. 1 Relation between active material content and strength activity index

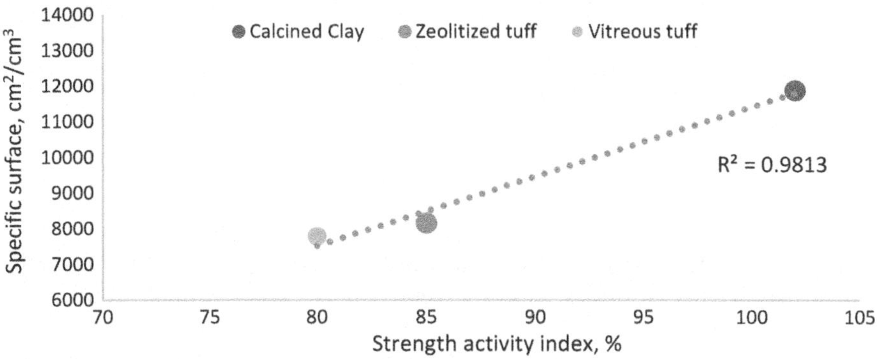

Fig. 2 Relation between specific surface and strength activity index

3.3 Performance in Ternary Blended Systems and Hollow Concrete Blocks

Figure 3 shows the evolution of the compressive strength of blended cements. In the first three days, the blended systems have lower values of compressive strength compared to the control. The values of strength at this age should not be taken as indicative of reactivity because at this time, the effects associated with physical transformations prevail. After 7 days, there is a slight increase in the compressive strength for cement containing calcined clay. The blended cements with both volcanic tuffs are still below the reference. This trend is maintained at 28 days. The substitution of cement by calcined clay resulted in a substantial increase in strength from 28 days compared to the control mortar.

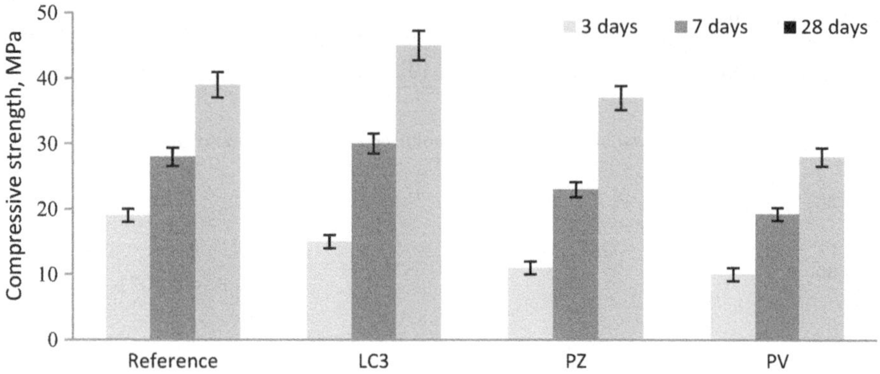

Fig. 3 Compressive strength of ternary blends

Table 5 Compressive strength and absorption of hollow concrete blocks

Hollow blocks	Average compressive strength at 7 days (MPa)	Average compressive strength at 28 days (MPa)	Absorption (%)
Block LC^3	5.93	6.70	6.90
Block PZ	4.08	6.02	7.10
Block PV	4.03	5.83	7.66
Block reference	4.64	5.47	6.80
Specification	4.00	5.00	≤10.00

In particular, LC^3 system is the cementitious material with promising potential, which suggests that the differences observed are in agreement with the reactivity of the materials assessed.

Compressive strength and water absorption of hollow concrete blocks are presented in Table 5. A significant increase in the compressive strength of the hollow concrete blocks formulated with LC^3 compared to the reference blocks and the specifications was observed. Performance of the blocks made with cements containing vitreous and zeolitized tuffs were comparable to control blocks and specifications. The results obtained were consistent with the reactivity and the compressive strength of blended cements.

4 Conclusions

Three supplementary cementitious materials was characterized, assessed and compared yours performance in cements and hallow concretes blocks. The highest pozzolanic reactivity is obtained for the calcined clay, followed by the zeolitized tuff and the vitreous tuff, with moderate reactivity, which seems to be more related to the specific surface than to the content of active material. The ternary blended cement and the hollow blocks of concrete with better performance are those that contain calcined clay, which is in agreement with the same behavior observed in the assessment of the pozzolanic reactivity. According to the results, the most effective material is calcined clay, however the performance of the two tufa materials should not be discarded for use as supplementary cementitious materials.

Acknowledgements The authors would like to acknowledge to the Cuban Geological Services and Ministry of Construction for the financial support. The authors would also like to acknowledge to the ECOPP of Moa for the technical and material support.

References

1. IEA/CSI-WBCSD, Technology Roadmap: Low-carbon transition in the cement industry 2060, París y Ginebra, Francia y Suiza, 2018. www.iea.org/technology-roadmap/cement
2. Juenger, M.C.G., Provis, J.L., Elsen, J., Matthes, W., Hooton, R.D., Duchesne, J., et al.: Supplementary cementitious materials for concrete: characterization needs. In: Materials Reserch Society Symposium Proceedings. Cambridge University Press, Cambridge (2012). https://doi.org/10.1557/opl.2012.1536
3. ONRM, Balance Nacional de Recursos y Reservas Minerales No Metálicos, La Habana, Cuba (2016)
4. Alujas Díaz, A., Almenares Reyes, R.S., Arcial Carratalá, F., Martirena Hernández, J.F.: Proposal of a methodology for the preliminary assessment of kaolinitic clay deposits as a source of SCMs. In: Martirena, F., Favier, A., Scrivener, K. (eds.) Calcined Clays for Sustainable Concrete. Springer, Netherlands, pp. 29–34 (2018). https://doi.org/10.1007/978-94-024-1207-9_5
5. NC TS 527:2013, Cemento hidráulico-Métodos de ensayo-Evaluación de las puzolanas (2013) 20. www.nc.cubaindustria.cu
6. NC 506: 2013, Cemento hidráulico. Método de ensayo. Determinación de la resistencia mecánica, (2013) 26. www.nc.cubaindustria.cu

Production of Limestone-Calcined Clay Cement in Guatemala

S. Torres, E. García, L. Velásquez, R. Díaz and Jose Fernando Martirena-Hernandez

Abstract This work describes the production process for the first formulation of a limestone-calcined clay cement in Guatemala, with a total production of 160 metric tons of cement. After reviewing more than 10 possible sites for extraction of clays at several regions, that could accomplish the criteria for usage as an activated clay material, a 67% of kaolin content material was finally selected, on a location around 50 km of our cement factory. With the selected material, a theoretical formulation by weight was defined of 48% clinker, 30% calcined clay, 17% limestone and 5% gypsum, the process began with the thermal activation of the clay, by passing 125 tons of raw material by a rotary kiln at 750–850 °C, temperature control was possible due to a series of control points using a thermal imaging camera, the activation process was controlled by X-Ray Diffraction, looking for the reduction of the signal for kaolin type clays, obtaining 90 tons of activated material. The cement produced had a 48% of clinker content verified by XRD, around 12% lower than any general use cement produced in the country. This cement was tested for strength development, setting time, hydration heat, and alkali-silica reactivity showing a very good performance. No critical difference was perceived between the concrete produced with the new cement and one with 83% of clinker with similar strength performance at 28 days evaluated by ASTM C39. The mortar, developed a superior strength compared to the national normative for a general usage cement ASTM C1157.

Keywords Limestone-calcined clay cement · Kaolin · Thermal activation · Strength

S. Torres (✉) · E. García · L. Velásquez · R. Díaz
Centro de Investigación y Desarrollo, Cementos Progreso, S.A., Guatemala City, Guatemala
e-mail: storres@cempro.com

J. F. Martirena-Hernandez
CIDEM, Universidad Central de las Villas, Santa Clara, Cuba

© RILEM 2020

J. F. Martirena-Hernandez et al. (eds.), *Proceedings of the International Conference of Sustainable Production and Use of Cement and Concrete*, RILEM Bookseries 22,
https://doi.org/10.1007/978-3-030-22034-1_10

1 Introduction

The cement industry in Guatemala over the past decades it has been using pozzolanic materials for general use cement, with a content of clinker around 60–65%, because there is a great amount of this materials in the country, due to volcanic activity. Recently we are working with the LC3 initiative [1], headed by Prof. Karen Scrivener from EPFL, Switzerland and Prof. Dr. Jose Fernando Martirena-Hernandez from the Universidad Central de Las Villas, Cuba, for the development of a new type of cement in the country with a lower clinker content, that could perform as a general use cement.

The limestone-calcined clay cement (LC3) is a mixture of this materials plus gypsum and clinker, the target is producing a cement with similar characteristics of an OPC cement, but also with an increased capability of ASR resistance [2], compressive strength resistance and other benefits regarding concrete durability [3].

This article presents the results of an industrial trial carried out in Guatemala City, using the concept of a producing a LC3 cement, and evaluating its properties of strength development and durability. In Guatemala the housing problem is one of the most difficult problems to solve for the population, not sufficient infrastructure and poverty has made this problem even worst; Cementos Progreso is the largest cement company in the region, so it has a responsibility of producing the amount of cement necessary to supply the market for the developing economies in Central America. Plus, the LC3 cement, is a way of doing this by producing a low carbon cement, with better performance using a less amount of clinker that traditional cement.

2 Industrial Trial: Selection, Thermal Activation, Grinding

First, we did an extensive search for the kaolinitic clays that could be used for its thermal activation in the LC3 cement; three regions of the country were studied for the clays: central north, east and southeast. From this areas were collected 69 samples and it was analyzed for thermal losses by TGA, chemical composition by XRF and XRD. We looked for clays that could be within the group of the kaolin (clays 1:1), from a geology point of view, using this as a guide in Fig. 1 shows the samples that were considered as a good raw material in one of the most promising deposits (C1) for the production of a low carbon cement.

The material was taken to the cement plant, with an original humidity content of 20%, it was dried to less than 1% in a rotary kiln especially designed for the drying of raw materials. The material was with a small particle size that was able to getting into the mill directly without a previous preparation (Fig. 2).

Figure 3 shows the XRD of the materials, seeing the changes on the signals at 12°, 26° and 32° 2θ, after the completion of the thermal activation of the clay. This was made at a 750-850 °C temperature, in a rotary kiln of 20 m long, with a speed of 9 tons per hour. In the last X-Ray diffraction, could see the clay completely activated.

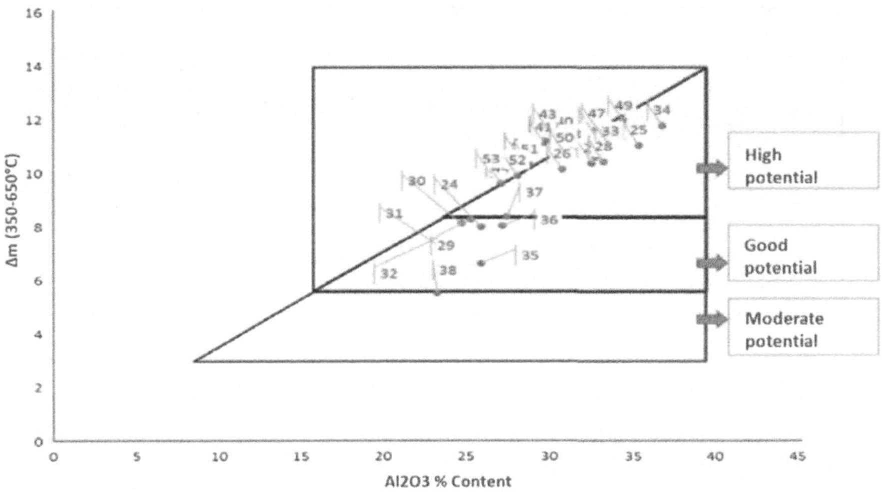

Fig. 1 Potentiality for samples in C1

Fig. 2 **a** The selected kaolinitic clay dried, 90 ton. **b** The kaolinitic clay thermally activated

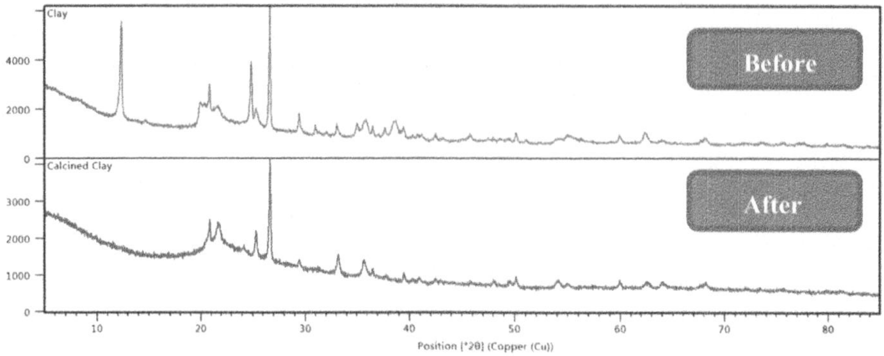

Fig. 3 Reduction of the kaolin signals in the clay after the thermal activation

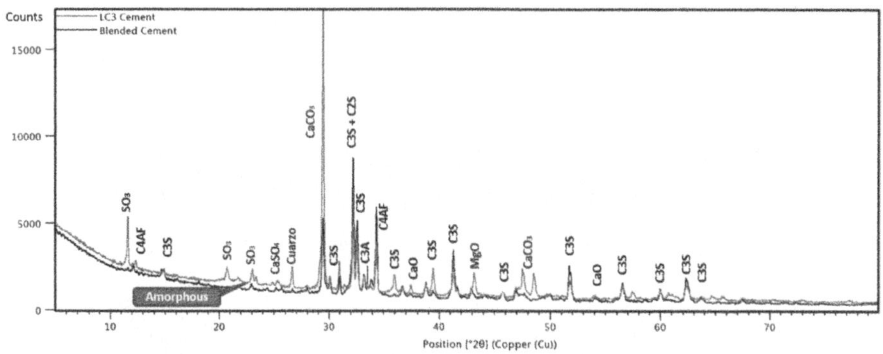

Fig. 4 Comparison between the mineral phases between LC3 cement and a blended cement commercially produced

After the thermal activation of the material, the grinding process begin in a ball mill for raw meal adapted for the production of cement, with a capacity of 25 ton per hour, with the mixture of 48% clinker, 30% calcined clay, 17% limestone and 5% gypsum, reaching a fineness of the product of 95% <45 μm. In Fig. 4 are presented the chemical phases, identified by X-Ray Diffraction, of the final low carbon cement produced, in this we could see the difference between the LC3 cement and a comparison with a blended cement produced with a clinker factor over 80%, the major differences are the amorphous content that is bigger in the low carbon cement, the content of limestone by the signal at 29.5° and 47°–49° 2θ and the signals corresponding to the alite (32°–33° 2θ) that indicates the reduced clinker content.

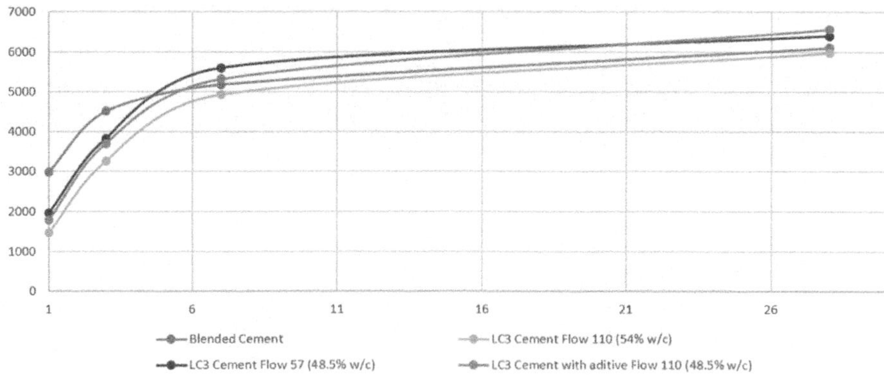

Fig. 5 Strength development for mortars of LC3 cement versus blended cement

3 Properties of the LC3 Cement

The cement produced showed an over grinding, the Blaine fineness was over 10,000 g/cm^3, that was caused do to the co-grinding of all the materials, the clinker reached the target granulometry, but the clay was over grinded in this process, the particle size caused that the demand of water in the mortars was greater than any other cement previously produced in our factory, that in average has a Blaine fineness of 3800 g/cm^3. The LC3 cement showed an initial vicat at 154 min and a final at 259 min, that is a similar to our blended cements.

In Fig. 5, it presents the strength development showed for different essays for the low carbon cement, with 54% of water and a flow of 110, with 48.5% of water and a flow of 57 and with 48.5% of water with a flow of 110 reached with the addition of 1.5 mL/kg of cement of a superplasticizer, showing the best results. The mortars of LC3 cement, had a slow early strength development, but the mortars with 48.5% w/c, surpassed the blended cement at an age of 7 days by an average of 380 PSI.

In Fig. 6, we could see that the behaviour is similar to the mortars, the strength at 7 days is superior to the cement with more than 30% of clinker, and shows that the influence of the superplasticizer for the improvement of the workability of the concrete that leads to a better distribution of the aggregates and lower air content that improves the strength development in the concrete.

We took samples of the concrete for microscopic evaluation by SEM, looking for any indication of a major difference with the blended cements normally produced. The major differences were the least amount of micro-fractures found in the cement paste that could be linked to the minor content of clinker that diminishes the contraction and expansion in the matrix and the major content of aluminum in the paste of 6-7% that in the blended cement is around 1.8% (Fig. 7).

The ready mix concrete was made using a mixture of additives for the improvement of the workability and trying to increase the fluidity of the mixture for better pumping; the concrete showed good behavior using this mixture and reached a strength

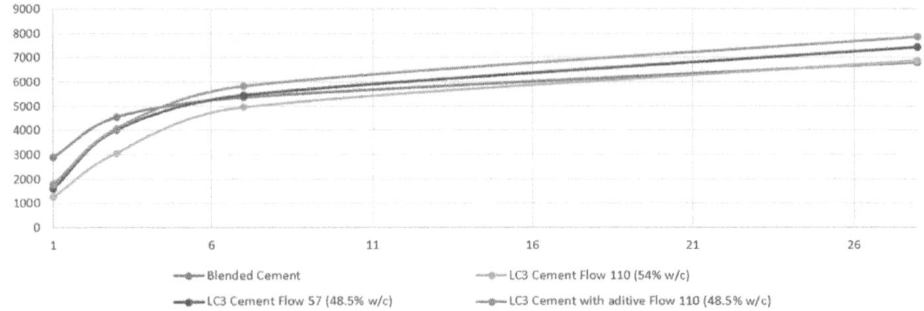

Fig. 6 Strength development for concrete of LC3 cement versus blended cement

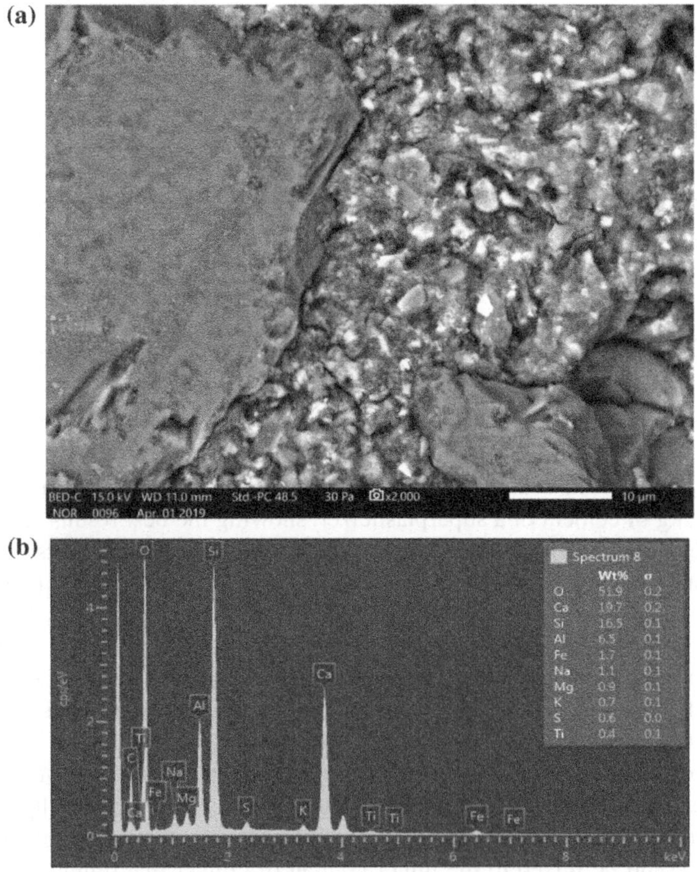

Fig. 7 **a** SEM and **b** EDS analysis of a concrete sample of LC3 cement

Fig. 8 Application of the ready mix of LC3 concrete

development at 28 days of 5000 PSI for a concrete design of 4000 PSI. Again in this test, the early strength development was slow in comparison with traditional concretes (Fig. 8).

4 Conclusions

The LC3 cement, in mortar and concrete, showed promising results for the development of a new type of cement in Guatemala. Even with a low early strength, the concrete was workable and performed well during setting. The reduction in the use of clinker gives this type of cement, less reactive properties that could be useful when reactive aggregates are used, which is very important in Guatemala, due to the vast amount of minerals from volcanic activity. For us, is necessary to improve the early strength in the concrete for better results and being able to make a product that solves the issues in infrastructure that impact our country.

References

1. Scrivener, K., Favier, A. (eds.): Calcined clays for sustainable concrete. In: Proceedings of the 1st International Conference on Calcined Clays for Sustainable Concrete. Springer Book Series, vol. 10 (2017)
2. ASTM Annual Book of Standards. Section 4, vol. 04.01 and 04.02. ASTM International, USA (2010)
3. Scrivener, K., Snellings, R., Lothenbach, B. (eds.): A Practical Guide to Microstructural Analysis of Cementitious Materials. CRC Press, Boca Raton (2015)

Processing of Calcined Clays for Applications in Cementitious Materials: The Use of Grinding Aids and Particle Classification After Grinding

Franco Zunino and Karen L. Scrivener

Abstract Calcined clays provide a promising opportunity to lower clinker levels in cements because of their widespread availability and their excellent reactivity in blended cements. On the other hand, one of the main drawbacks of calcined clay utilization for blended cement manufacture is the negative impact on workability. This project explores the use of grinding aids to control the resulting particle size distribution of calcined clay and limestone. It was observed that after grinding, calcined clays exhibit a strongly bimodal particle size distribution, where the clay minerals concentrate mainly in the finer particle population. Particle classification (air separation) techniques were applied to remove the impurities (mainly quartz and iron oxides) and therefore, increase the amount of kaolinite in the resulting material. An increase in the kaolinite content from 29 to 45% by mass was achieved in one step and without pre-dispersion of the particles.

Keywords Grinding aids · Agglomeration · Reactivity

1 Introduction

Calcined clays are a promising opportunity to lower clinker levels in cements because of their widespread availability and their excellent reactivity in blended cements. The combination of metakaolin and limestone in OPC-based systems produces a synergy that enables the production of high performance cement with a significantly lower clinker factor. Clays are mixtures of clay minerals (such as kaolinite, illite and montmorillonite) and other impurities, such as quartz, iron oxide and other rock forming minerals. Due to this inherent heterogeneity of the material, the grinding of clays results in a characteristic bimodal particle size distribution. This characteristic distribution is observed when grinding both raw and calcined clay, and also at industrial

F. Zunino (✉) · K. L. Scrivener
Laboratory of Construction Materials, IMX, École Polytechnique Fédérale de Lausanne (EPFL), 1015 Lausanne, Switzerland
e-mail: franco.zunino@epfl.ch

© RILEM 2020

J. F. Martirena-Hernandez et al. (eds.), *Proceedings of the International Conference of Sustainable Production and Use of Cement and Concrete*, RILEM Bookseries 22,
https://doi.org/10.1007/978-3-030-22034-1_11

scale grinding setups. This opens the possibility of applying particle classification processes to increase the kaolinite content in low grade clays.

Grinding aids (GAs) are incorporated during comminution of clinker to reduce electrostatic forces and minimize agglomeration of clinker and SCM grains [1]. Such additions are commonly used to increase cement fineness and compressive strength for given specific energy consumption (Ec) of the grinding mill [1–3]. After the grinding process, GAs may not preserve their original molecule structures. However, they do remain adsorbed onto the cement particles to entail variations of cement properties whether in the fresh or hardened state [1]. Thus, grinding aids may have beneficial effects on rheology and hydration of limestone calcined clay cements (LC^3).

2 Materials and Methods

The effect of grinding aids on the agglomeration of calcined clay during grinding was studied using unground calcined clay passing #8 sieve, and a lab scale rotary jar mill. The clay had a calcined kaolinite content of 62%. The grinding was performed in controlled conditions of time and load of the mill, incorporating different dosages of commercial grinding aids based on poly carboxylate ether (PCE), glycol or amines. As the clay was calcined before grinding, the workability of LC^3 cast using these clays was assessed using the mini-cone slump test, while hydration kinetics were studied using isothermal calorimetry.

To assess the potential of particle classification as a mean to increase the kaolinite content of ground clays, a natural clay with 30% kaolinite content was selected. Two different particle classification techniques, gravimetric precipitation and air separation, were applied over the same material. The first one is a classic and accurate method normally use by geologists on soil science, while the second one has the biggest potential for industrial scalability. The procedure was applied to a previously ground and dried batch of raw clay. A Hermle Z 206 lab centrifuge was used with 50 mL tubes. The clay dispersion was composed of raw ground clay as solute and sodium metaphosphate (0.1%) aqueous solution in a water to solid ratio of 6. At the end of the centrifugation, the liquid solution was placed in plastic container. The containers were placed in a 60 °C oven to speed-up the evaporation and to allow the retrieval of the dry samples. The gravimetric separation was performed to obtain 5 batches with their own target separation size (the limit size between particles sedimented and in suspension): 0.8, 3, 5, 11 and 50 microns. The time required to achieve the desired separation limit was established using the Stoke's law adapted for centrifugation [4].

Cement plants normally perform grinding in close circuit configuration, combining a mill with an air classifier that controls the particle size of the output. In order to demonstrate the potential for clay concentration using this technology, a lab-scale air separator was used. The speed of the classifier was initially adjusted using a limestone calibration curve. Afterwards, a clay curve was constructed based on experimental results. The reactivity of the fine and rejected (coarse) fractions of clay were analysed by means of the R^3 test [5], while their physical (particle size distribution by laser diffraction and specific surface area by nitrogen adsorption) and chemical (kaolinite content by TGA) properties were also assessed.

3 Results and Discussion

3.1 Effect of Grinding Aids on Grinding

Calcined clay was ground in the conditions described above for 60 min in order to observe the effects of grinding aids incorporation on the strong agglomeration observed in clays without the addition of these molecules. The three commercial grinding aids were included in the dosages recommended by the manufacturer. As observed in Fig. 1, the incorporation of grinding aids a strong effect on reducing the clay covering layer of the mill walls and grinding media, which impacts the efficiency of the process. Particle size distribution of samples collected every 15 min during the grinding process allowed to observe that fine clay is obtained faster with the use of grinding aids. However, the ultimate fineness is similar as this is mainly controlled by the mill geometry and the grinding media load.

Fig. 1 Photographs of ground clay after 60 min incorporating (from left to right) no grinding aid and 0.15% PCE-based grinding aid

Particle size distribution and specific surface area measurements of the samples were collected every 15 min and compared. It was observed that the increase in the surface area as a function of particle size is almost negligible (2% increase of surface area, for a D_{V50} reduction from 50 to 10 μm), in contrast to what is normally observed in cement and other SCMs. This is explained as the main source of surface area from clays is their internal porosity, which is independent of the particle size over the range explored in this study.

3.2 Effect of the Incorporation of Grinding Aids on Workability and Hydration of LC^3

The effect of the grinding aids addition on the workability of LC^3 is illustrated on Fig. 2. Results are presented as slump versus addition of a commercial PCE superplasticizer, and compared against a reference OPC w/c 0.5 mixture. As observed, the incorporation of some grinding aids can improve the flowability of LC^3 to levels approaching the behaviour of OPC. The most effective grinding aid is the PCE based. For this reason, another batch of clay was ground with an increased dosage of this product to see if the effects could be further increased. Thus, the addition of PCE both in the solids or during mixing as a regular superplasticizer is effective to increase the slump of LC^3 based mixtures.

Regarding the effects of grinding aids on LC^3 hydration, it was observed that alkanolamines (TEA, TIPA and DEIPA) have a strong effect on the intensity of the aluminate (second) peak of hydration. The effect seem to be more intense in the

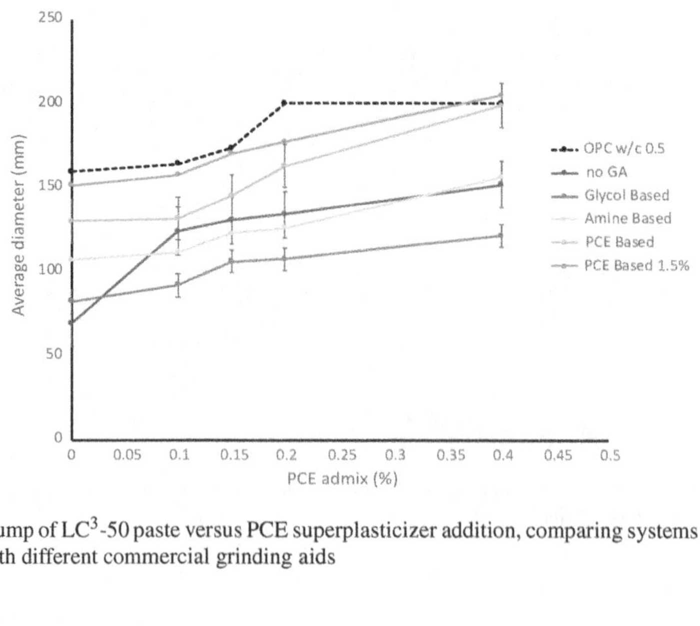

Fig. 2 Slump of LC^3-50 paste versus PCE superplasticizer addition, comparing systems with clays ground with different commercial grinding aids

system containing DEIPA, followed by TIPA and finally TEA. Previous research has suggested that this heat increase is linked to the complexation of iron contained in calcined clay, preventing the formation of a solid ferrous hydroxide phase that prevents further dissolution of clay. In order to confirm this hypothesis, model systems with synthetic iron free clay (prepared by mixing quartz and pure metakaolin to match the grade of the natural clay used) and also incorporating white OPC instead of grey cement were tested. Each system was intended to test the effect of alkanolamines on the aluminate peak in LC^3 mixtures with iron free clays and cement. Results shows that in the iron-free clay system, the effects of TEA, TIPA and DEIPA is comparable as the one observed in the natural (iron rich) clay LC^3. On the other hand, the complete iron free system (synthetic clay and white OPC) showed a very minor effect of the molecules studied. This confirms that the interaction is indeed related to the presence of iron, but not sourced from calcined clay. In contrast, the effect appears to be linked to the iron contained in cement, in particular, C_4AF.

3.3 Particle Classification of Ground Clay to Increase the Kaolinite Content

The measurements of kaolinite content on the fine fraction collected after gravimetric separation showed that it is concentrated in the fine fraction of the particle size distribution, corresponding typical first bump observed between about 0.1 and 1 μm. This is linked to the different grindability of kaolinite and the impurities normally found on clays (quartz and rock forming minerals). In this regard, clay minerals become finer quicker generating the characteristic bimodal distribution, while the impurities remain concentrated in the coarser fraction. Thus, particle classification techniques based on size difference could provide a technically suitable way to concentrate kaolinite in low-grade natural clays.

A consolidated plot containing the results of characterization of the fine fraction obtained by air separation is shown on Fig. 3 for a speed of the classifier of 12,000 rpm. It can be observed that, as gravimetric separation, air separation is effecting in removing the impurities (mainly quartz) from the raw materials and con centrating the kaolinite in the remaining one. The amount of kaolinite was increased from 30% in the raw clay to about 45% under the most restrictive conditions. However, the reactivity as measured by the R^3 test shows an increase on the heat release at 24 h from 70 to 125 Joules per gram of solids. This is explained due to the simultaneous increase of kaolinite content and specific surface area of the final material as compared to the initial clay. Furthermore, it can be seen that repeated processing of the rejected fraction do not further reduce the amount of kaolinite in the coarse material. Thus, the process appear to be limited by the efficiency of the separator and the agglomeration of the material.

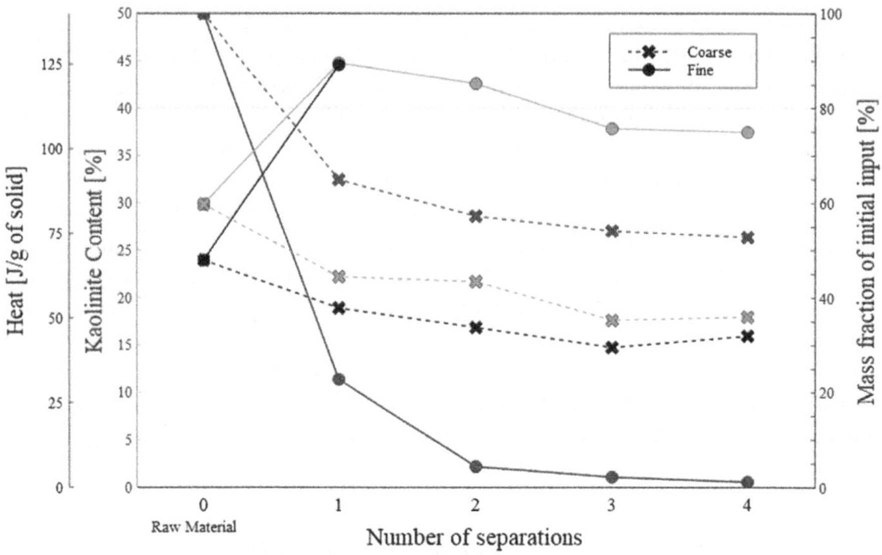

Fig. 3 Consolidated results of particle separation trials using a lab scale air classifier. Results shown represent values measured over the coarse (reject) and fine fraction of the classified material

4 Conclusions

Based on the presented results, the following conclusions can be drawn:

- The incorporation of grinding aids effectively reduces the agglomeration of clays upon grinding. This leads to a reduction of the time required to reach a fine material suitable for LC^3 manufacture under the same grinding energy conditions.
- The incorporation of grinding aids in calcined clay can lead in some cases to significant improvements in the flowability of fresh LC^3.
- Alkanolamines have a strong impact on the aluminate reaction during LC3 hydration. This effect is linked to the interaction between alkanolamines and iron-bearing phases of OPC.
- Kaolinite is concentrated in the finer portion of the particle size distribution of ground clays. Particle classification techniques can be used to remove part of the (coarser) impurities in order to increase the kaolinite content of the final material.

Acknowledgements The authors would like to acknowledge financial support by the Swiss Agency of Development and Cooperation (SDC) grant 81026665. The Swiss federal commission for scholarships for foreign students (FCS) is acknowledged for supporting Franco Zunino's studies through scholarship 2016.0719.

References

1. Assaad, J.J., Issa, C.A.: Effect of clinker grinding aids on flow of cement-based materials. Cem. Concr. Res. **63**, 1–11 (2014). https://doi.org/10.1016/j.cemconres.2014.04.006
2. Katsioti, M., Tsakiridis, P.E., Giannatos, P., Tsibouki, Z., Marinos, J.: Characterization of various cement grinding aids and their impact on grindability and cement performance. Constr. Build. Mater. **23**, 1954–1959 (2009). https://doi.org/10.1016/j.conbuildmat.2008.09.003
3. Sohoni, S., Sridhar, R., Mandal, G.: The effect of grinding aids on the fine grinding of limestone, quartz and Portland cement clinker. Powder Technol. **67**, 277–286 (1991). https://doi.org/10.1016/0032-5910(91)80109-v
4. Hathaway, J.C.: Procedure for clay mineral analyses used in the sedimentary petrology laboratory of the USGS, US. Geol. Surv. Methods, 8–13 (1955)
5. Avet, F., Snellings, R., Alujas Diaz, A., Ben Haha, M., Scrivener, K.: Development of a new rapid, relevant and reliable (R3) test method to evaluate the pozzolanic reactivity of calcined kaolinitic clays. Cem. Concr. Res. **85**, 1–11 (2016). https://doi.org/10.1016/j.cemconres.2016.02.015

First Experiences with Geopolymeric Mortars of Alkaline Activation Based on Natural Pozzolans

Joaquín Raúl Cuetara Ricardo and Rolando Batista Gonzales

Abstract The geopolymers are a new class of inorganic polymers that are obtained from the alkaline activation of natural minerals or industrial wastes and by-products. In the present investigation we describe the destructive and non-destructive tests carried out on the mortars made from the alkaline activation of the natural pozzolans from the Las Carolinas deposit to measure mechanical and durability properties. For this purpose, sodium hydroxide (NaOH) and sodium silicate (Na_2SiO_3) were used as alkaline agents. The results were analyzed through an experimental methodology where variables such as alkaline solution and curing time were determined. Finally, the mortars were analyzed mechanically, where their compressive strength was determined at 7 and 28 days of age, and their durability was also evaluated from the measurements of electrical resistivity and ultrasonic pulse velocity. The results obtained were compared with the values specified by the Cuban standard regarding the properties of mortars, also revealed that the pozzolanic material used as source of pure aluminosilicate has low alumina content, which affects the low mechanical resistance of the mortars The final product presents a series of properties that demonstrate the potential of natural pozzolans as a cementing material, which does not have any type of industrial process, or additives, or CO_2 emissions during its use, the main reason for this research.

1 Introduction

Portland cement, is the undeniable result of the historical development of humanity, today is the most used material after water [1]; its production, except the last two years, is been keeping ascendant [2]. Its use in practically all construction works, its relatively low cost, the possibility of its massive industrial production and the good results obtained in its applications have been the reason why this binder has displaced all those who preceded it, who have been relegated to minor applications in masonry work. Paradoxically, it has also turned out to be one of the main responsible

J. R. C. Ricardo (✉) · R. B. Gonzales
Construction Department, Technological University of Havana (CUJAE), Havana, Cuba
e-mail: jrcuetara@ceinpet.cupet.cu

© RILEM 2020

J. F. Martirena-Hernandez et al. (eds.), *Proceedings of the International Conference of Sustainable Production and Use of Cement and Concrete*, RILEM Bookseries 22,
https://doi.org/10.1007/978-3-030-22034-1_12

for the environmental degradation of the planet because its production process is fundamentally based on the intensive exploitation of non-renewable resources (raw materials and fuels) and the emission of significant volumes of greenhouse gases [3].

The development and economic growth of the construction industry has promoted a deep interest in the production and valuation of materials of lower environmental impact, economically competitive and with performances equivalent to or even superior to traditional materials. Geopolymers can serve as an alternative to cover environmental, technical and economic problems and promote sustainable development of building materials.

The term geopolymer was coined by Joseph Davidovits [4] in the decade of the 1980s to designate inorganic synthetic polymers of aluminosilicates that come from the chemical reaction known as geopolymerization. The geopolymers have a high potential to be used in many fields, but the use as substitutes for portland cement predominates, field in which most of the research has been directed. These upstarts have the advantage of emitting low CO_2 contents during their production, great chemical and thermal resistance, and good mechanical properties, both at room temperature and at extreme temperatures.

In this sense, the factors that intervene in the properties of the final product have been identified, being the physicochemical characteristics and dosage criteria of the raw materials the most relevant parameters. Among the precursor raw materials (source of aluminosilicate) used to activate alkaline are the natural pozzolans, which are basically composed of silica and alumina oxide [5], and in general, their vitreous nature makes them suitable for be activated alkaline.

Our country is characterized by the abundance of resources identified from natural pozzolans, since it has more than 300 million tons of this material with less than 70% zeolitization [6]. At the moment four deposits are exploited (Castilla-La Pita, Piojillo-Tasajeras, El Chorrillo and San Andrés) for their benefit and others that are used for the manufacture of cement (Los Congos, Las Carolinas, Siguaney, Las Margaritas-El Rubio and Palmarito del Cauto) [7]. The subject is novel and encouraging since it has the basic resources to induce the development of geopolitical mortar production for construction from our natural pozzolans. The investigation starts from using as a precursor (source of aluminosilicate) the natural pozzolans from the Las Carolinas deposit, located in the municipality of Cienfuegos, to manufacture geopolitical mortars without using the CPO, which improves its environmental profile and contributes to the sustainable development of the national industry of construction materials.

2 Materials and Methods

In the investigation, a natural pozzolan was used as a precursor, located approximately 12 km northwest of the municipality of Cienfuegos. The deposit is known by the name Las Carolinas, and the alterations observed are zeolitization, sericitisation, argilation and carbonization. It predominates, in more than 90%, the clinoptilolite

Table 1 Composition of natural pozzolan

Ensayos	Resultados	ASTM C 618
Silicon dioxide (SiO_2) (%)	59.92	>70%
Ferric oxide (Fe_2O_3) (%)	2.77	
Aluminum oxide (Al_2O_3) (%)	12.59	
Calcium oxide (CaO) (%)	3.95	–
Magnesium oxide (MgO) (%)	1.80	–
Sodium oxide (Na_2O) (%)	0.70	>1.5
Sulfur trioxide (SO_3) (%)	0.0319	>4.0

*Resistant Activity Index

Table 2 Chemical requirements

Properties	Values	ASTM C 618
By Frattini (mg/g)	636	–
Specific surface area (cm^2/g)	3882.9	–
Real specific weight (kg/dm^3)	2.24	–
RAI*-7 días (MPa)	77.4	≥75
RAI*-28 días (MPa)	81.0	≥75
Adsorption (%)	22.25	–
Porosity (%)	31.84	–

Table 3 Physical properties of aggregates

Properties	Sand	Gravell
Current specific weight (g/m^3)	2.61	2.63
Specific saturated weight (g/m^3)	2.65	2.66
Apparent specific weight (g/m^3)	2.72	2.70
Absorption percent (%)	1.8	1.0
Loose volumetric mass (kg/m^3)	1477	1371
Compact volumetric mass (kg/m^3)	1663	1510
Percent of holes (%)	37.0	45.0
Finest material sieve 200 (%)	9.9	1.2

and mordenite minerals, in addition, it is accompanied by quartz, montmorillonite and chlorite [7]. The characterization of this material is shown in Tables 1 and 2, where the main components are silica and alumina in a ratio of 4.76 times.

The aggregates used in the study are of natural origin, from the La Molina quarry, located in the Mariel municipality, in the province of Artemisa. The tests were carried out according to the Cuban norms NC 186; 181; 177 and 182, all of the year 2002. The fine aggregate fineness module is 3.58; Tables 3 shows its main physical properties.

The presentation of the NaOH used to make the alkaline solution is in the form of flakes with a purity of the order of 98%, and is used in the Dairy Company for the cleaning of pipes once the process of making ice cream, cheese and yogurt is

Table 4 Characteristics of Na_2SiO_3

Composition	Quantity
Dióxido de Silicio (SiO_2) (%)	2.9
Óxido de Sodio (Na_2O) (%)	9.9
Water (H_2O) (%)	87.2
Bulk density (g/cm^3)	1.42

completed. The Na_2SiO_3 was supplied by the company Electroquímica de Sagua, in Villa Clara, and its composition is shown in Table 4. For the alkaline solution, a NaOH solution with a molarity of 10, and another of NaOH and Na_2SiO_3 with a molarity of 10 was used and with relation Na_2SiO_3/NaOH of 0.4.

2.1 Experimental Method

The objective of the investigation is to know the influence of the curing time and the type of alkaline activating solution on the response variables (bulk density, electrical resistivity, ultrasonic pulse velocity and compression strength). The tests to be carried out are based on the methodology, within our reach, proposed by the bibliography used to carry out the research. To infer statistically in the results obtained during the work carried out, a 2^k factorial design was considered in which two factors with two levels each will be taken into account.

The factors to be studied during the investigation are the curing time and the activating solution, and they will be measured numerically and categorically, respectively. The curing time has two levels (10 and 28 days) and the activating solution two others (sodium hydroxide and sodium hydroxide and sodium silicate).

2.2 Procedure Description

For the dosage, a Type V masonry mortar was considered in accordance with the Cuban standard (NC-175, 2002), which specifies the properties of the mortars according to the need and proposes the quantity of mortar materials for it. In the case in question the dosage in weight of the cement is the following: a part of cement, three parts of fine aggregate and a part of hydrate of lime.

The amount of water is a function of the required settlement. For the investigation a consistency was conceived in the range of 18–20 cm. During the realization of the test mixtures it was perceived that, for the amount of materials proposed by the Cuban standard and without the use of chemical additives, a stony mortar is obtained. In order to correct this situation, it was decided to use less amount of fine aggregate,

Table 5 Quantity of materials

Materials	Solution of NaOH	Solution of NaOH + Na_2SiO_3
	(1 experiment + 1 replica)	1 experiment + 1 replica
Pozzolan (kg)	0.454	0.457
NaOH (ml)	181.6	131.1
Na_2SiO_3 (ml)	–	54.9
Fine (kg)	1.136	1.143
Extra water (ml)	110	110

so the final dosage was as follows: one part of cement, 2.5 parts of fine aggregate. The amount of materials to be used for each type of mortar is shown in Table 5.

The curing of the mortars was carried out in the stove at a temperature similar to the climate of our country. For this, the temperature was set in the range between 35 and 40 °C, where the mortars were introduced once they were demolded after the first 24 h.

3 Results and Discussion

Among the objectives of the research is the need to determine the influence of the alkaline solution and the curing time on the response variables, which are: density in hardened state, electrical resistivity (ER), ultrasonic pulse velocity (UPV) and compressive strength (CS). To do this, mortars of 4 × 4 × 16 cm were manufactured and tested according to the methodology established in the standards and techniques [8].

Figure 1 shows the analysis of the standardized effects performed from the StatGraphics Centurion XV software [9]. In it is reflected as in the case of the density in hardened state the time factor cured has a negative effect on the results, however, in the case of the ER and the UPV the same factor positively influences the results obtained. The values of CS are conditioned by the alkaline solution, logically, this factor and the amount of aluminosilicates give rise to the formation of polymer chains once the ions are reoriented in solution, therefore, the hardness of the mortar is influenced, among other factors, for the quality of it.

3.1 Analysis of the Hardened State Bulk Density

It was determined that there is a tendency to decrease the density of mortars over time, although in those where the alkaline solution contains sodium silicate the values

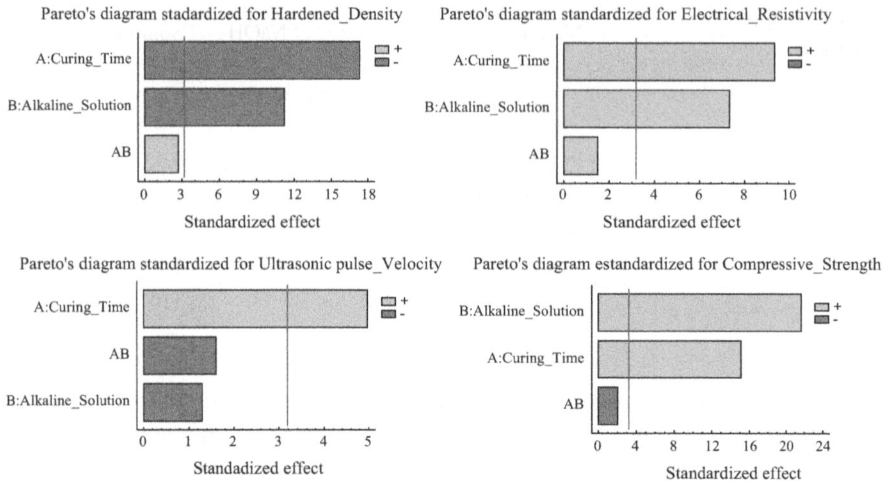

Fig. 1 Analysis of standardized effects

are slightly higher than mortars where the alkaline solution contains only sodium hydroxide. This is because, as the curing was carried out in the stove at a temperature between 35 and 40 °C, all the extra water used to improve the workability of the mortar evaporated, hence the decrease in the densities of the mortars.

3.2 Analysis of the Ultrasonic Pulse Velocity

The ultrasonic pulse velocity (VPU) values obtained were very low, furthermore, during the tests it was perceived that the receiver recognized between 5 and 10% of the wave emitted by the transducer. According to the RED DURAR [10] and based on a hypothesis test, where a VPU of 2000 m/s was taken as a null hypothesis, it was determined that in all cases the quality of the mortar is "poor".

3.3 Analysis of Electrical Resistivity

The test was performed on mortars cured at 10 and 28 days. Taking into account that the propagation of the current is through the water contained in the pores and empty spaces inside the concrete, the measurements were made after being wetted. Figures 2 and 3 show the values obtained during the lab work.

According to the TC-154 of the RILEM (Polder et al., 2000), the RE values show a mortar with a risk of "moderate" corrosion. However, it is noteworthy that with the passage of time the values of RE tend to increase. Mortars made from alkaline

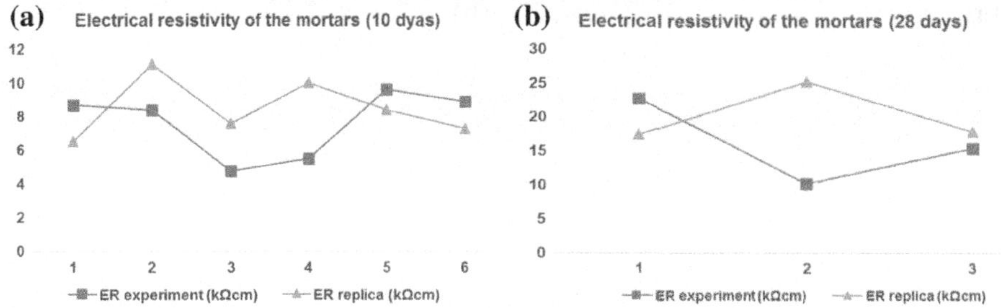

Fig. 2 Geopolymeric mortar with hydroxide solution **a** after 10 days of curing and **b** after 28 days of curing

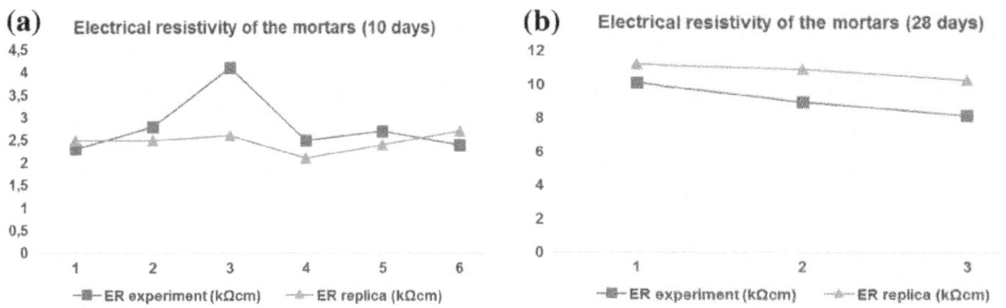

Fig. 3 Geopolymeric mortar with sodium hydroxide and sodium silicate solution **a** after 10 days of curing and **b** after 28 days of curing

solution based on sodium hydroxide have higher values compared to the alkaline solution of sodium hydroxide and silicate.

3.4 Analysis of the Compressive Strength

The results obtained at the different ages evaluated and the different solutions are shown in Figs. 4 and 5. In addition, the results obtained were compared, based on a hypothesis test, with the value specified by the Cuban standard for mortars (NC 175, 2002), which is 2.4 MPa (Type I mortar).

To consummate the comparison from a hypothesis test, the equality of the CS value of 2.4 MPa for a Type I mortar was defined as null and, as an alternative, that all the CS values are unequal at 2.4 MPa. In this case, the P-value is greater than 0.05, so the null hypothesis is accepted, which raises the equality between the CS values between the geopolymeric mortars tested and those specified by the Cuban standard for Type I mortars.

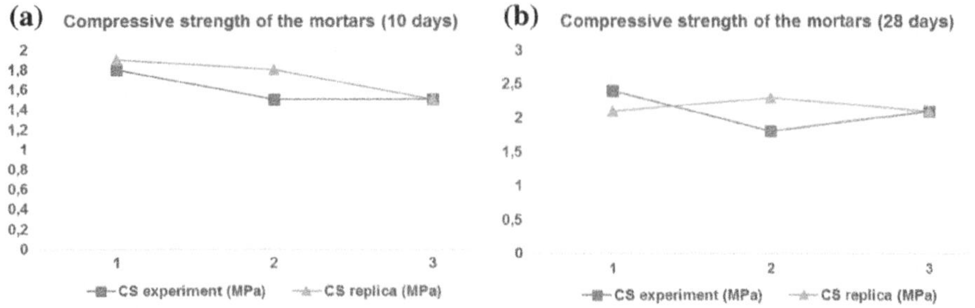

Fig. 4 Geopolymeric mortar with hydroxide solution **a** after 10 days of curing and **b** after 28 days of curing

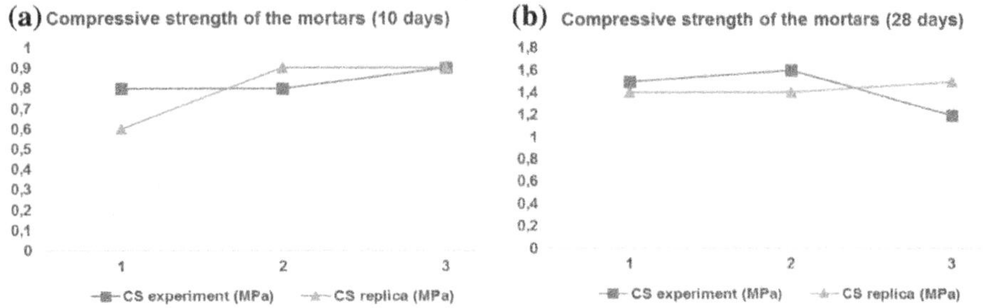

Fig. 5 Geopolymeric mortar with sodium hydroxide and sodium silicate solution **a** after 10 days of curing and **b** after 28 days of curing

3.5 Comparison Between Alkaline Solutions

With the intention of knowing if geopolymeric mortars can be manufactured from the natural pozzolans with the materials available to us, simple comparisons were made between the data obtained from the activated mortars based on an alkaline solution with sodium hydroxide and sodium hydroxide and a part of sodium silicate. As it was conceived from the design phase of the experiment, the type of pozzolan, the concentration of the alkaline solution and the fine aggregate used were kept constant in order to rule out variables that could influence the final result.

In all cases (bulk density in hardened state, ER, UPV and CS) comparisons were made from hypothesis tests, where the approach of the test is as follows:

- Null hypothesis (Ho): mean 1 (Na) = mean 2 (Si_Na)
- Alternative hypothesis (H_1): mean 1 (Na) > mean 2 (Si_Na)

All in order to demonstrate whether the replacement of a part of the sodium hydroxide used by sodium silicate influences the yields that were measured in the course of the investigation. The docima raised by the software shows that the density in hardened state and the VPU of the mortars made with the alkaline solution containing part of sodium silicate are greater than that of the mortars made with the alkaline solution that only contains sodium hydroxide, and in the case of RE and RC, the opposite happens.

4 Conclusions

- As a first experience in the manufacture of geopolymeric mortars, it is not possible to define whether the precursor (natural pozzolan from the Las Carolinas deposit) is suitable for the manufacture of this type of mortar, as the influence of the fine aggregate content employed is unknown. and its fineness, the concentration of the alkaline activator and the curing temperature.
- According to the results of ER and the statistical analysis, the influence of the curing time on the response variables can be inferred, although both factors influence the ER measurements. In addition, it was found that with the curing time the values tend to increase, and from the simple comparison between the data of the response variable it was shown that the values of ER taken in the geopolymer mortars activated with the sodium hydroxide solution they are greater than those activated with the solution of sodium hydroxide and sodium silicate with a 95% reliability.
- Both, the curing time and the type of alkaline solution influence the CS, however, the interaction between them does not. Logically, the curing time influences more than the type of alkaline solution, in addition, according to the bibliography reviewed for the realization of the present study and the results obtained, it was demonstrated that the sodium silicate used to catalyze the mortar hardening process It is not adequate, because as the main characteristic, this compound is not reactive.
- The Si ratio: Al = 4.76 which presents the material used as a precursor (natural pozzolan) is much higher than the one recommended by the international bibliography, which must be less than two. This element directly influenced that the results obtained of resistance to compression were only similar to those specified by the Cuban standard for masonry mortars.

Acknowledgements The authors wish to thank the investigation done by the recently graduated Ing. Dayanis Gorina Perdomo during the realization of her thesis, and the help provided by the technicians of the CECAT laboratory.

References

1. Aitcin, P.-C.: Cements of yesterday and today concrete of tomorrow. Cem. Concr. Res. **30**, 11 (2000)
2. WBCSD-Cement. Cement sustainability initiative, Getting the numbers right (GNR). Available from: http://www.wbcsdcement.org/GNR-2016/world (2016)
3. García, I.G., et al.: Estudio del efecto sumidero de CO_2 de los materiales de base cemento. Revista Técnica Cemento-Hormigón **939**, 14 (2010)

4. Davidovits, J.: Properties of geopolymer cement. In: Proceeding 1 International Conference on Alkali Cements and Concretes. Kiev-Ucrania (1994)
5. Malhotra, V.M., Mehta, P.K.: Pozzolanic and cementituos materials. In: Malhotra, V.M. (ed.) Advances in Concrete Technology, vol. 3, p. 191. British Library Cataloguing, 1075 AW Amsterdam, The Netherlands (1996)
6. Batista, R., et al.: Valoración del potencial de las rocas y minerales industriales para el desarrollo municipal en la República de Cuba. First ed., p. 249 Centro nacional de información geológica, La Habana (2011)
7. Rodríguez, G., Brito-Rojas, A., Coutín-Correa, D.: Sedimentary zeolite deposits in Cuba. In: Zeolites, p. 13. International Zeolite Association, I.Z. Association, Editor, La Habana (2014)
8. NC-EN-1015-10. Mortero de albañilería. Determinación de la densidad aparente en seco del mortero endurecido. Oficina Nacional de Normalización, La Habana (2008)
9. Polhemus, N.W.: Statistical analysis using StatGraphics Plus. In: Quality Control and Experimental Design. S.G. Corporation, Editor, Englewood Cliffs, NJ (1999)
10. RED-DURAR. Manual de inspección, evaluación y diagnóstico de corrosión en estructuras de hormigón armado, p. 208 (2000)

Surfaces of Response. An Effective Methodology to Estimate the Optimal Factor k of the Supplementary Cementitious Materials

Joaquín Raúl Cuetara Ricardo

Abstract Pozzolanas of natural origin have been widely used in the manufacture of mortars and concrete for millennia, from the ancient Greek and Roman civilizations to the present. In the present work, the k-factor of the natural pozzolans of the Western region of Cuba is studied and its influence on the quality of the concretes is evaluated. The response surface methodology was used to locate the optimal replacement values. In the design of the experimental program, the determination of the resistance to compression at 7 and 28 days, and the effective porosity at 28 days were conceived. The study showed that, according to Cuban regulations, the factor k of these pozzolans is 0.2, however, with the use of the proposed methodology it was possible to optimize the replacement levels and the curing time.

Keywords Natural pozzolans · Compressive strength · Porosity · Durability

1 Introduction

Concrete is the most widely used construction material in the world, and after water is the most exploited. As a result of its high demand, the cement industry is considered among the most polluting industrial sub-sectors, together with the steel industry, chemistry, oil refining and paper production.

Without question, Portland cement has been one of the materials that has most favored the development of humanity. Its manufacture is practiced in more than 150 countries, mainly in Asia, Europe and the Middle East [1]. Today is associated with the level of development of a country, however, has also been contradictory one of the main responsible for the degradation of the environment because its production process is based on the intensive exploitation of non-renewable resources (raw materials and fossil fuels) and it emits significant volumes of greenhouse gases.

In the area of construction materials, most of the research work is aimed at finding sustainable alternatives where impacts on the environment are reduced. Having

J. R. Cuetara Ricardo (✉)
Construction Department, Technological University of Havana (CUJAE), Havana, Cuba
e-mail: jrcuetara@ceinpet.cupet.cu

© RILEM 2020

J. F. Martirena-Hernandez et al. (eds.), *Proceedings of the International Conference of Sustainable Production and Use of Cement and Concrete*, RILEM Bookseries 22,
https://doi.org/10.1007/978-3-030-22034-1_13

knowledge of the replacement coefficients of the supplementary cement materials is fundamental in the manufacture of concrete from these materials. This work is aimed at providing sufficient tools to counteract the negative effects caused by cement production and, in unison, obtain concretes of similar or better quality, all from the method proposed by the author.

The supplementary cementing materials (SCM) have both hydraulic and pozzolanic activity, and both are chemical nature. The former, in the presence of water harden, similar to what happens with ordinary portland cement (OPC) and the latter, are reactions of silica oxides in the presence of calcium hydroxides (Portlandite) and water to form hydrated calcium silicates [2].

The formation of calcium silicates hydrated from the pozzolanic reaction favors the densification of the concrete microstructure, decreases its porosity and increases its resistance [3]. The effect of SCM on the pore structure is strongly related to its composition and the water-cement ratio of the mixture [4]. Also, for the pozzolanic reaction to occur it is necessary to cure enough to ensure the presence of water [5].

Many of the properties of concrete made with natural pozzolans (NP) have been studied, López and Castro [6], for example, determined the electrical resistance to chloride ion penetration of concrete made with cements added with NP. They determined that there are differences around the 2900 °C with respect to concrete made with OPC and the results coincided with those obtained by Metha and Monteiro [7]. They concluded that a concrete with OPC at the age of 90 days or more usually has an electrical resistance to penetration of 2500 °C chloride ion more than a concrete with SCM replacement.

Making concrete with a OPC part replaced by NP helps reduce the effects of sulfate attack. Meridaa and Kharchib [8], showed that the expansion decreases 65.21% in comparison with the control concrete. Other researchers have evaluated how MCS influence electrical conductivity, compressive strength and pore size distribution in lime-pozzolan binder pastes [9]. In this study, non-zeolitized NP, fly ash and microsilica were used as SCM, and concluded that direct tests of resistance in pulps better indicate the contribution potential of pozzolan-based SCM than conventional methods for estimating pozzolanic activity. The properties evaluated showed that pozzolans with a higher reaction rate are not always the best performers, hence the importance of conducting the tests directly on mortars and concrete, however, they all influenced the measurements made, although some more than others.

The factor k or cementing efficiency is related to the part of cementing material in a special concrete that is considered equivalent to Portland cement. The Cuban norm that specifies the hydraulic concrete (NC-120, 2014) establishes that the k factor of the pozzolans is the value obtained in practice that indicates the part of OPC that can be replaced with a part of SCM without affecting the resistance to the compression of a cured concrete until 28 days.

Based on statistics, this paper aims to determine the k factor of natural pozzolans in the Western region of Cuba and its influence on the quality of concrete. In order to obtain this, concretes with different levels of replacement will be manufactured and the importance of the factors of curing time and replacement level on the response variable (compressive strength) will be verified. Likewise, the performance for durability of the concretes manufactured from the capillary absorption test will be evaluated, and they will be statistically compared with the standard concrete.

2 Materials and Methods

For the manufacture of the concretes, a OPC was used that corresponds to the P-35 denomination (type I) of the Cementos Curasao factory, of the Mariel municipality, in the province of Artemisa. Table 1 shows its main characteristics and the values are compared with those established by the Cuban norm (NC 95: 2002).

The aggregates used in the study are of natural origin, from the La Molina quarry, located in the Mariel municipality, in the province of Artemisa. The tests were carried out according to the Cuban norms NC 186; 181; 177 and 182, all of the year 2002. Table 2 shows its main physical properties. The maximum size of the coarse aggregate is 19.1 mm and the fine aggregate fineness module is 3.58.

In the present work, a natural pozzolan was used as a precursor, located approximately 12 km northwest of the municipality of Cienfuegos. The deposit is known by the name Las Carolinas, and the alterations observed are zeolitization, sericitisation, argilation and carbonization. It predominates, in more than 90%, the clinoptilolite

Table 1 Physical properties of cement (P-35)

Properties	Values	NC 95: 2011 [10]
Specific surface area (cm^2/g)	3317	$\geq$2800
Fineness (%)	1.7	$\leq$10
Initial setting time (h)	102:00:00	$\geq$0.75
Final setting time (h)	3:02:00	$\leq$10
Volumetric weight (kg/m^3)	1168	
Bulk density (g/cm^3)	3.15	
Normal consistency (%)	24.5	

Table 2 Physical properties of aggregates

Properties	Sand	Gravell
Current specific weight (g/m^3)	2.61	2.63
Specific saturated weight (g/m^3)	2.65	2.66
Apparent specific weight (g/m^3)	2.72	2.70
Absorption percent (%)	1.8	1.0
Loose volumetric mass (kg/m^3)	1477	1371
Compact volumetric mass (kg/m^3)	1663	1510
Percent of holes (%)	37.0	45.0
Finest material sieve 200 (%)	9.9	1.2

Table 3 Composition of natural pozzolan

Ensayos	Resultados	ASTM C 618
Silicon dioxide (SiO_2) (%)	59.92	>70%
Ferric oxide (Fe_2O_3) (%)	2.77	
Aluminum oxide (Al_2O_3) (%)	12.59	
Calcium oxide (CaO) (%)	3.95	–
Magnesium oxide (MgO) (%)	1.80	–
Sodium oxide (Na_2O) (%)	0.70	>1.5
Sulfur trioxide (SO_3) (%)	0.0319	>4.0

*Resistant Activity Index

Table 4 Chemical requirements

Properties	Values	ASTM C 618
By Frattini (mg/g)	636	–
Specific surface area (cm^2/g)	3882.9	–
Real specific weight (kg/dm^3)	2.24	–
RAI*-7 días (MPa)	77.4	≥75
RAI*-28 días (MPa)	81.0	≥75
Adsorption (%)	22.25	–
Porosity (%)	31.84	–

and mordenite minerals, in addition, it is accompanied by quartz, montmorillonite and chlorite [11]. The characterization of this material is shown in Tables 3 and 4, where the main components are silica and alumina in a ratio of 4.76 times.

2.1 Experimental Method

The interest of the study is to determine the k-factor of the NP from the Las Carolinas quarry, from the Western region of Cuba, and measure the influence of the replacement level and the curing time on the compressive strength. The effective porosity values of the replacement concrete, selected at random, will also be compared with the standard.

It was decided to use a multilevel factorial design for the study of the compressive strength of concrete. In the Replacement_Puzolana factor, five levels were included, the low and high values as well as other levels equally spaced between them, and in the case of the Factor_Days, two levels, low and high, were included. Table 5 specifies the factors and the number of levels for each of them. Each treatment was assigned three blocks to be tested, that is, one experiment and two replications were performed for each treatment.

Table 5 Factors under study and their levels

Factors	Low (−1)	Medium-Low (−0.5)	Medium (0)	Medium-High (0.5)	High (1)
Replacement_Puzolana (%)	0	10	15	20	25
Days_Cured (days)	7	–	–	–	28

The behavior of these factors on the resistance to compression based on the response surface methodology (RSM) was analyzed. This design is supported by a set of mathematical and statistical techniques that are useful for modeling and analyzing problems in which a response of interest is influenced by several variables, and the goal is to optimize this response [12]. The difference between RSM designs and factorial designs is the inclusion of quadratic effects. According to Walpole and others [13], the quadratic effect causes the estimation of the response surface by exhibiting a curvature. In the present work, the levels of the factors that lead to an optimal response were determined experimentally with the help of statistical software StatGraphics Centurion XV [14].

In addition, the effective porosity was measured by the method established in the Cuban standard (NC-345, 2011). The measurements were made in the standard concrete and with 10 and 20% replacement of Portland cement by natural pozzolan. For the study in question the witnesses were extracted from the test pieces of 15 cm in diameter and 30 cm high to avoid the wall effect that occurs in the area close to the surface. In order to determine the effective porosity, three witnesses were analyzed for each type of concrete. In this case, no experiment was designed, only the type of concrete was taken at random and tested at the laboratory level, determining if there is a statistically significant difference between the results obtained.

2.2 Procedure Description

A standard concrete with a characteristic resistance to compression of 30 MPa at 28 days was considered, and depending on the degree of control in the laboratory, its average resistance to compression will be 32 MPa. The mixtures considered in the study have the same type of aggregate and dosed in the same proportion in order to obtain a constant granulometric curve. A proportion of fine and coarse aggregate of 45/55% respectively was used. The water-cement ratio (0.45) was considered for an environment where aggressiveness is high, as established by the Cuban norm (NC-120, 2014).

To improve the workability of fresh concrete, the Dynamon SRC-20 additive belonging to the new MAPEI system was used. It is an acrylic-based superfluidifying additive (second generation advanced) modified for premixed concretes character-

Table 6 Dosage and quantity of materials

Materials	PP	P10%	P15%	P20%	P25%
Relationship a/c	0.45	0.45	0.45	0.45	0.45
Cement P-35 (kg)	21.10	18.99	17.94	16.88	15.83
Natural pozzolan (kg)	–	2.11	3.17	4.22	5.28
Wate (L)	9.51	9.51	9.51	9.51	9.51
Coarse (kg)	48.25	47.85	47.65	47.45	47.25
Fine (kg)	39.95	39.65	39.50	39.35	39.15
Additive (%-ml)	0.35–65	0.71–120	0.94–150	1.26–190	1.63–230

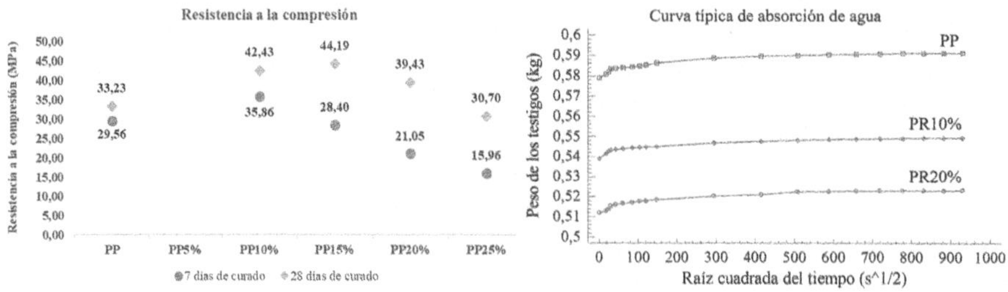

Fig. 1 Typical curves of water absorption by capillarity and compressive strength

ized by a low water-cement ratio, high mechanical resistance and long maintenance of the workability. The desired settlement measured by the Abrams cone will be between approximately 14 and 16 cm. The dose used varied according to the replacement content of OPC by NP. The amount of materials to be used for each type of concrete to be processed is shown in Table 6.

3 Results and Discussion

The water absorption test by capillarity and compressive strength of the concrete was applied according to the methodology established by Cuban standards (NC-345, 2011) and (NC-244, 2005) respectively, using witnesses with a diameter of 10 cm and cylindrical test tubes of 15 × 30 cm respectively. Figure 1 shows the results obtained from the aforementioned tests, and in both cases it is observed how the replacement of OPC by NP influences the reduction of the porosity of the concrete and the increase of the compressive strength of the concrete. the same.

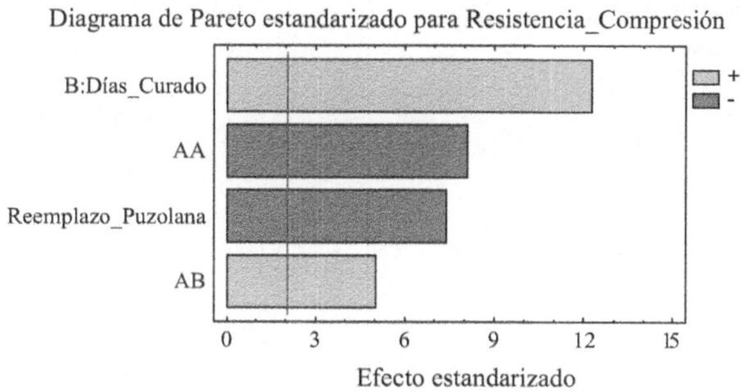

Fig. 2 Influence of the factors under study in the resistance to compression

3.1 Analysis of Resistance to Compression and Porosity

As part of the investigation, the results obtained were statistically processed to know the influence of each factor and its interaction on performance (compressive strength of concrete). The used statistic is the analysis of variance (ANOVA) which divides the variability of the compressive strength in separate pieces for each of the effects, and in this way proves the statistical significance of each one of the factors comparing its average square against an estimate of experimental error [13].

According to Montgomery [12], the term statistical significance is used to show that there are differences or not, in other words, it is defined as the probability of making the decision to reject the null hypothesis when it is true (decision known as error of type I). This decision is often made using the P-value that is nothing more than the probability of obtaining a result at least as extreme as the one obtained. The lower this is, the more significant the result will be.

In the study in question, the main effects, the effect of the interaction and the pure quadratic term have a P-value less than 0.05 which indicates that they are significantly different from zero with a confidence level of 95.0%. Figure 2 represents the Pareto diagram and the length of each bar is proportional to the value of the statistic calculated for the corresponding effect. Any bar beyond the noise line is statistically significant at the selected level of significance, set by default by 5%. For this reason it can be said that all the effects influence the resistance to compression, but the curing time is who else does it in the final result.

In the case of effective porosity, The Cuban standard establishes the determination of water absorption by capillarity in hardened concretes by means of the test method proposed by Fagerlund [15]. In our country this method is widely used to establish the durability requirements in the design of concrete and to check the capillarity of these in their hardened state.

The results were analyzed in the same software used in the compression resistance analysis. An analysis of variance was carried out to compare if there is a statistically significant difference between the results obtained for concrete with 10 and 20% replacement with respect to the standard concrete. In Fig. 3 the diagram of sample

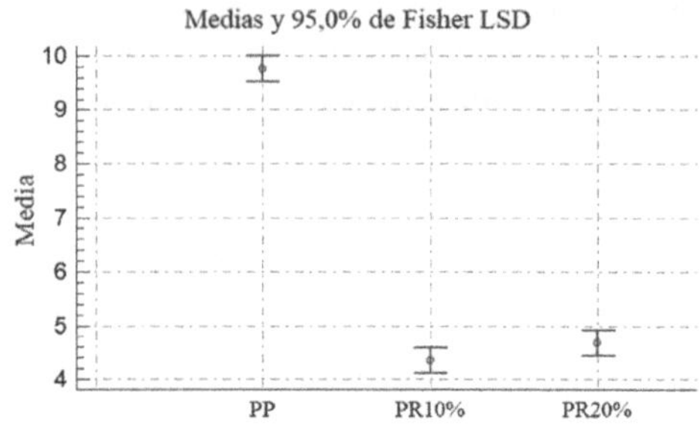

Fig. 3 Diagram of sample means

means is shown. In contrast to the standard concrete, those with 10% and 20% replacement have a reduction in effective porosity of 55 and 52% respectively.

The multi-rank test was also performed. There are two homogeneous groups, concretes with 10 and 20% replacement, for this reason there is no statistically significant difference between these levels. The method currently used to discriminate between the means is Fisher's minimum significant difference procedure (LSD). With this method there is a 5.0% risk when saying that each pair of means is significantly different.

The truth is that the replacement of OPC by NP in the manufacture of concrete positively influences the compressive strength and porosity of them. Its fineness, similar to that of cement, has a filler effect that makes up for the lack of fineness of the Cuban aggregates. Also, the pozzolanic reaction as a secondary process provides better distributed resistant compounds that also fill pores.

3.2 Determination of the K Factor Based on the Criteria of the Cuban Norm

The values obtained of compressive strength for each replacement level were compared with those of the standard concrete. It was identified that there is a significant statistical difference between them except for concrete with 25% replacement. The compressive strength for a 20% replacement of OPC by NP is increased by 16% with respect to the standard concrete, which is why it can be said that the cementing efficiency of this NP is $k = 0.2$.

Once this factor is known, it is necessary to analyze other aspects. If this level of replacement is energetically and ecologically significant, it is necessary to add to the achievement of saving 20% of OPC, the added value that implies to cure during 28 days.

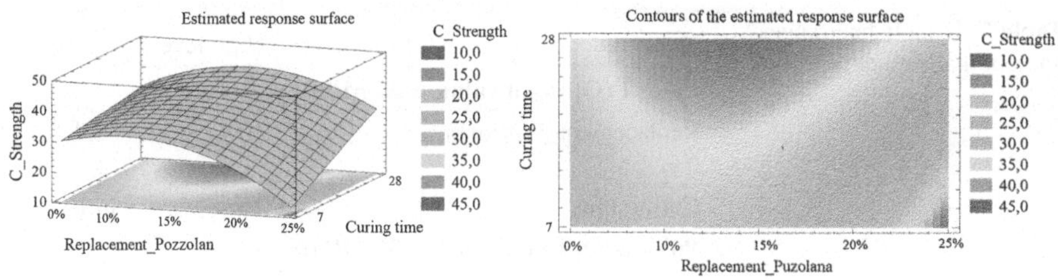

Fig. 4 Surfaces and estimated response contours

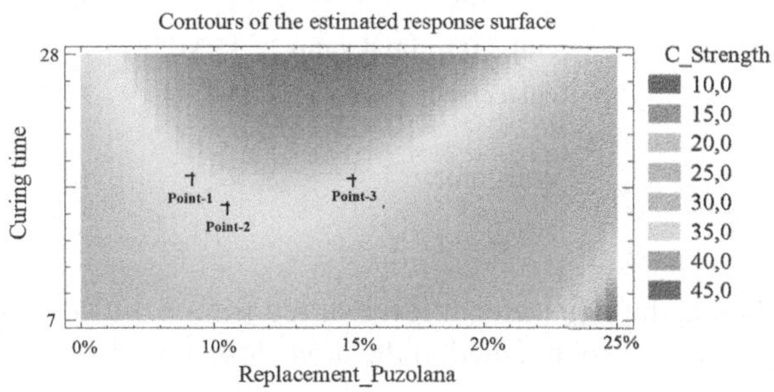

Fig. 5 Optimal points in the established replacement ranges

3.3 *Estimation of the Factor K from the Proposed Methodology*

Based on the previous analysis, the response surface methodology is proposed to optimize replacement levels based on the value of compressive strength desired by the researcher. This methodology provides efficient ways to locate a set of experimental conditions that facilitate a maximum or minimum response. Figure 4 details the hill-like curvature generated by the software and its contours.

In it, it is perceived that with the increase of the curing time the compression resistance also increases. Logically, the pozzolanic reaction depends on the hydration of the cement particles, therefore, there must be enough moisture for it to hydrate and then react the pozzolanic material with the consequent formation of resistant compounds.

In order to know the best points at a known value of compressive strength, it was decided to optimize the response for 35 MPa. A curing time range was established that was maintained between 7 and 15 days and then the values of the replacement level were moved. In Fig. 5 and Table 7 the optimal points and the coordinates of the same are detailed.

Each pair of coordinates indicates the best adjustment to obtain the desired compressive strength, in this case 35 MPa. The intention is to reduce the curing time to logical values only with the use of the additional cementing material in question.

Table 7 Established ranges and optimal values

Factor	Low level	High level	Optimum
Point 1 (Optimal value = 35 MPa)			
Replacement_Pozzolan (%)	0	10	8.4
Curing time	7	15	14.6
Point 2 (Optimal value = 35 MPa)			
Replacement_Pozzolan (%)	10	15	10.7
Curing time	7	15	12.4
Point 3 (Optimal value = 35 MPa)			
Replacement_Pozzolan (%)	15	20	15.2
Curing time	7	15	14.7

In each selected area, the replacement level varied from 8.4 to 15.2%. On the other hand, the curing time was maintained in the area closest to 15 days.

4 Conclusions

From the analysis of the results of this study that determined and estimated the k factor of the NP in question from the specification of the Cuban standard and the response surfaces methodology respectively, the influence of this on the quality of the concretes was also evaluated. It can be concluded that:

- The k factor of natural pozzolans in the Western region of Cuba is 0.2. For this replacement level the improvements would mean a 16% increase in the compressive strength with respect to the standard concrete.
- In order to obtain compressive strength values of 35 MPa and cure the shortest possible time, the response surface methodology identified three points that reflect the optimal values of cure time and replacement level. In no case was the desired compressive strength achieved without cure for less than 12 days, even for low replacement levels. Hence the importance of curing concrete when natural pozzolans are used in their manufacture.
- All values of effective porosity are less than 10%, although those obtained for replacement levels of 10 and 20% are significantly lower than those obtained for the standard concrete. For these replacement levels the porosity was reduced by 55 and 52% respectively.
- The study agrees on the importance of direct resistance tests in concrete. The use of statistics showed the ability of natural pozzolans to improve the quality of concrete. You have to recognize the role that curing time plays even over the replacement level.

Acknowledgements The author wishes to thank the work performed by the CECAT laboratory technicians and the tutorial of Professor Juan José Howland Albear.

References

1. Aitcin, P.-C.: Cements of yesterday and today concrete of tomorrow. Cem. Concr. Res. **30**, 11 (2000)
2. Hewlett, P.C.: Lea's Chemistry of Cement and Concrete, 4 edn. Elsevier, China, 1035 (2004)
3. Meddah, M.S., Tagnit-Hamou, A.: Pore structure of concrete with mineral admixtures and its effect on self-desiccation shrinkage. ACI Mater. J. **106**(3), 11 (2009)
4. Diamond, S.: Mercury porosimetry. An inappropriate method for the measurement of pore size distributions in cement-based materials. Cem. Concr. Res. **30**, 9 (2000)
5. Mehta, P.K.: High-performance, high volume fly ash concrete for sustainable development. In: International Workshop on Sustainable Development and Concrete Technology. Beijing, China (2004)
6. López, M., Castro, J.T.: Efecto de las puzolanas naturales en la porosidad y conectividad de poros del hormigón con el tiempo. Revista Ingeniería de Construcción **25**(3), 13 (2010)
7. Mehta, K., Monteiro, P.J.: Concrete: Microstructure, Properties and Materials, 3rd edn, p. Xxi. McGraw-Hill, New York (2005)
8. Meridaa, A., Kharchib, F.: Pozzolan concrete durability on sulphate attack. Proc. Eng. **114**, 6 (2015)
9. Uzal, B., et al.: Pozzolanic activity of clinoptilolite: a comparative study with silica fume, fly ash and a non-zeolitic natural pozzolan. Cem. Concr. Res. **40**, 7 (2010)
10. NC-95. Cemento Portland. Especificaciones. Pficina Nacional de Normalización, La Habana (2001)
11. Rodríguez, G., Brito-Rojas, A., Coutín-Correa, D.: Sedimentary zeolite deposits in Cuba. In: Zeolites. International Zeolite Association, p. 13. I.Z. Association, La Habana (2014)
12. Montgomery, D.C.: Diseño y análisis de experimentos, 3 edn, p. 589. G.E.I. S.A. México (1991)
13. Walpole, R., et al.: Probabilidad y estadística para ingeniería y ciencias, 9 edn, p. 792. Pearson Education Inc., México (2012)
14. Polhemus, N.W.: Statistical analysis using StatGraphics Plus. In: Quality Control and Experimental Design. S.G. Corporation, Englewood Cliffs, NJ (1999)
15. Fagerlund, G.: On the Capillarity of Concrete, p. 20. N.C. Research (1982)

Determination by Ultraviolet-Visible Spectroscopy of Adsorption of the Superplasticising Admixtures Dynamon SX 32 and Dynamon SRC 20 in Calcined Clay of Layers A, B and C of the Yaguajay Deposit and the Pontezuela Deposit

Alina Hereira Díaz, Roger Castillo Delgado and Adrián Alujas Díaz

Abstract This work proposes the ultraviolet–visible molecular absorption spectroscopy as method of chemical–physical analysis, simple and economically viable for Cuba, for the sake of quantifying the superplasticizers adsorption in calcined clay. The superplasticizers Dynamon SX 32 and Dynamon SRC 20 were rehearsed (commercial name of the signature MAPEI), used in the industry of the construction in the country and four roasted clays of the locations of Yaguajay (3) and Pontezuela (1). Firstly the spectra were obtained for both superplasticizers and the longitude of wave for work was selected; then the calibration straight line were elaborated for the quantitative analysis and lastly, the method was proven in the four clays. For both superplasticizers the longitude of selected wave was 259 nm. The calibration straight line presented superior determination's coefficients to 0.995 and variation's coefficients for point inferior to 3%. The adsorption of the superplasticizers in the roasted clays oscillated between 40 and 60%. The results demonstrate the feasibility of using the method proposed for the analytic determination as well as the strong adsorption that these superplasticizers present in the roasted clays for the studied systems.

Keywords Calcined clays · UV-Vis spectrophotometry · Adsorption

1 Introduction

Clinker production emits enormous amounts of carbon dioxide into the atmosphere, contributing to the deterioration of the health of the environment. An alternative that has produced good results is the use of supplementary cementitious materials,

A. H. Díaz (✉) · R. C. Delgado · A. A. Díaz
Facultad Química Farmacia, Universidad Central de las Villas. CEQA,
Carretera de Camajuani km 5, Santa Clara, Villa Clara, Cuba
e-mail: alinahd@uclv.edu.cu

© RILEM 2020

J. F. Martirena-Hernandez et al. (eds.), *Proceedings of the International Conference of Sustainable Production and Use of Cement and Concrete*, RILEM Bookseries 22,
https://doi.org/10.1007/978-3-030-22034-1_14

such as calcined clay, to partially replace Clinker [1]. This is an advantage for Cuba because it has a fundamentally karst geology and clays are frequent materials.

However, for this new type of low-carbon cement [2, 3], the addition of superplasticizers to improve its flow properties has not had the expected effects. It is assumed that, due to the high content of calcined clay in the mixture, a portion of these added additives are preferably adsorbed on the mixture rather than on the cement grains, not efficiently fulfilling their dispersant purpose.

In addition, the quantification of plasticizers is done internationally, in a general way, using techniques such as Total Organic Carbon, HPLC Chromatography and Gas Chromatography, among others, which are much more expensive.

2 Materials and Methods

The tests were carried out in the laboratories of the Centro de Estudios de Química Aplicada (CEQA) of the Faculty of Pharmacy-Chemistry of the Universidad Central "Marta Abreu" de Las Villas (UCLV).

2.1 Materials, Equipment and Reagents

- Water: distilled once with POBEL distiller
- Calcium hydroxide: in powder form, p.a., ACRO-S ORGANICS
- Sodium hydroxide: in the form of pearls, p.a., UNI-CHEM
- Dynamon SX 32 (SX-32) superplasticising additive: 22,1% amber aqueous solution of non-sulphonated acrylic polymers, free of formaldehyde, produced by MAPEI.
- Dynamon SRC 20 (SRC-20) superplasticising additive: 22% amber aqueous solution of modified acrylic polymers, free of formaldehyde. It belongs to the MAPEI Dynamon SR system.
- Pontezuela clays and technological types A, B and C of Yaguajay. Dryed with Binder stove at 200 °C during 24 h. Calcinating with Nabertherm muffle at 850 °C for 1 h. Grinded and pulverized with a MB-600 ball mill and a 90% fineness criterion passed through a 90% sieve μm [4].
- Commonly used glassware: beaker (VWR), flask and pipette (POBEL).
- UV-Vis spectrophotometer (L6S), with quartz cuvettes with 1 cm optical path and 2 nm resolution (Inesa Analytical Instrument, China).
- pH meter (JENWAY 3020) (JENWAY Scientific Equipment Suppliers, UK)
- Analytical balance (DENVER SI-234) (Denver Instrument, USA)
- Orbital shaker (digital KS 501) (IKA-Werke GmbH & Co. KG, Germany)
- Centrifuge (Neofuge 15) (Heal Force Bio-Meditech Holdings Limited, China).
- Water: distilled once with POBEL distiller
- Calcium hydroxide: in powder form, p.a., ACRO-S ORGANICS

- Sodium hydroxide: in the form of pearls, p.a., UNI-CHEM
- Dynamon SX 32 (SX-32) superplasticising additive: 22.1% amber aqueous solution of non-sulphonated acrylic polymers, free of formaldehyde, produced by MAPEI.
- Dynamon SRC 20 (SRC-20) superplasticising additive: 22% amber aqueous solution of modified acrylic polymers, free of formaldehyde. It belongs to the MAPEI Dynamon SR system.
- Pontezuela clays and technological types A, B and C of Yaguajay. Dryed with Binder stove at 200 °C during 24 h. Calcinating with Nabertherm muffle at 850 °C for 1 h. Grinded and pulverized with a MB-600 ball mill and a 90% fineness criterion passed through a 90% sieve μm [4].
- Commonly used glassware: beaker (VWR), flask and pipette (POBEL).
- UV-Vis spectrophotometer (L6S), with quartz cuvettes with 1 cm optical path and 2 nm resolution (Inesa Analytical Instrument, China).
- pH meter (JENWAY 3020) (JENWAY Scientific Equipment Suppliers, UK)
- Analytical balance (DENVER SI-234) (Denver Instrument, USA)
- Orbital shaker (digital KS 501) (IKA-Werke GmbH & Co. KG, Germany)
- Centrifuge (Neofuge 15) (Heal Force Bio-Meditech Holdings Limited, China).

2.2 *Experimental Procedure*

2.2.1 Preparation of Aqueous Solution with pH Between 11 and 12

Calcium hydroxide was dissolved in distilled water to saturation, then vacuum filtered with quantitative filter paper, and finally a pH between 11 and 12 was reached by dissolving sodium hydroxide in it. This pH value was quantified using a JENWAY 3020 potentiometer with a glass electrode.

2.2.2 Ultraviolet-Visible Molecular Absorption Spectra

The ultraviolet-visible molecular absorption spectra of Dynamon SX 32 and Dynamon SRC 20 superplasticisers were obtained from their solutions in distilled water, at a ratio of 1/99 v/v, and in an aqueous solution of pH 11.77.

UV-Vis L6S spectrophotometer was used, with deuterium and tungsten lamps, and quartz cuvettes of 1 cm optical passage. Wavelengths from 190 to 800 nm were recorded. The data were processed with UVWin8 software version 1.17.

2.2.3 Stability of Superplasticisers in Basic Solutions

The ultraviolet-visible molecular absorption spectra were obtained from Dynamon SX 32 and Dynamon SRC 20 superplasticisers in aqueous solution of pH 11.77 at different times. The times were: immediately after preparation, and every 15 min for 2 h, for a total of 9 spectra. For the evaluation the statistical parameters were taken into account: standard deviation (S) and coefficient of variation in percent (VC%), which were calculated with Microsoft Excel 2013 software.

2.2.4 Obtaining Calibration Lines for Superplasticisers

The calibration lines (absorbance vs. mass concentration) were prepared with five dilutions of the superplasticizer of industrial quality, in an aqueous solution of pH 11.77. The aqueous solution of pH 11.77 was used as a reference.

The IBM SPSS Statistics 23 software was used for the statistical analysis of the data.

2.2.5 Adsorption of Superplasticisers in Calcined Clay in Basic Solutions

For this test, the proportion of calcined clay/superplasticiser that is actually used in LC3 cement was considered.

5 g of calcined clay were weighed with a DENVER SI-234 analytical balance, and then mixed with 8.5 ml of aqueous solution of pH 11.77 and 0.3 ml of superplasticizer. The mixture was kept with circular agitation in an orbital agitator KS 501 digital, to 150 min^{-1} during 90 min. Later, the supernatant solution of the solid was separated with a centrifuge Neofuge 15 to 4000 min^{-1} during 35 min. Finally the absorbance of the transparent liquid was measured with a spectrophotometer L6S. Three replicates were made. The reference or white solution had all the procedure described above but without the superplasticizer.

For the analysis of the three replicas, the following statistical parameters were taken into account: standard deviation (S) and coefficient of variation in percent (VC%), which were calculated with Microsoft Excel 2013 software.

Adsorption was expressed in percent, and calculated using the formula:

$$\%\,\text{adsorption} = 100 * \frac{c_0(\text{g/ml}) - c_f(\text{g/ml})}{c_0(\text{g/ml})}$$

where: c_0 (mg/ml) is the concentration of the industrial quality superplasticiser and c_f (mg/ml) is the concentration of the superplasticiser in the solution collected after the adsorption process, measured with the calibration line.

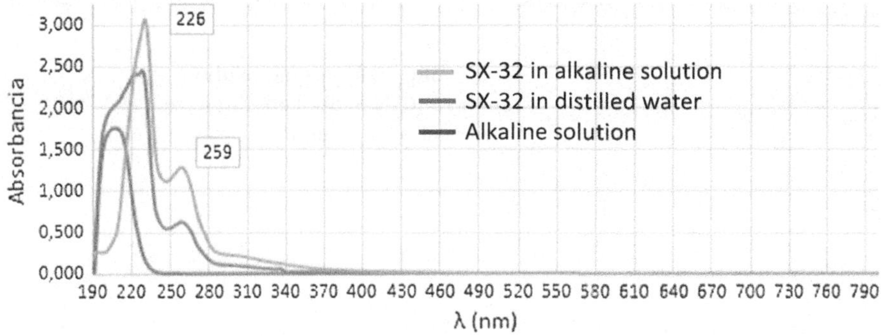

Fig. 1 UV-Vis spectra of Dynamon SX-32 superplasticizer

3 Discussion of Results

3.1 *Obtaining UV-Vis Spectra from Superplasticisers*

The ultraviolet-visible molecular absorption spectra were obtained from Dynamon SX 32 and Dynamon SRC 20 superplasticisers, in neutral and basic solution, as well as from the basic solution itself.

The spectrum in neutral aqueous solution showed that both superplasticisers present very similar spectra, which is due to their similar chemical structure.

In each of them two bands are distinguished. The first one is very intense, and its maximum absorbance is at 226 nm (Dynamon SX 32) and 218 nm (Dynamon SRC 20). This band may be due to an electronic transition between $\pi \rightarrow \pi^*$ molecular orbitals in a poorly conjugated system. The second band is less intense and in both cases its maximum absorbance is at 259 nm; it may be associated with an electronic transition between molecular orbitals $n \rightarrow \pi^*$ [5].

The comparison of the spectra of superplasticisers in neutral aqueous solution with the spectrum of the basic aqueous solution showed that the first most intense band of superplasticisers overlaps partly with the band of the basic aqueous medium and is very sensitive to changes in pH, so it is not useful for quantitative chemical analysis. Therefore, it was decided to work with this second band, see Figs. 1 and 2.

On the other hand, the spectrum of ultraviolet-visible absorption of the calcined clay in basic aqueous solution, i.e. of its dissolution products, does not present considerable absorption in the area selected for the quantitative chemical analysis, see Fig. 3.

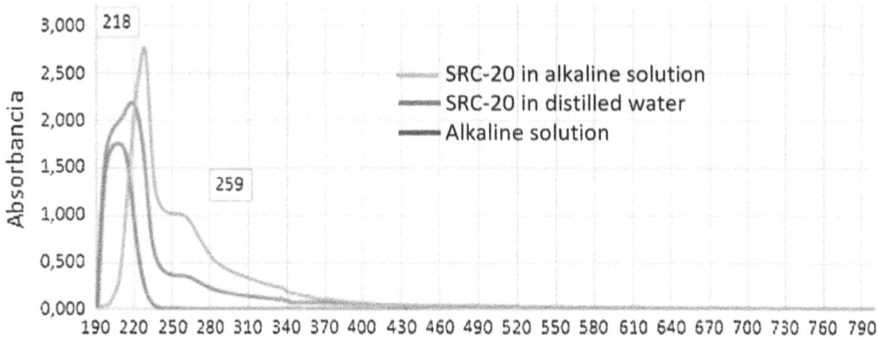

Fig. 2 UV-Vis spectra of Dynamon SRC-20 superplasticizer

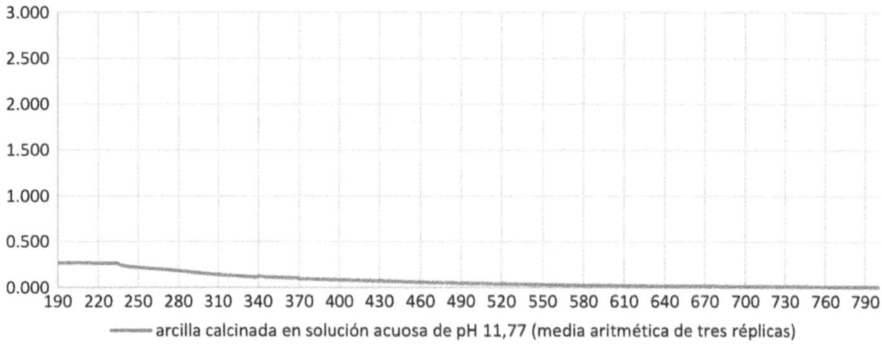

Fig. 3 UV-Vis spectrum of calcined clay solution products in basic solution

3.2 Stability of Superplasticisers in Basic Solution

The analysis of the spectra showed that they overlap perfectly (Dynamon SX 32) or almost perfectly (Dynamon SRC 20) in the region of the weak band and the surrounding band (from 228 to 288 nm).

Numerical analysis of absorbances at 259 nm showed randomness and little variation in the data. For the Dynamon SX 32 superplasticizer a standard deviation of 0.002 and a coefficient of variation in percent of 0.25 were obtained; while for the Dynamon SRC 20 superplasticizer they were 0.032 and 3.08 respectively Table 1.

3.3 Obtaining Calibration Lines

The Shapiro-Wilk test showed statistical significance higher than 0.05 for each one of the five data trios that made up the calibration line, for both superplasticisers;

Table 1 Absorbances measured at different times for Dynamon SX-32 and SRC-20 in basic solution

Dynamon SX-32	**t (min)**	**0**	**15**	**30**	**45**	**60**	**75**	**90**	**105**	**120**	**S**	**CV%**
	A	0.778	0.778	0.780	0.779	0.780	0.781	0.782	0.783	0.783	**0.002**	**0.25**
Dynamon SRC-20	**t (min)**	**0**	**15**	**30**	**45**	**60**	**75**	**90**	**105**	**120**	**S**	**CV%**
	A	1.075	1.004	1.068	0.990	1.048	1.059	1.036	1.021	1.081	**0.032**	**3.08**

Table 2 Variance analysis

	Sum of squares	gl	Mean square	F	Sig
Between groups	242.772	3	80,924	2.363	0.212
Inside groups	137.009	4	34.252		
Total	379.780	7			

Table 3 Adsorption in clays Yaguajay layer A and B

			SX 32	SRC 20
Yaguajay	**Layer A**	A1	1.040	1.244
		A2	1.020	1.232
		A3	1.058	1.234
		Media	1.039	1.237
		S	0.019	0.006
		CV%	1.83	0.52
		c (g/ml)	0.0038	0.0046
		% adsorp.	**60.60 ± 0.73**	**52.27 ± 0.21**
	Layer B	A1	1.268	1.551
		A2	1.233	1.537
		A3	1.224	1.490
		Media	1.242	1.526
		S	0.023	0.032
		CV%	1.87	2.09
		c (g/ml)	0.0046	0.0057
		% adsorp.	**52.33 ± 0.70**	**40.37 ± 0.92**

therefore we can affirm that all have normal distribution and in agreement parametric tests were applied, in addition all the normal deviations were lower than 0.02 and all the coefficients of variation lower than 3% (Table 2). The absorbance values are presented in Tables 3 and 4.

Table 4 Adsorption in clays Yaguajay layer C and Pontezuela

			SX 32	SRC 20
Yaguajay	**Layer C**	A1	1.074	1.235
		A2	1.039	1.237
		A3	1.090	1.277
		Media	1.068	1.250
		S	0.026	0.024
		CV%	2.44	1.90
		c (g/ml)	0.0039	0.0046
		% adsorp.	**59.45 ± 0.96**	**51.73 ± 0.92**
Pontezuela		A1	1.006	1.038
		A2	1.002	1.024
		A3	0.982	1.015
		Media	0.997	1.026
		S	0.013	0.012
		CV%	1.29	1.13
		c (mg/ml)	0.0036	0.0038
		% adsorp.	**62.± 0.33**	**60.94 ± 0.31**

4 Conclusions

1. The ultraviolet-visible molecular absorption spectra of superplasticisers SX 32 and SRC 20 present a band whose maximum absorption is at 259 nm, and is not interfered with by the basic aqueous medium or by the clay dissolution products in this medium, and therefore can be used for quantitative chemical analysis.
2. Calibration lines were obtained for both superplasticisers, with determination coefficients higher than 0.995 for the simple linear regression model, the intercepts obtained are statistically insignificant, therefore the Bouguer–Lambert–Beer law is complied with for the concentration range used.
3. The calcined clays studied adsorb 52–63% (Dynamon SX 32) and 40–60% (Dynamon SRC 20) of the amount of superplasticizer added.
4. There are no statistically significant differences between the adsorption values in percent, taking as grouping factors the type of superplasticizer and the type of clay.

References

1. Suraneni, P., Weiss, J.: Examining the pozzolanicity of supplementary cementitious materials using isothermal calorimetry and thermogravimetric analysis. Cem. Concr. Compos. (2017)

2. Fernandez, R., Martirena, F., Scrivener, K.L.: The origin of the pozzolanic activity of calcined clay minerals: a comparison between kaolinite, illite and montmorillonite. Cem. Concr. Res. **41**(1), 113–122 (2011)
3. Antoni, M., Rossen, J., Martirena, F., Scrivener, K.: Cement substitution by a combination of metakaolin and limestone. Cem. Concr. Res. **42**(12), 1579–1589 (2012)
4. Duverger, K.: Pozzolanic reactivity of the calcination products of the clays of Yaguajay deposit. Universidad Central "Marta Abreu" de Las Villas, Tesis de grado (2018)
5. Pérez, C., Ortiz, P.: Espectroscopy, vol. I (2010)

Effect of Gypsum Content on the Compressive Strength of LC^3 Cement

Christian Moreira and João Henrique Silva Rêgo

Abstract Due to environmental damage caused by the Portland cement manufacturing process, industry and academia have been required to reduce greenhouse gas emissions from the production process. The massive use of supplementary cimentitious materials (SCM's) is one of the most viable options for reducing clinker factor and pollutant emission. LC^3 cement is a type of cement that uses high additions contents, counting on the synergistic reaction between calcined clay and limestone to obtain mechanical strengths similar to those of traditional cements. This research intends to investigate the influence of the gypsum content on the compressive strength of a LC^3 cement. For the accomplishment of this study six mortars were produced. In three of them the cement used was produced with 3, 5, 7% gypsum, without SCM's. In three mortars, a 45% replacement cement was used (30% calcined clay and 15% limestone). The proportion of gypsum in these cements was 3, 5 and 7% in relation to the total mass. For the hardened state analysis, the compressive strengths were compared at the ages of 1, 3, 7, 28 and 91 days. A statistical analysis was performed on the results. From the results obtained it was possible to verify that the gypsum content influences the compressive strength of the LC^3 cement. The highest mechanical strengths were obtained with the use of larger quantities of gypsum.

Keywords Cement LC^3 · Compressive strength · Gypsum

1 Introduction

Portland cement is one of the most consumed products in the world. Concretes and mortars account for an immense share of the materials market and are vital to the development of the building and infrastructure industry. So great is its importance

C. Moreira · J. H. S. Rêgo (✉)
Programa de Pós-Graduação em Estruturas e Construção Civil – PECC/UNB, Brasília, Brazil
e-mail: jhenriquerego@unb.br

C. Moreira
e-mail: concretoleve.eng@gmail.com

© RILEM 2020

J. F. Martirena-Hernandez et al. (eds.), *Proceedings of the International Conference of Sustainable Production and Use of Cement and Concrete*, RILEM Bookseries 22,
https://doi.org/10.1007/978-3-030-22034-1_15

that the production and consumption indexes of portland cement usually serve as indicators of growth or retraction of the national economy.

Portland cement is based on clinker. The clinker manufacturing process involves high CO_2 emissions due to the burning of fuels and the decarbonation process of the rock. For the production of one tonne of cement, approximately 900 kg of CO_2 is emitted, which makes the production of the product responsible for 5–8% of all anthropogenic emissions of the gas [1].

Therefore, the cement production process is considered invasive and harmful to the environment. It has become a key point in the cement industry to develop products that meet ever more complex technological and commercial demands while being less aggressive in the middle without compromising the essential qualities of the product. Several solutions have been tried. One of the most viable is the large scale use of supplementary cementitious materials (SCM) in composition with clinker.

LC^3 is the acronym for limestone/calcined clay cement or calcareous filler cement and calcined clay. The hydration characteristics of the components of the LC^3 cement (calcined clay and limestone) are largely known. It is currently accepted that limestone filler reacts with clinker components when added in small amounts and accelerates the hydration of cement at the earliest ages. The filler creates extra points of heterogeneous nucleation, favors the reactions and improves the packaging of the system. It provides, however, the dilution effect, which limits its percentage of use.

The calcined clays act as pozzolans, reacting with the calcium hydroxide generated by the clinker hydration and forming extra C-A-S-H. The clays also act as nucleation points in the early ages, due to their fineness and also collaborate in the packaging of the system [2].

Several authors [1–8] state that there is a synergistic reaction between the calcined clay and the limestone filler. This reaction would further improve the microstructural characteristics of the paste, leading to an increase in mechanical strength and improvement in the pore system.

Several factors influence the performance of LC^3 cements and are important parameters in their formulation. Substitution content is one such factor [9]. In this research, the authors present a prognosis based on thermodynamic modeling that shows that up to 60% of clinker can be replaced without compromising performance compared to a normal cement.

In their paper, [10] worked the effect of the fineness of the different components in a mixture containing 55% of clinker, 30% of calcined clay and 15% of limestone in the mechanical properties and microstructure. The calcined clay contained 50% kaolinite. The increased fineness of calcined clay and clinker improves mechanical strength at all ages and the increase in finer filler has only effects at the earliest ages.

Since the work of [11] the ratio of calcined clay to limestone is the most used is 2:1, that is, two parts of calcined clay and one of filler. The same results were found by other authors [3, 12, 13]. The optimum ratio depends on the amount of kaolinite present in the clay [14].

In a study using "poor" kaolinite clays (44% kaolinite), calcined at 750 °C in combination with limestone, it was demonstrated that the hydrated phases found correspond to the pozzolanic reaction (contribution of the calcined clay) and to the stabilization of phases, modifying the porous structure and developing interesting mechanical properties with large reduction of energy consumption and CO_2 emissions. The authors concluded that composite cements with calcined clay and limestone exhibit mechanical strength similar to that at 28 days (although calcined clay has a low degree of purity) due to the interaction between these two components. The hydrated phases found correspond to the pozzolanic reaction (contribution of the calcined clay) and to the stabilization of phases, modifying the porous structure. The addition of calcined clay and filler increases the volume of pores with smaller diameter [12].

Very few papers focus on the influence of gypsum content on LC^3 cement behavior, although it is stated that the system is sensitive to this parameter. The optimum gypsum content in LC^3 cements remains a gap that demands further research [13].

2 Materials and Methods

For the accomplishment of the experiments, the cements were manufactured in laboratory. The reference cement was of type CP-I, made only with clinker and gypsum, without mineral additions. The clinker was ground together with 3, 5 and 7% gypsite, by mass, giving rise to three types of CP-I. Clinker and gypsum were introduced in a ball mill with 25 kg of steel and capacity of 7 kg of material. The grinding time was set to obtain a standard interval that would result in a cement thin enough to comply with the limits established by Brazilian standarts. The fineness was verified through its surface area obtained with the Blaine method. By trial and error 180 min was defined as best milling time.

To obtain the LC^3 cement, the 7 kg of ground materials represented 55% of the total cement mass. The other 45% were composed of clay material (30%) and limestone (15%). To perform a qualitative and quantitative analysis of the chemical constituents of the SCM's, the X-ray fluorescence assay was performed. The results obtained are set forth in Table 1.

The clinker X-ray fluorescence analysis showed that the requirements of [15] were met: the percentage of magnesium oxide less than or equal to 6.5% and of sulfur trioxide (not identified in the test) less than or equal to 4.0%.

The BET test measures the surface area of the materials, and it is important to note that this test considers the roughness of the material. Table 2 shows the surface area results of the additions that will be used in the cements.

Laser granulometry is important for checking the fineness of the material. Figure 1 shows the result of the test on the calcined clays and on the filler. To facilitate analysis, Table 3 demonstrates the diameters in which 10, 50 and 90% of the material passes, in addition to the average diameters.

Table 1 Result of X-ray fluorescence of calcined clay, clinker and filler

	Calc. clay (%)	Clinker (%)	Filler (%)
SiO_2	61.13	17.60	4.01
Al_2O_3	27.9	4.20	0.35
MgO	0.4	6.40	3.9
Fe_2O_3	2.2	3.70	0.13
CaO	3.45	65.52	48.73
Na_2O	0.04	0.30	0.009
K_2O	0.51	0.90	0.012
TiO^2	1.9	0.24	0.015
P_2O_5	0.04	0.14	0.009
MnO (%)	0	0.03	0.002
LOI (%)	2.2	0.67	42.37
Total (%)	99.77	99.7	99.53

Table 2 BET results material surface area BET (m^2/g)

Calc. Clay	13.979
Filler	3.774

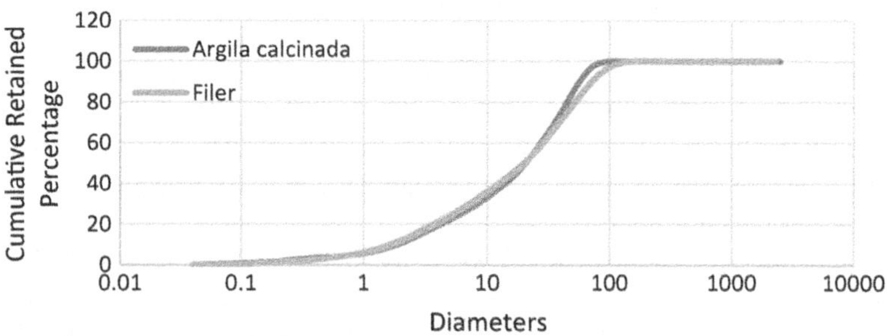

Fig. 1 Graphs of granulometry laser of calcined clay and limestone filler

Table 3 Diameters referring to 10, 50 and 90% of the passing material and average diameter

Diameters (μm)	Calcined clay	Filler
D10	1.86	1.66
D50	20.33	19.74
D90	56.59	71.48
Aver. diameter	25.14	29.16

Table 4 Percent formulations of reference cements and LC^3

	CODE	CL (%)	GP (%)	AC (%)	MK (%)	FA (%)
1	REF3	97	3	–	–	–
2	REF5	95	5	–	–	–
3	REF7	93	7	–	–	–
4	AC345A	52	3	30	–	15
5	AC545A	50	5	30	–	15
6	AC745A	48	7	30	–	15

CL—Clinker content, AR—calcined clay content, GP—gypsum content, FA—limestone content

Table 5 Dosage of superplasticizer additive and scattering obtained

Mortar	Spreading (mm)	Superplasticizer (%)
REF3	211.20	–
REF5	210.80	–
REF7	210.50	–
AC345A	209.90	0.75
AC545A	210.60	0.73
AC745A	211.50	0.70

2.1 Methods

At this stage of the research six formulations were investigated, according to Table 4.

The molded cylinders were ruptured at 1, 3, 7, 14, 28 and 91 days and their mechanical strength was evaluated, and 04 test pieces were used for each age. The final result for each age will be the average of the four test bodies. If the mean deviation is greater than 6% a new mean should be made by removing the most discrepant value. The tests were carried out according to the Brazilian compression standard ABNT - NBR 7215-1996 [17]. The mortars were prepared by varying the superplasticizer content to achieve the same consistency of the reference mortar.

3 Results

The spreading results and the superplasticizer additive content used in each mortar are shown in Table 5.

There was a loss of mortar workability with the addition of additional materials. This fact corroborates the observations made by [8], which warn of a considerable increase in the demand for water in mixtures containing mineral additions. It should be noted that none of the mortars required a superplasticizer content higher than that recommended by the manufacturer (1.5%).

Table 6 Classification by groups of resistance to compression at 28 days

Age	Mortar	Comp. strength (MPa)	St. dev (MPa)	Group 1	Group 2	Group 3
28 days	REF3	34.1	2.6479	X	X	
	REF5	38.4	1.7087		X	X
	REF7	40.4	1.0776			X
	AC345	32.4	5.5463	X		
	AC545	33.8	4.1153	X	X	
	AC745	36.2	1.1833	X	X	X

The results of compressive strength are presented in Table 6, divided into groups of statistical similarity. The groups were obtained by the Duncan test.

At 28 days the LC^3 cement with the highest gypsum content was the one that presented the highest results of compressive strength, comparable to the reference cements. The results confirm the adequate mechanical performance of the LC^3 cement, shown by the research [13–15].

It is observed that there was a homogeneity in the results. It was stated that each cement had an optimum content of SO_3, and that the lack or excess of the compound could negatively impact the mechanical strength of the cement, mainly because of its influence on the porosity of the paste. The ettringite is stabilized in the LC3 cements, not reacting further with the aluminates. Thus, hemicarboaluminates or monocarboluminates are formed, replacing the monosulphates. The free ettringite in the cement, present in larger quantities with higher gypsum indexes, can cause cracking in the late ages [16].

4 Conclusion

The ideal gypsum content, leading to the larger mechanical strength results, appears to be 7% for the LC^3 cement, although all it are adequate and in some cases even superior to the reference cements in terms of mechanical strength.

References

1. Scrivener, K.L.: Options for the future of cement. Indian Concr. J. **88**(7), 11–21 (2014)
2. Tironi, A., Scian, A.N., Irassar, E.F.: Ternary blended cement with limestone filler and kaolinitic calcined clay. In: Calcined Clays for Sustainable Concrete, pp. 195–201. Springer, Dordrecht (2015)
3. Antoni, M., Rossen, J., Martirena, F., Scrivener, K.: Cement substitution by a combination of metakaolin and limestone. Cem. Concr. Res. 1579–1589 (2012)

4. Tironi, A., Castellano, C.C., Bonavetti, V.L., Trezza, M.A., Scian, A.N., Irassar, E.F.: Kaolinitic calcined clays—Portland cement system: hydration and properties. Constr. Build. Mater. 215–221 (2014)
5. Maity, S., Bishnoi, S., Kumar, A.: Field application of limestone-calcined clay cement in India. In: Calcined Clays for Sustainable Concrete, p. 435–441. Springer, Dordrecht (2015)
6. Bishnoi, S., Maity, S., Kumar, M., Saxena, S.K., Wali, S.K.: Pilot scale production of limestone calcined clay cement. In: Calcined Clays for Sustainable Concrete, pp. 69–74. Springer, Dordrecht (2017)
7. Medepalli, S., Shah, V., Bishnoi, S.: Production of lab scale limestone calcined clay cements using low grade limestone. In: The 7th International Conference on Sustainable Built Environment (2016)
8. Bishnoi, S., Maity, S.: Limestone calcined clay cement: the experience in India this far. In: Calcined Clays for Sustainable Concrete, pp. 64–68. Springer, Dordrecht (2017)
9. Martirena, F., Scrivener, K.: Ecomaterials in low-cost housing. Connecting cutting-edge science with the grassroots. In: Technologies and Innovations for Development, pp. 101–111. Springer, Paris (2012)
10. Vizcaino Andrés, L.M., et al.: Industrial trial to produce a low clinker, low carbon cement. Materiales de Construcción **65**(317), 045 (2015)
11. Damidot, D., Lothenbach, Herfort, D., Glasser F.P.: Thermodynamic sand cement science. Cem. Concr. Res. **41**(7), 679–695 (2011)
12. Tironi, A., Scian, A.N., Irassar, E.F.: Ternary blended cement with limestone filler and kaolinitic calcined clay. In: Calcined Clays for Sustainable Concrete, pp. 195–201. Springer, Dordrecht (2015)
13. Antoni, M.: Investigation of cement substitution by blends of calcined clays and limestone. Tese de Doutorado, 223p. École Polytechnique Fédérale de Lausanne, Lausane (2013)
14. Rocchi, A.C.N.L., Oliveira, J.M.: Evaluation of the compressive strength of mortars with the use of LC3 cement with different proportions of limestone and calcined clay. Final Project Monograph, Department of Civil and Environmental Engineering, University of Brasília (2017)
15. Lins, D.: Analysis of Compressive Strength of LC^3 Cement and the Influence of Mineral Additions on its Performance 2017. Dissertation, University of Brasília, Brazil
16. Mohammed, S., Safiullah, O.: Optimization of the SO3 content of an Algerian Portland cement: study on the effect of various amounts of gypsum on cement properties. Constr. Build. Mater. **164**, 362–370 (2018)
17. ASSOCIAÇÃO BRASILEIRA DE NORMAS TÉCNICAS. NBR 7215: Cimento Portland - Determinação da resistência à compressão. Rio de Janeiro, 1996

Experimental Pilot Plant for Low Carbon Cements Development: The Cuban Innovative Project

G. Esperanza, Ivan Machado-Lopez and Jose Fernando Martirena-Hernandez

Abstract A new type of pilot plant for research and production of supplementary cementitious materials (SCMs) and others highly reactive pozzolan blends, has being constructed at the Central University of "Las Villas", with Swiss cooperation and technical support of IPIAC Nery. The raw materials: low grade kaolinite clays, limestone, clinker and gypsum are available in many geological settings, at low cost, and have the quality required for building admixtures production. Technological flow is quite simple, with eleven steps, and great level of mechanization and automation according to process needs. Pilot scale equipment was designed and constructed on the basis of European standards. Drying and calcination of clays, with a feed rate of 100–300 kg/h, is carried out in a rotary horizontal diesel fired kiln, at temperatures between 600 and 850 °C. Grinding, blending and homogenization of cement components is carried out at a rotary balls mill, and a centrifugal blower fan send the final product to a centrifugal solid gas separator, getting ready for packing and storage. The technical and operational facilities of this pilot plant offers a great innovative potential that will be used for academic, investigative and educational purposes, contributing additionally to materials production local programs development, a national priority.

Keywords Calcined clay · Cementitious materials · Low carbon cement · Pilot plant · LC3

G. Esperanza (✉)
Centro de Estudios de Química Aplicada, Universidad Central "Marta Abreu de Las Villas", Carretera a Camajuaní km 5.5, 54830 Santa Clara, Cuba
e-mail: guillermoep@uclv.edu.cu

I. Machado-Lopez · J. F. Martirena-Hernandez
Centro de Investigación y Desarrollo de Estructuras y Materiales, Universidad Central "Marta Abreu de Las Villas", Carretera a Camajuaní km 5.5, 54830 Santa Clara, Cuba

© RILEM 2020

J. F. Martirena-Hernandez et al. (eds.), *Proceedings of the International Conference of Sustainable Production and Use of Cement and Concrete*, RILEM Bookseries 22,
https://doi.org/10.1007/978-3-030-22034-1_16

1 Introduction

Ordinary Portland Cement (OPC) is recognized as a major construction material throughout the world and its production is responsible for about 5–8% of global CO_2 emission, a great environmental problem that will most likely be increased due to its exponential demand. Strategies to lower these environmental impacts are already applied by the industry and their future potential has been evaluated in detail by the IEA (International Energy Agency).

Several researchers at universities, institutes and even the cement industry are investigating alternatives to produce green and low cost building materials, mainly at emerging and develop economies.

Changes in the Cuban economy will cause a significant increase in demand for cement in the short term (2018–2020), and the current productive capacity of the Cuban cement industry cannot supply that demand. This will produce a deficit of estimated cement between 0.5 and 1 MM tons per year, which in the absence of alternatives will have to be imported at high costs.

The potential for ternary blends of limestone, calcined clay and clinker (LC3), to give good performance at 50% clinker content or less has been demonstrated in the collaborative research between the Laboratory of Construction Materials (LMC) at EPFL, Switzerland, and CIDEM in Cuba.

The introduction of LC3 cement could contribute to increase the capacity of cement production in the country by 17–45% in the short term, with small investments of short cycle, from using the current levels of industrial production of clinker.

Further important advantages of LC3 technology are: cheaper or similar production costs; the LC3 can by produced using the existing equipment in a cement plant; and the use of LC3 also does not need any major changes in concrete technology.

2 Project's Genesis

This pilot plant was acquired entirely with the collaboration of Habitat II (International Project), auspiced by COSUDE (Swiss cooperation). Equipment was manufactured by IPIAC-Nery Enterprises, based in Portugal. Unique of its kind in the region, located in a university, this semi-industrial facility will be used to carry out different scaling studies of the research related to the development of new construction materials, especially those aimed at the clinker replacement in cement industry, which is recognized as one of the strategic problems in this field [9].

3 Raw Materials

The basic raw materials studied for the production of SCMs (LC3 blends) in Cuba are kaolinite clays, limestone, gypsum and clinker [2]. On the other hand, is possible to use in these formulations, another type of compounds considered industrials and agricultural wastes: blast furnace slag, coal fly ash, silica fumes and sugar cane ash [5], or low grade natural minerals and pozzolanic materials (calcined clays from common clays). Always, the purpose is to reduce clinker content in OPC, and the emissions of CO_2 to atmosphere.

3.1 Kaolinite Clays

Clays are the weathering product of all rock types and as such are abundantly available close to the earth's surface in all geological settings. They are made up of silicon and aluminium oxides, which together constitute around three quarters of the earth's crust [8]. Cuba has many geographical sites and regions where kaolinite clays and limestone may be easily found and extracted at low cost. Pinar del Rio, Villa Clara, Ciego de Avila, Santiago de Cuba, and Holguin are some of these provinces [6]. These clayey materials with 40% of kaolinite content or higher when properly calcined, a highly reactive pozzolan can be obtained, practically as reactive as metakaolin (MK) [4, 10].

The substitution of clinker employing calcined clay in cement formulations, improve several properties of mortars and concretes, for example: compressive and flexural strength, pores microstructure, and durability. Also was demonstrated that, mortars and concretes strengths and its rheological properties, are strongly dependent on the kaolinite content of the calcined clay [7].

Clays are traditionally calcined in rotary kilns at temperatures around 600–850 °C, approximately half of the temperature needed for clinker production. Scaling up tests carried out at lab scale to pilot or industrial production in rotary kilns is extremely complex because the clays reactivity depends on the kaolinite content, the activation temperature, residence time and specific surface. A high grade of mineral's particles interaction, blending, and mass and heat transfer is needed to ensure the optimal conditions in clays calcination process. The burners and parameters such as rotation speed and fuel pressure should be adjusted to the calcining conditions in order to produce an optimized product according to quality requirements [1].

3.2 Limestone ($CaCO_3$)

Sedimentary rock composed mainly of calcium carbonate ($CaCO_3$), usually in the form of calcite or aragonite. It may contain considerable amounts of magnesium

Table 1 Average chemical composition of clays samples from Guayos, Manicaragua and Pontezuela

Clay	SiO_2	Al_2O_3	Fe_2O_3	CaO	MgO	SO_3	Na_2O	K_2O	TiO_2	LOI
Guay.	54.63	27.33	12.60	1.73	0.93	0.70	0.30	1.57	0.80	1.8
Manic	48.33	26.63	11.31	1.48	2.9	0.06	2.22	1.2	0	2.2
Pontez	45.71	28.04	21.49	0.14	0.82	0.27	0.17	0.54	0.7	1.8

Table 2 Properties of limestone from Guayos, Santi Spíritus

Compound	SiO_2	Al_2O_3	Fe_2O_3	CaO	MgO	SO_3	$CaCO_3$
(%) of mass	3.02	0.94	0.49	51.59	0.71	4.17	94.4

Table 3 Chemical analysis of clinker and gypsum from Siguaney cement plant

Compound	SiO_2	Al_2O_3	Fe_2O_3	CaO	MgO	SO_3	LOI	IW
Clinker	20.97	4.77	4.42	65.42	1.23	0.26	0.40	0.17
Gypsum	8.10	2.03	1.97	30.27	2.81	31.39	20.91	9.91

Table 4 Pilot plant technological equipment

Main	Auxiliary
Rotary kiln, Balls mill Heat Exchanger, S-G separators (cyclones)	Conveyor belt, Additive storage silos Blowers, Endless Screw, Burner

carbonate (dolomite) as well; minor constituents also commonly present include clay, iron carbonate, feldspar, pyrite, and quartz. Its reaction with alumine, present in clays, conduces to the formation of monocarboaluminate hydrate $C_4A\underline{C}H_{11}$ [3].

The typical chemical composition (weight %) of a low grade clays, limestone, clinker, and gypsum to be used in LC3 or LC2 manufacture, are shown in Tables 1, 2, and 3.

4 Brief Description of Technological Process

Technological flowsheet of pilot plant for low carbon cement and others highly reactive pozzolan blends manufacture is quite simple (Fig. 1), with eleven well-known operational stage and a great level of mechanization and automation according to process needs and technical specifications. The main and auxiliary equipments, installed and ready to use, are presented in Table 4. These pilot scale equipment was entirely designed and constructed by IPIAC-Nery, on the basis of European standards, making possible its operation under several technical conditions (Table 5; Fig. 2).

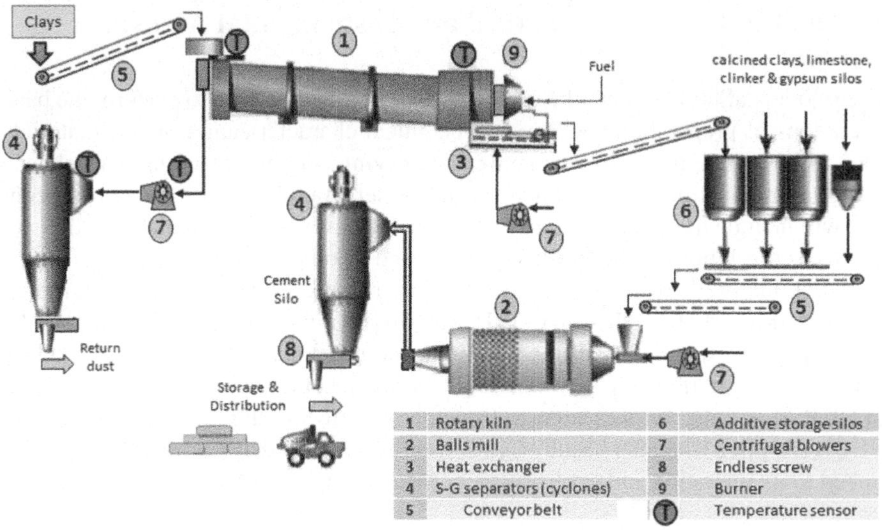

Fig. 1 Low carbon cement manufacture process flow chart

Table 5 Technical parameters of the rotary kiln

Parameter	Pilot scale rotary kiln
Inside diameter	0.42 m
Total length	3.75 m
Rotation speed	1.8–3.5 rpm
Throughput average	100–300 kg/h
Temperature range	600–900 °C
Heating	Direct diesel
Kiln atmosphere	Air

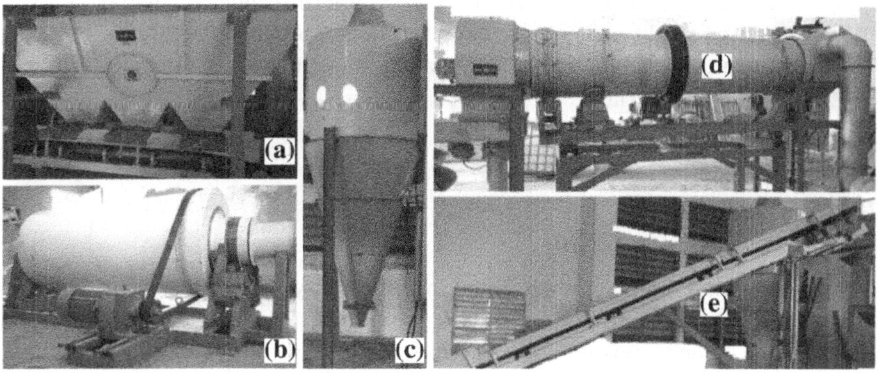

Fig. 2 Silos (**a**), balls mill (**b**), cyclone (**c**), rotary kiln (**d**), and conveyor belt (**e**)

5 Quality Control. Analytical and Instrumental Assurance

The pilot installation has two laboratories. The first will be used to perform the basic operations of preparation and physical chemical characterization of raw materials, such as: grinding, mixing, homogenization, drying, sieving, blending, particle size distribution, specific density, fineness, and other granulometric tests. The second lab, will include the basic instrumental equipment for chemical characterization and quality control, both of the finished product and of other materials.

Additionally, there is another group of more specific tests, such as: compressive and flexural strength, DTA, TGA (thermogravimetry), AAS, FTIR, UV-VIS spectrophotometry, X-Ray diffraction, chloride resistance, isothermal calorimetry, porous pH, hydration stop, carbonation experiments, etc., which can be performed in CIDEM (Centre for Research & Development of Structures and Materials) and CEQA (Applied Chemistry Study Centre), entities attached to the pilot plant, directly linked to its operation and exploitation of its technical possibilities.

6 Pilot Plant Potentialities

Once in full operation regime, the technical and operational facilities installed in this pilot plant, offers a great innovative potential that could be used for several purposes. Some actions that may be executed rapidly are the followings:

- Semi-industrial trials, with national or international customers.
- Scale up test for new building blends, additives or cement formulations.
- Optimization studies of several technical parameters.
- Assessment of new ternary cement blends; training of technical personal.
- Raw materials preparation (grinding, blending, drying, homogenization).
- Raw materials and final products characterization (instrumental analytics).
- Assessment of final products according to Cuban normative (NC).
- Development, production and commercialization of innovative products in the area of importations materials.

7 Conclusions

Unique of its kind in the region, located in a university, this semi-industrial facility will be used to carry out different scaling studies of the researches related to the development of new construction materials, especially those aimed at the clinker replacement in cement industry, which is recognized as one of the strategic problems in this field.

It might possible to produce in our country supplementary cementitious materials (SCMs) at pilot scale, as clinker replacement material to improve quality and reduce the cost of industrial cement manufacture and many others construction materials.

The technical and operational facilities of this pilot plant offers a great innovative potential that will be used for academic, investigative and educational purposes, contributing additionally to materials production local programs development, a national priority.

References

1. Almenares, R., Vizcaino, L.M., Damas, S., Antoni, M., Alujas, A., Martirena, F.: Industrial calcination of kaolinitic clays to make reactive pozzolans. Case Stud. Constr. Mat. (2017). https://dx.doi.org/10.1016/j.cscm.2017.03.005
2. Alujas, A., Fernández, R., Quintana, R., Scrivener, K.L., Martirena, F.: Pozzolanic reactivity of low grade kaolinitic clays: Influence of calcination temperature and impact of calcination products on OPC hydration. Appl. Clay Sci. **108**, 94–101 (2015)
3. Antoni, M., Rossen, J., Martirena, F., Scrivener, K.L.: Cement substitution by a combination of metakaolin and limestone. Cem. Concr. Res. **42**(12), 1579–1589 (2012)
4. Avet, F., Scrivener, K.: Investigation of the calcined kaolinite content on the hydration of Limestone Calcined Clay Cement (LC3). Cem. Concr. Res. **107**, 124–135 (2018)
5. Jing, Yu., Cong, Lu, Leung, Christopher K.Y., Li, Gengying: Mechanical properties of green structural concrete with ultrahigh-volume fly ash. Constr. Build. Mater. **147**, 510–518 (2017). https://doi.org/10.1016/j.conbuildmat.2017.04.1880950-0618
6. Poll-Legrá, L., Almenares, R., Romero, Y., Alujas, A., Leyva, C., Martirena, J. (2016). Evaluación de la actividad puzzolánica del material arcilloso del depósito La Delta Moa, Cuba. *Minería y Geología*. V 32, No. 1, enero-marzo, p. 15–27 (2016). ISSN 1993 8012
7. Scrivener, K., Avet, F., Maraghechi, H.: Impacting factors and properties of limestone calcined clay cements (LC3). Green Mat. **7**(1), 3–14 (2019). https://doi.org/10.1680/jgrma.18.00029
8. Scrivener, K.: Options for the future of cement. Indian Concr. J. **88**(7), 11–21 (2014)
9. Shashank, B., Soumen, M., Amit, M., Shiju, J., Sreejith, K.: Pilot scale manufacture of limestone calcined clay cement: The Indian experience. Indian Concr. J. **88**(7), 22–28 (2014)
10. Vizcaíno, L., Sánchez, S., Pérez, A., Damas, S., Scrivener, K., Martirena, F.: Industrial trial to produce low clinker, low carbon cement. Press. J. Mater. Construcción, IIETC, Spain. **65**(2015) e045 (2015). ISSN-L: 0465-2746. https://dx.doi.org/10.3989/mc.2015.00614

Ultrasonication Effect of Silica Fume on Compressive Strength of Cement Pastes

Yuri Sotero Bomfim Fraga, João Henrique da Silva Rêgo and Valdirene Maria Silva Capuzzo

Abstract Portland cement production is responsible for about 6–8% of the world's CO_2 emissions. To reduce this emission, several researchers have studied the replacement of clinker by highly reactive supplementary cementitious materials such as rice husk ash, metakaolin and silica fume in the production of high performance cementitious materials. In these studies, it was verified that the silica fume agglomerates due to the small size of its particles and with that its reactive potential is reduced. Thus, the objective of this research is to verify the ultrasonication effect of silica fume on the mechanical performance of cement pastes. Three cement pastes were studied, a reference with 100% Portland cement CP V-ARI and two with 10% substitution of the Portland cement by silica fume, one in its natural state and the other ultrasonicated. It was prepared 12 test specimens for each trait by setting the scattering at 140 ± 10 mm in the mini slump test. Three specimens were ruptured at 1, 3, 7 and 28 days of hydration. With the results, a statistical analysis ANOVA was performed through the Duncan's test in Statistica v10 software. Through the results, it was verified that the ultrasonication of the silica fume increased its reactive potential and consequently its mechanical performance. Thus, by ultrasonication of silica fume it is possible to produce high-performance cementitious materials with lower consumption of Portland cement and consequently reduce the emission of CO_2 into the atmosphere.

Keywords Silica fume · Ultrasonication · Mechanical performance · Statistical analysis

1 Introduction

Cementitious materials are the main raw material used in civil construction. Several researches have been carried out with the objective of improving the mechanical

Y. S. B. Fraga (✉) · J. H. da Silva Rêgo · V. M. S. Capuzzo
University of Brasilia, Brasilia, Distrito Federal, Brazil
e-mail: yurisotero.engcivil@gmail.com

© RILEM 2020

J. F. Martirena-Hernandez et al. (eds.), *Proceedings of the International Conference of Sustainable Production and Use of Cement and Concrete*, RILEM Bookseries 22,
https://doi.org/10.1007/978-3-030-22034-1_17

performance of these materials with lower consumption of cement. With high performance concrete (HPC), for example, it becomes possible to produce slimmer structures. This reduces considerably the consumption of cement and the emission of CO_2 in the atmosphere [16].

One of the most studied materials for the production of high performance cementitious materials is silica fume (SF). The SF is a highly reactive pozzolanic material due to its high SiO_2 content, generally higher than 90%, its high degree of amorphism and its very fine grain size. It acts in two ways on the microstructure of cementitious materials: the pozzolanic (chemical) effect in which the silica fume reacts with the CH released during the hydration of the cement, forming additional C-S-H that acts to fill the capillary voids and the filler effect which densifies the microstructure, since the small particles fill the spaces between the grains of Portland cement, reducing its porosity [8].

In the research conducted by Zhang et al. [19], by replacing 10% of Portland cement by SF in cement pastes with w/b ratio = 0.50, an increase in compressive strength was observed in relation to the paste with 100% Portland cement. The increase in mechanical strength was observed in the three evaluated ages: 3, 7 and 28 days. This increase in the mechanical performance of cementitious materials with silica fume in the up to 10% Portland cement substitution content was also observed by several authors [1, 9, 10, 13, 15].

However, the silica fume particles have a high agglomeration tendency, which reduces their pozzolanic effect in cementitious materials [10, 21]. One of the most researched and promising methods for the dispersion of small particles is through ultrasonicators. In addition to dispersing microparticles such as silica fume, these devices can also disperse nanoparticles such as nanosilica [12, 14, 21].

Nevertheless, there are still some gaps in the literature on the subject such as the ultrasonication parameters and its influence on the mechanical properties of cementitious materials. Thus, this research has as objective to evaluate the ultrasonication effect of silica fume used in substitution to Portland cement in the compressive strength of cement pastes.

2 Methodology

2.1 Materials

To carry out this study was used Portland cement CP V-ARI, meeting the Brazilian standard, brand Cauê, Intercement group, with fineness index of 0.9% obtained according to the methodology of ABNT NBT 11579 [3] and specific gravity equal to 3.00 g/cm^3 according to ABNT NBR 16605 [4]. The start and end of setting time were equal to 2 h00 min and 2 h55 min, respectively, according to ABNT NBR 16607 [6] and compressive strength of 25.4 MPa at 1 day, 36.1 MPa at 3 days and 42.7 MPa at 7 days of hydration obtained according to ABNT NBR 16697 [7].

Table 1 Chemical composition of the materials

–	Chemical composition (%)										
	SiO_2	Al_2O_3	MgO	Fe_2O_3	CaO	Na_2O	K_2O	TiO_2	P_2O_5	MnO	LOI
CP V-ARI	19.91	4.20	1.78	2.20	63.61	0.38	0.36	0.23	0.17	0.07	3.36
SF	94.33	<0.01	0.49	0.19	0.78	0.39	1.04	0.02	0.14	0.06	2.28

Table 2 Nomenclature of cement pastes

Paste	Composition
REF	100% CP V-ARI
10SF	90% CP V-ARI +10% SF (natural state)
10SFU	90% CP V-ARI +10% SF (ultrasonicated)

The densified silica fume (SF) used was of the brand Tecnosil with specific gravity equal to 2.16 g/cm^3 according to ABNT NBR NM 16605 [4], apparent density of 595.07 kg/m^3 and fineness in the sieve 45 μm equal to 9.1%. The superplasticizer (SP) used was Grace's Tecflow 50 N with specific mass determined with the pycnometer equal to 1.095 in relation to the specific gravity of the water, the pH value was 3.8 and the solids content was 40%, determined according to the procedure of ABNT NBR 10908 [2]. The chemical composition of CP V-ARI and SF was shown in Table 1.

2.2 Production of Cement Pastes

The ultrasonication process of the SF consisted of placing it in a glass beaker with the superplasticizer and with 50% of water of the pastes. Sonics VCX 750 ultrasonicator was used with ultrasonic probe model CV334. The parameter of ultrasonication were adapted from Barkoula et al. [12]: amplitude of 75%, effective time of ultrasonication of 10 min in cycles of 50 s with application of ultrasonic pulses and 10 s without application of pulses. Around the beaker with the mixture to be ultrasonicated, ice was used to avoid evaporation of water.

Three pastes with water/binder ratio of 0.40 were prepared according to the methodology of ABNT NBR 16606 [5]. The 10% content of CP V-ARI was replaced by SF in its natural state and after the ultrasonication process. The nomenclature of the pastes was shown in Table 2.

For the production of the pastes the superplasticizer content was varied until the consistency of 140 ± 10 mm in the mini slump test described by Kantro [18]. To not change the water/binder ratio, the amount of water of the superplasticizer was discounted at 60% of its mass. The quantity of materials used to prepare seven specimens was shown in Table 3.

Table 3 Quantitative materials for molding 7 specimens of cement pastes

PASTE	MATERIAL				
	CP V-ARI (g)	SF (g)	Superplasticizer		Water (g)
			(g)	(%)	
REF	2100.0	0.0	10.5	0.50	833.7
10SF	1890.0	210.0	12.6	0.60	832.5
10SFU	1890.0	210.0	14.7	0.70	831.2

Twelve cylindrical specimens 50 × 100 mm were molded for each paste. After molding, the specimens were placed in a humid chamber for 24 h and then under submerged cure to 1, 3, 7 or 28 days.

2.3 Analysis of the Compressive Strength

The specimens were rectified and the compressive strength of 3 test specimens was checked for each trait by age studied: 1, 3, 7 and 28 dias. To perform the compressive strength of the pastes, an ANOVA variance analysis was performed to verify the significance of the compressive strength between the pastes with SF in its natural state or SF after ultrasonication process. This test was performed in Microsoft Office Excel 2010 software, in the "Anova: single factor" option, with 5% of significance.

Then the Duncan's test was performed to verify the heterogeneity between the groups of compressive strength taking into account the mean and the standard deviation of the samples at 1, 3, 7 and 28 days. This test was performed in Statistica software, version 10, with a 95% confidence interval.

3 Results and Discussion

At all ages the results of ANOVA were significant for the compressive strength of the pastes. To verify the differences in mechanical performance, the Duncan's test was performed for all ages, as shown in Table 4.

At 1 day of hydration, the REF and 10SFU pastes were classified in group 2, so statistically presented higher compressive strength than the 10SF paste that was classified in group 1. In this age the ultrasonicated SF contributed to the filler effect to compensate the reduction of 10% of Portland cement that was replaced [10]. Thus, due to the agglomeration of SF, there was a decrease in mechanical performance of the 10SF paste compared to the pastes REF and 10SFU [21].

At 3 and 7 days the paste 10SFU resulted in higher compressive strength than the paste 10SF. The paste REF showed a tendency to decrease the compressive strength

Table 4 Duncan's test for the compressive strength of pastes at 1, 3, 7 and 28 days

Age	Paste	Mean of compressive strength (MPa)	Standard deviation (MPa)	Group 1	Group 2
1 day	REF	37.0	1.0693		X
	10SF	31.2	0.9609	X	
	10SFU	36.3	0.2000		X
3 days	REF	55.3	2.7875	X	X
	10SF	53.0	2.5502	X	
	10SFU	62.4	4.9238	X	X
7 days	REF	58.8	5.2482	X	X
	10SF	53.2	5.0501	X	
	10SFU	68.0	4.6000		X
28 days	REF	66.8	5.1209		X
	10SF	55.7	4.0079	X	
	10SFU	71.9	3.4530		X

Note Groups are independent between ages. Thus, groups of one age are not related to groups of the other ages

compared to the paste 10SFU and increase the compressive strength compared to the paste 10SF.

As with 1 day, at 28 days the 10SFU and REF pastes presented a statistically compressive strength higher than the 10SF paste. This demonstrates that the use of SF in its natural state (agglomerated) reduces its effectiveness of the pozzolanic reaction with calcium hydroxide (CH), as reported by Barbhuiya et al. [11].

In addition, according to Juenger and Ostertag [17], the high degree of agglomeration of the SF generates fissures originating in the center of the SF that can contribute to the non-effectiveness of the SF in the gain of the mechanical performance of the cementitious materials.

4 Conclusions

Through this research, it was possible to observe that the ultrasonication process was efficient to disperse the SF particles. At all ages the paste 10SFU resulted in a higher compressive strength compared to the paste 10SF.

It was also possible to infer that the ultrasonication of the silica fume contributed to the increase in the compressive strength of the paste 10SFU in relation to the paste REF at 3 and 7 days due to the filler effect and the pozzolanic effect at these ages.

It is estimated that ultrasonication of SF can further improve the mechanical performance and durability of cementitious materials. Thus, it is recommended to study the ultrasonication of SF with different parameters and at more advanced ages such as 91 days and 180 days of cementitious materials. It is suggested that, in addition to the compressive strength, the rheological properties, the heat of hydration, the microstructure and the durability of the cementitious materials are studied.

References

1. Alonso-Domínguez, D., Álvarez-Serrano, I., Reyes, E., Moragues, A.: New mortars fabricated by electrostatic dry deposition of nano and microsilica additions: Enhanced properties. Constr. Build. Mater. **135**, 186–193 (2017)
2. ABNT, Associação Brasileira de Normas Técnicas. NBR 10908: Aditivos para argamassa e concreto - Ensaios de caracterização. Rio de Janeiro (2008)
3. ABNT, Associação Brasileira de Normas Técnicas. NBR 11579: Cimento Portland — Determinação do índice de finura por meio da peneira 75 μm (n° 200). Rio de Janeiro (2013)
4. ABNT, Associação Brasileira de Normas Técnicas. NBR 16605: Cimento Portland e outros materiais em pó – Determinação da massa específica. Rio de Janeiro (2017)
5. ABNT, Associação Brasileira de Normas Técnicas. NBR 16606: Cimento Portland — Determinação da pasta de consistência normal. Rio de Janeiro (2017)
6. ABNT, Associação Brasileira de Normas Técnicas. NBR 16607: Cimento Portland — Determinação dos tempos de pega. Rio de Janeiro (2017)
7. ABNT, Associação Brasileira de Normas Técnicas. NBR 16697: Cimento Portland — Requisitos. Rio de Janeiro (2018)
8. Aïtcin, P.C.: Concreto de alto desempenho. PINI, São Paulo (2000)
9. Amin, M.S., Abo-El-Enein, S.A., Rahman, A.A., Alfalous, K.A.: Artificial pozzolanic cement pastes containing burnt clay with and without silica fume. J. Therm. Anal. Calorim. **107**, 1105–1115 (2012)
10. Bajja, Z., Dridi, W., Darquennes, A., Bennacer, R., Bescop, P.L., Rahim, M.: Influence of slurried silica fume on microstructure and tritiated water diffusivity of cement pastes. Constr. Build. Mater. **132**, 85–93 (2017)
11. Barbhuiya, S., Chow, P., Das, A.: Nanomechanical properties of cement paste containing silica fume. In: Proceedings of International Conference on Architecture and Civil Engineering (ICAACE'14). Anais [...]. Dubai, pp. 92–97 (2014)
12. Barkoula, N.M., Ioannou, C., Aggelis, D.G., Matikas, T.E.: Optimization of nano-silica's addition in cement mortars and assessment of the failure process using acoustic emission monitoring. Constr. Build. Mater. **125**, 546–552 (2016)
13. Bhattacharya, M., Harish, K.V.: An integrated approach for studying the hydration of Portland cement systems containing silica fume. Constr. Build. Mater. **188**, 1179–1192 (2018)
14. Cai, Y., Hou, P., Cheng, X., Du, P., Ye, Z.: The effects of nanoSiO2 on the properties of fresh and hardened cement-based materials through its dispersion with silica fume. Constr. Build. Mater. **148**, 770–780 (2017)
15. Chen, Y., Deng, Y., Li, M.: Influence of nano-SiO_2 on the consistency, setting time, early-age strength, and shrinkage of composite cement pastes. Adv. Mat. Sci. Eng. 1–8 (2016)
16. Citek, D., Rydval, M., Rehacek, S., Kolisko, J.: Material properties of ultra-high performance concrete in extreme conditions. Key Eng. Mater. **711**, 157–162 (2016)
17. Juenger, M.C.G., Ostertag, C.P.: Alkali–silica reactivity of large silica fume-derived Particles. Cem. Concr. Res. **34**, 1389–1402 (2004)
18. Kantro, D.L.: Influence of water-reducing admixtures on properties of cement paste—A miniature slump test. Cem. Concr. Aggr. **2**, 95–102 (1980)
19. Zhang, X., Du, X., Zhao, X., Zhang, R., Hou, P., Zhou, Z., Cheng, X.: The synergistic effect of nano-SiO_2 with silica fume in cement-based material. J. Sustain. Cement-Based Mat. **6**, 267–279 (2016)

20. Zhang, Z., Zhang, B., Yan, P.: Hydration and microstructures of concrete containing raw or densified sılica fume at different curing temperatures. Constr. Build. Mater. **121**, 483–490 (2016)
21. Zhang, Z., Zhang, B., Yan, P.: Comparative study of effect of raw and densified silica fume in paste, mortar and concrete. Constr. Build. Mater. **105**, 82–93 (2016)

Influence of Fixation of Consistency or Superplasticizer Content on Strength of Cement Pastes with Silica Fume or Nanosilica

Yuri Sotero Bomfim Fraga, Lucas de Paula Vasques, João Henrique da Silva Rêgo and Valdirene Maria Silva Capuzzo

Abstract Currently, the production of pastes, mortars and concretes with replacement of Portland cement by highly reactive cementitious materials to reduce the environmental impact caused by cement production is studied. Among these materials, silica fume and nanosilica stand out due to their properties that improve the mechanical performance and durability of pastes, mortars and concretes. Most of the research involving these materials uses superplasticizer to fix the consistency of the researched material. Despite this, there are researches that fix the superplasticizer content in their research. Thus, the objective of this research is to verify the influence of the fixation of the consistency or the superplasticizer content on the mechanical performance of cement pastes with silica fume or nanosilica. For this, five specimens were studied replacing Portland cement by 2% of colloidal nanosilica or 10% of silica fume. The specimens were molded by fixing the superplasticizer content or fixing the consistency of the pastes by the mini slump test. The compressive strength of three specimens at 1, 3 and 7 days was verified. The results showed that when incorporating silica fume or nanosilica in Portland cement, it became difficult to mold the pastes when the superplasticizer content was fixed, which influenced its mechanical performance. In order to carry out research with replacement of Portland cement by highly reactive cementitious materials for the production of pastes, mortars and concretes, it is recommended to fix the consistency of these materials.

Keywords Silica fume · Nanosilica · Consistency · Superplasticizer content · Mechanical performance

1 Introduction

Currently, the academic community aims to develop cementitious materials with decreasing amounts of clinker, due to the fact that this component is the main responsible for the high energy consumption and CO_2 release in the atmosphere during the

Y. S. B. Fraga (✉) · L. de Paula Vasques · J. H. da Silva Rêgo · V. M. S. Capuzzo
University of Brasilia, Brasilia, Distrito Federal, Brazil
e-mail: yurisotero.engcivil@gmail.com

© RILEM 2020

J. F. Martirena-Hernandez et al. (eds.), *Proceedings of the International Conference of Sustainable Production and Use of Cement and Concrete*, RILEM Bookseries 22,
https://doi.org/10.1007/978-3-030-22034-1_18

Portland cement production [7]. According to Scrivener [9], the most viable option to obtain more sustainable cements involves mixing Portland cement clinker with increasing levels of Supplementary Cement Materials (SCMs).

Among the pozzolanic SCMs, silica fume (SF) and nanosilica (NS) can be highlighted. These materials have been the object of several studies due to their high reactive potential and gain of strength. Kupwade-Patil et al. [4] demonstrates that silica fume can act in favor of the durability of the composites. The use of SF resulted in a considerable decrease in the porosity of the cementitious materials. However, according to Li et al. [6], the use of SF demands elevation of the w/b ratio or use of superplasticizer in order to maintain the workability of the paste, mortar or concrete.

According to Ghafari et al. [3] due to its smaller particle size, the use of NS generates a higher level of particle packing (filer effect) resulting in a denser microstructure and consequently leading to a better mechanical performance of the material. In addition, the high specific surface makes the material more reactive than the silica fume, besides generating higher demand of water to maintain the workability of the cementitious mixture [2].

Most of the researches involving cement pastes with SF and/or NS fix the consistency in the fresh state of the samples by means of the variation of the superplasticizer content, for example, in the research of Narattha et al. [8] and Sharkawi et al. [10]. However, there are studies that fix the superplasticizer content to evaluate the properties of the material, which results in the variation of the consistency of the samples, as exemplified by the work of Vasques et al. [11]. Therefore, this study aims to verify the influence of the fixation of the superplasticizer content or the consistency in the mechanical performance of cement pastes with SF or NS.

2 Methodology

2.1 Materials

In this research it was used the Portland cement CP V-ARI, meeting the Brazilian standard, Cauê—Intercement group, with a fineness index equal to 0.9% and specific gravity equal to 3.00 g/cm^3. The start and finish setting times were equal to 2 h00 min and 2 h55 min, respectively, and the compressive strength of 25.4 MPa at 1 day, 36.1 MPa at 3 days and 42.7 MPa at 7 days of hydration.

The densified silica fume (SF) used was of the brand Tecnosil with specific gravity equal to 2.16 g/cm^3, apparent density equal to 595.07 kg/m^3 and fineness in the sieve 45 μm equal to 9.1%. The colloidal nanosilica (NS) used was the AkzoNobel Cembinder 30, with solids content of 30% and mean diameter of 22.75 nm obtained by the DLS test. The superplasticizer (SP) used was Grace's Tecflow 50 N with specific gravity, determined with the pycnometer, equal to 1.095 in relation to the specific gravity of the water, the pH value was 3.8 and the solids content was 40%. The chemical composition of CP V-ARI, SF and NS was shown in Table 1.

Table 1 Chemical composition of the materials

–	Chemical composition (%)										
	SiO_2	Al_2O_3	MgO	Fe_2O_3	CaO	Na_2O	K_2O	TiO_2	P_2O_5	MnO	LOI
CP V-ARI	19.91	4.20	1.78	2.20	63.61	0.38	0.36	0.23	0.17	0.07	3.36
SF	94.33	<0.01	0.49	0.19	0.78	0.39	1.04	0.02	0.14	0.06	2.28
NS	90.61	<0.01	0.10	0.11	0.02	2.03	0.03	0.03	<0.01	<0.01	6.19

Table 2 Composition of cement pastes

Paste	Composition
REF	100% CP V-ARI
2NS-FC	98% CP V-ARI + 2% NS (fixing the consistency)
2NS-FS	98% CP V-ARI + 2% NS (fixing the superplasticezer content)
10SF-FC	90% CP V-ARI + 10% SF (fixing the consistency)
10SF-FS	90% CP V-ARI + 10% SF (fixing the superplasticezer content)

2.2 *Production of Cement Pastes*

Five cement pastes were prepared according to the methodology of ABNT NBR 16606 [1], all with w/b ratio = 0.40. Portland cement was partially replaced by SF or NS and during the production of the pastes the consistency or superplasticizer content was fixed. The consistency of the pastes was measured by the mini slump test described by Kantro [5]. The spreading established for the reference paste was equal to 140 ± 10 mm, the same scattering for the 2NS-FC and 10SF-FC pastes. In the 2NS-FS and 10SF-FS pastes the superplasticizer content was the same added to the REF paste. To maintain the exact w/b ratio, the amount of water of the colloidal nanosilica (70% of its mass) and the amount of water of the superplasticizer additive (60% of its mass) were discounted. Table 2 shows the nomenclatures of the pastes studied.

Nine cylindrical 50 × 100 mm specimens were molded for each one of the pastes mentioned in Table 2. After molding, the specimens were placed in a humid chamber for 24 h and then submitted to submerged cure up to 1, 3 or 7 days. The quantity of materials used to prepare seven specimens is shown in Table 3.

Table 3 Quantitative materials for molding 7 specimens of cement pastes

PASTE	MATERIAL						
	CP V-ARI (g)	SF (g)	Colloidal NS 30% (g)	Superplasticizer		Water (g)	Consistency (mm)
				(g)	(%)		
REF	2100.0	0.0	0.0	10.5	0.50	833.7	145
2NS-FC	2058.0	0.0	140.0	23.1	1.10	728.2	139
2NS-FS	2058.0	0.0	140.0	10.5	0.50	735.7	–
10SF-FC	1890.0	210.0	0.0	12.6	0.60	832.5	136
10SF-FS	1890.0	210.0	0.0	10.5	0.50	833.7	135

2.3 *Analysis of the Compressive Strength*

An ANOVA variance analysis was performed to verify the difference between the compressive strength of the pastes. This test was performed in Microsoft Excel software, in the ANOVA: single factor option, with a significance of 5%.

After that, the Duncan's test was performed to verify the heterogeneity between the groups of compressive strength taking into account the mean and the standard deviation of the samples at 1, 3 and 7 days. This test was performed in Statistica software, version 10, with a 95% confidence interval.

3 Results and Discussion

3.1 *Consistency*

During the production of the pastes with fixed superplasticizer content it was observed that this fixation directly influenced the workability of the mixture. As shown in Table 3, the inclusion of SF or NS generates a demand for water or the use of superplasticizer in the mixture, especially in the case of NS. However, when the w/b ratio and the SP content were maintained, the pastes had their workability affected. While the workability of the 10SF-FS paste was lower than that of 10SF-FC, the effect on 2NS-FS was more intense and resulted in a paste with very low workability, as demonstrated by the mini slump test that couldn´t even be performed. This effect on the FS pastes generated difficulties to mold the specimens, affecting the quality and resulting in several voids in these (mainly in 2NS-FC). The consistencies of 2NS-FS and 10SF-FS are shown in Fig. 1a, b, respectively.

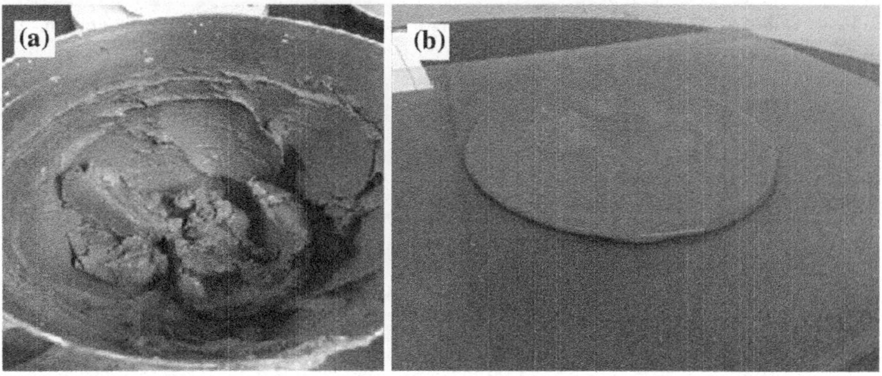

Fig. 1 Consistency of **a** 2NS-SF; **b** 10SF-FS

3.2 Compressive Strength

A statistical analysis was performed with the results of compressive strength at 1, 3 and 7 days of hydration. First, an analysis of variance ANOVA was performed, which showed that the results of compressive strength were statistically different. With this, the Duncan's test was performed, as shown in Table 4.

At 1 day, there is no relevant difference between the pastes containing SF. This probably occurs due to the fact that SF particles exhibit their effects only at later ages. On the other hand, the pastes with NS presented significant difference. The 2NS-FC paste resulted in higher compressive strength than the 2NS-FS paste.

At 3 days by comparing the pastes with NS it was observed that 2NS-FS had lower performance in compared to the paste 2NS-FC. The low workability of the paste and the difficulty to molding it culminated in specimens with voids that decreased the mechanical performance of 2NS-FS. It is also noted that the performance of 10SF-FC paste was higher than the 2NS-FS paste. It can be affirmed that the effect of the low workability mentioned previously affected the mechanical performance in such a way that the effect of higher reactivity of NS in relation to the SF mentioned by Ghafari et al. [3] was not relevant.

At 7 days, the 2NS-FS presented higher compressive strength in compared to the REF, although still lower than 2NS-FC paste. The 10SF-FS paste was the lower performance among the pastes analyzed, but the difference between its compressive strength and 10SF-FC paste decreases, due to the later effect of SF in the pastes.

4 Conclusions

It was observed that the negative effect of fixing the superplasticizer content becomes more relevant at 3 days, although at the age of 1 day it can already be observed in pastes with NS. Due to this, with the advancement of hydration of the paste,

Table 4 Duncan's test for the compressive strength of pastes at 1, 3 and 7 days of hydration

Age	Paste	Mean of compressive strength (MPa)	Standard deviation (MPa)	Group 1	Group 2	Group 3	Group 4
1 day	REF	37.0	1.0693	X	X	–	–
	2NS-FC	41.2	4.5092		X	–	–
	2NS-FS	30.8	7.5302	X		–	–
	10SF-FC	31.2	0.9609	X		–	–
	10SF-FS	30.5	1.0017	X		–	–
3 days	REF	55.3	2.7875		X		–
	2NS-FC	70.3	1.1533			X	–
	2NS-FS	46.2	5.8398	X			–
	10SF-FC	53.0	2.5502		X		–
	10SF-FS	43.4	1.1676	X			–
7 days	REF	58.8	5.2482		X	X	
	2NS-FC	74.1	0.6351				X
	2NS-FS	64.9	4.3108			X	
	10SF-FC	53.2	5.0501		X		
	10SF-FS	46.1	0.5132	X			

Note Groups are independent between ages. Thus, groups of one age are not related to groups of the other ages

the presented results became increasingly heterogeneous, resulting from the more intense effect of the molding problems that occurred due to the low workability of the pastes with fixed content of superplasticizer.

At 7 days, the pastes with fixed consistency and variation of superplasticizer content presented a considerably higher compressive strength than those with the fixed superplasticizer content. In addition, among all the analyzed pastes the higher compressive strength was presented by 2NS-FC. The effect of NS was more intense than SF, as expected, as well as the fixation of consistency that provided adequate workability and easy molding, ensuring that the SCM used achieved a higher potential for compressive strength of the cement paste. Finally, it is suggested to fix the consistency of the pastes, mortars and concretes when used SCM as SF or NS to maintain its workability.

References

1. Associação Brasileira De Normas Técnicas. NBR 16606: Cimento Portland – Determinação da pasta de consistência normal. Rio de Janeiro (2017)
2. Andrade, D.S., Rêgo, J.H.S., Morais, P.C., Lopes, A.N.M., Rojas, M.F.: Investigation of C-S-H in ternary cement pastes containing nanosilica and highly-reactive supplementary cementitious materials (SCMs): microstructure and strength. Constr. Build. Mater. **198**, 445–455 (2019)
3. Ghafari, E., Costa, H., Júlio, E., Portugal, A., Durães, L.: The effect of nanosilica addition on flowability, strength and transport properties of ultra high performance concrete. Mater. Des. **59**, 1–9 (2014)
4. Kupwade-Patil, K., Palkovic, S.D., Bumajdad, A., Soriano, C., Büyüköztürk, O.: Use of silica fume and natural volcanic ash as a replacement to Portland cement: Micro and pore structural investigation using NMR, XRD, FTIR and X-ray microtomography. Constr. Build. Mater. **158**, 574–590 (2018)
5. Kantro, D. L.: Influence of water-reducing admixtures on properties of cement paste—a miniature slump test. Cem. Concr. Aggr. **2**, 95–102 (1980)
6. Li, L.G., Zheng, J.Y., Zhu, J., Kwan, A.K.H.: Combined usage of micro-silica and nano-silica in concrete: SP demand, cementing efficiencies and synergistic effect. Constr. Build. Mater. **168**, 622–632 (2018)
7. Mehta, P.K., Monteiro, P.J.M.: Concreto: Microestrutura, Propriedades e Materiais, 2nd edn. IBRACON, São Paulo (2014)
8. Narattha, C., Thongsanitgarn, P., Chaipanich, A.: Thermogravimetry analysis, compressive strength and thermal conductivity tests of non-autoclaved aerated Portland cement–fly ash–silica fume concrete. J. Therm. Anal. Calorim. **122**, 11–20 (2015)
9. Scrivener, K.L.: Options for the future of cement. Indian Concr. J. **88**, 11–21 (2014)
10. Sharkawi, A.M., Abd-Elaty, M.A., Khalifa, O.H.: Synergistic influence of micro-nano silica mixture on durability performance of cementious materials. Constr. Build. Mater. **164**, 579–588 (2018)
11. Vasques, L.P., Fraga, Y.S.B., Rêgo, J.H.S., Capuzzo, V.M.S.: Efeito da sonicação da sílica ativa e da nanossílica no estado fresco de materiais cimentícios. In: 60° Congresso Brasileiro do Concreto. Anais […]. Foz do Iguaçu, pp. 1–15 (2018)

Part II
Concrete

The Effect of Various Superplasticizers on Ultra High Strength Concrete

Anthony Torres, Federico Aguayo, Srinivas Allena and Michael Ellis

Abstract Ultra-High Strength Concrete (UHSC) is a high-strength and highly ductile material formulated to provide compressive strengths exceeding 130 MPa. UHSC materials typically have a very low water-to-cementitious ratio (w/cm), which requires the use of superplasticizers, also known as High Range Water Reducing Admixtures (HRWRA), to disperse the fine particles and to make the material workable for placing, handling and consolidating. Common examples of superplasticizer compositions include Polynaphthalene Sulfonate (PNS), Polymelamine Sulfonate (PMS) and Polycarboxylate Ether (PCE) based polymers. This study focuses on assessing the impact of various superplasticizers on the performance of UHSC. Four different types of superplasticizers were used; two different PCE based superplasticizers from a leading manufacturer, one PNS superplasticizer, and one PCE superplasticizer, both of which were provided by a local chemical provider. Specific properties assessed were the superplasticizers' viscosity, concrete workability through the mortar-spread test, and 7, 14, and 28 day compressive strengths. Two mixtures were produced with two w/cm (0.20 and 0.15), which would subsequently increase the amount of HRWRA needed, from 34.7 to 44.5 L/m^3. The results show that both name bland PCE superplasticizers produce a higher spread and a higher compressive strength at all ages tested up to 28 days than the two local superplasticizers. Additionally, the rheology test demonstrated that the name brand PCE superplasticizers had a lower viscosity at all angular speeds than the local superplasticizers counterparts.

Keywords Superplasticizer · High strength concrete · HRWRA · Strength

1 Introduction

The technology development of concrete and demand for high strength construction materials gives momentum to the development of Ultra-High Strength Concrete

A. Torres (✉) · F. Aguayo · M. Ellis
Texas State University, San Marcos, TX, USA
e-mail: ast36@txtate.edu

S. Allena
Washington State University, Tri Cities, WA, USA
© RILEM 2020

J. F. Martirena-Hernandez et al. (eds.), *Proceedings of the International Conference of Sustainable Production and Use of Cement and Concrete*, RILEM Bookseries 22,
https://doi.org/10.1007/978-3-030-22034-1_19

(UHSC). UHSC is a high-strength and highly ductile material formulated to provide compressive strengths exceeding 130 MPa. However, some current UHSC preparation methods require costly materials and relatively sophisticated technology. UHSC mixtures consist of Portland cement, fine aggregate, and steel fibers. UHSC components also often require the addition of materials such as supplementary cementing material (e.g., silica fume and quartz flour) and superplasticizers for increased workability and high early-age strength. Superplasticizers are used in fresh concrete to modify the rheological properties. High range water reducing admixtures (HRWR) are a particular class of superplasticizers, which are used to manipulate the rheological profile of a concrete slurry. High range water reducing admixture requirements are outlined by ASTM C494 "Standard Specifications for Admixtures in Concrete" [1]. Several chemical additives can meet ASTM requirements for high range water reducing admixtures. Common examples of HRWR additives include Polynaphthalene Sulfonate (PNS), Polymelamine Sulfonate (PMS) and Polycarboxylate Ether (PCE) based polymers. These chemicals are all used to disperse cement particles and increase flowability of the slurry. The effectiveness of a water reducer is a result of the chemical structure; some variations in polymer based superplasticizers include charge density differentiation, side chain length, main backbone length, degree of backbone polymerization and composition of functional groups.

This study focuses on assessing the impact of various superplasticizers on the performance of UHSC. Four different types of superplasticizers were used; two different PCE based superplasticizers from a leading manufacturer (PCE Name Brand 1, PCE Name Brand 2), one PNS superplasticizer, and one PCE superplasticizer (PCE Local Provider), both of which were provided by a local chemical provider. The local provided products are believed to provide similar end results to the name brand products at a cheaper cost to the consumer. The comparison should help identify cost effective alternatives to the construction industries more commonly used products. The benefit to a construction project is improved placement time and reduction in cost. If this study is able to identify viable alternatives to the mainstream/name brand chemicals, then the viable cost solutions can be presented to UHSC concrete designers. The parameters, which were measured to determine acceptable performance were spread flow, strength development, and rheology properties of the superplasticizers.

2 Literature Review

2.1 Chemical Structures of PCE and the Effect on Concrete Rheology

The effect of different types of superplasticizer on fresh, rheological and strength properties of self-consolidating concrete (SCC) was investigated by Mardani-Aghabaglou et al. [2]. The admixtures used had the same main chain and the same polymer structure but different molecular weight and different side chain density of carboxylic acid groups. Molecular side chains of a polymer can be visualized as branches of a tree where the main chain is the tree trunk. The author summarizes how the dispersion mechanism of PCE (polycarboxylate)-based superplasticizers is more related to a steric hindrance effect produced by the presence of neutral sidelong graft chains. The dispersion from PNS and PMS type superplasticizers is shown to be caused by the presence of negatively charged anionic groups which are responsible for the adsorption of the polymers on the surface of cement particles. The study indicates better dispersion with high molecular weight PC with large side chains (similar to a tall tree with many large branches). Fresh concrete properties of the SCC mixtures were measured by using the slump flow, V-funnel and L-box tests as well as slump-flow loss after 15 and 30 min. The rheological parameters including apparent yield strength and plastic viscosity were measured by an impeller type rheometer suitable for SCC mixtures. These test were performed in accordance with European Federation of National Associations Representing for Concrete (EFNARC), which is the authoritative voice of contractors, manufacturers, raw material suppliers and consultants in the specialized construction and concrete systems industry. The hardened samples were cast in 150 mm cubic molds. The specimens were demolded after 24 h and were stored in a moist room in water at 23 ± 2 °C until the testing day. For each mixture, twelve specimens were prepared and tested at 1, 3, 7 and 28-day ages for compressive strength in accordance with EN 12,390-3 standard. The ultrasonic pulse velocity (UPV) values of the concrete mixtures were determined in accordance with the ASTM C597 Standard Test Method for Pulse Velocity Through Concrete at the same ages. The dynamic elastic modulus was calculated using the UPV assuming constant Poisson ratio. The data shows that the dynamic modulus increased with age due to lower mass and UPV values.

3 Materials and Mixture Design

Two UHCS mixtures were designed based on recommendations of the literature [3–6] with two different w/c ratios to determine the impact of the amount of superplasticizer on UHSC performance. The mixture designs can be seen in Table 1.

Table 1 Mixture proportions

Constituents	w/cm = 0.20 (kg/m^3)	w/cm = 0.15 (kg/m^3)
Type I/II Portland Cement [7]	889	889
Silica Fume	222	222
Fine Sand	833	1009
HRWRA (L/m^3)	34.7	44.6
Water	229	166

Compressive strength specimens were molded using 50-mm (2-in.) cube molds. All samples were compacted according to the mortar cube compaction method described in ASTM C109 [8] Compressive Strength of Hydraulic Cement Mortars. Cubes specimens were used to avoid problems with end preparation of cylindrical specimens [4]. After the specimens were properly cured, they were individually tested according to BS 12390-3-2009 [9]. The British Standard was used as it provides greater detail to testing hardened concrete cubes in compression than ASTM C 39-15a [9]. An average of three samples was tested per data point reported in the results section.

4 Results

The shear resistance of each superplasticizer was measured at 3 different rates (20, 50 and 100 rpm) using a Brookfield rheometer at room temperature, shown in Fig. 1. Observing the superplasticizer fluids response to shear at different rates helps to identify the effect of the superplasticizer on the flow and strength of the UHPC mixtures.

The spread flow results on each mixture can be seen in Table 2.

It can be seen in both Fig. 1 and Table 2 that the viscosity of the superplasticizers used has a direct effect on the flow results of the mixtures. In other words the PCE provided by a local provider had the highest viscosity produced in the Brookfield rheometer, and the UHSC mixtures produced with the locally provided PCE had the lowest spread. Although the superplasticizers are marketed as comparable products there appears to be a compositional difference that is having an effect on the spread of the UHSC. Although both manufactures do not publish their exact chemical com-

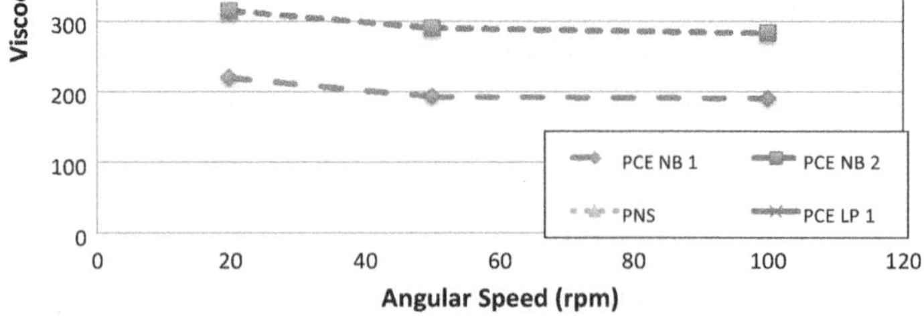

Fig. 1 Viscosity of Superplasticizers at various rotational speeds

Table 2 Spread flow results

Superplasticizer	w/cm = 0.20 (kg/m^3)	w/cm = 0.15 (kg/m^3)
PCE NB* 1	186.7	201.2
PCE NB* 2	181.6	197.9
PNS	159.3	167.1
PCE LP**	154.2	152.6

*NM = Name brand
**LP = Local provider

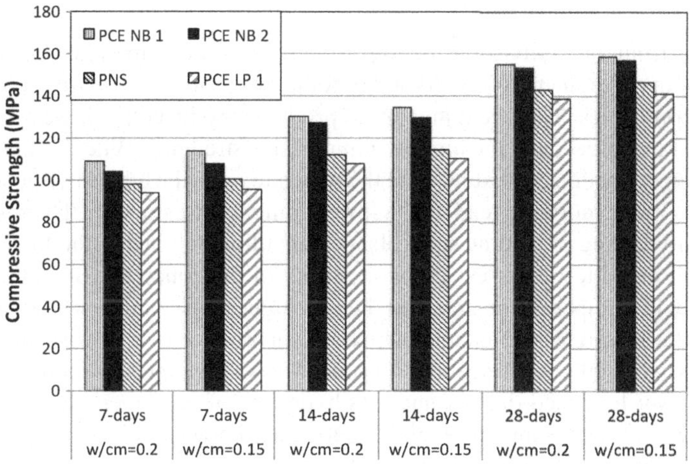

Fig. 2 Compressive strength results

position of percent dissolved solids in solutions for their superplasticizers, this study aims to simply elucidate the direct effect of each on the performance of two UHSC mixtures. All concrete sample compressive strength measurements were collected according to BS EN 12390-3-2009 [10] standard, which is a British standard used for 50-mm (2-in.) concrete cubes. Steel plates were slightly coated with a lubricant and each test sample was placed between the platens and loaded into the center of the load frame. The screeted face of the cube was not placed in contact with the platens during analysis. This produced a load on the more smooth cube surface and allowed more perpendicular contact with the plate. The results of the compression testing can be seen in Fig. 2.

The results of the compression strength testing demonstrate a few, as expected, results, that are consistent with the viscosity and spread tests previously discussed. Specifically, the compressive strengths increase with an increase in age. Additionally, the strengths increase with a decrease in w/cm. The maximum strength achieved was 158.9 MPa produced by the 28-day, 0.15 w/cm, PCE NB 1 mixture.

The lowest compressive strength obtained was the 7-day, 0.20 w/cm, PCE LB with a strength of 93.9 MPa. However, this should be noted as a sufficiently high compressive strength when compared to conventional concrete. It is also observed that the PCE NB 1 consistently produced the highest results with the PCE NB 2 shortly behind. These results were followed by the PNS and the PCE LP for the lowest compressive strengths.

5 Conclusions

The results from this study show the impact of four various superplasticizers on ultra high strength concrete mixtures. A total of four various mixtures were used from two different companies. The two name brand products produced the lowest viscosities, highest mortar spread, and the highest compressive strengths. Whereas the two local provider superplasticizers results were the opposite, with the highest viscosity, which produced a lower mortar spread, and a lower compressive strength. The results from the viscosity testing of just the superplasticizers ultimately led to the final results of the mortar spread test to assess the workability of the concretes. Such that the two locally provided superplasticizers had the two highest viscosities, which produced a lower mortar spread in the concrete, which in turn produced a less workable concrete. These less workable concretes also produced lower compressive strengths, likely due to a lower hardened density and less hydration of all cementing particles. The results from this study can ultimately lead to better knowledge regarding UHSC and specifically the impact of various superplasticizers.

References

1. ASTM C494: Standard Specifications for Admixtures in Concrete, American Society for Testing and Materials, USA (2009)
2. Mardani-Aghabaglou, A., Tuyan, M., Yılmaz, G., Arıöz, Ö., Ramyar, K.: Effect of different types of superplasticizer on fresh, rheological and strength properties of self-consolidating concrete. Constr. Build. Mater. **47**, 1020–1025 (2013). https://doi.org/10.1016/j.conbuildmat.2013.05.105
3. Collepardi, M., Monosi, S., Valenti, M.: Optimization of superplasticizer type and dosage in fly ash and silica fume concretes. Am. Concr. Inst. **119**, 425–444 (1989)
4. Collepardi, S., Coppola, L., Toli, R., Colepardi M.: Mechanical Properties of Modified Reactive Powder Concrete (2007)
5. Dili, A.S., Santhanam, M.: Investigationsonreactivepowder concrete: a developing ultra high-strength technology. Indian Concr. J. **78**(4), 33–38 (2004)
6. Shaheen, E., Shrive, N.G.: Optimization of mechanical properties and durability of reactive powder concrete. ACI Mater. J. **103**(6), 444–451 (2006)
7. ASTM C150: Standard Specification for Portland Cement, American Society for Testing and Materials USA (2009)

8. NF EN12390-3: Testing hardened concrete—Part 3: Compressive strength of test specimens, 18 p (2003)
9. Standard Test Method for Compressive Strength of Hydraulic Cement Mortars (Using 2-in. or [50-mm] Cube Specimens), ASTM Standard C 109-16a, West Conshohocken, Pa, USA (2016). https://www.astm.org
10. Standard Test Method for Compressive Strength of Cylindrical Concrete Specimens ASTM International, ASTM Standard C 39-15a, West Conshohocken, Pa, USA (2015). https://www.astm.org

Can Sustainability of Concrete Construction Be Improved Through a Better Understanding of Field Practices? Lessons from Haiti

Elise Berodier, Loic Aron, Judith Princeton and Ivan Bartolini

Abstract This work is the result of the first collaborative investigation between the National Laboratory for Building and Public Works in Haiti, the Ecole Polytechnique Fédérale de Lausanne and the Swiss Agency for Development and Cooperation on the construction practices with concrete. We have set up the first practical method combining material science to socio-economics to quantify and analyze the actual practices with concrete in the field. We have collected and analyzed real concrete samples randomly selected in ordinary environment in the urban area of Port-au-Prince and recorded qualitative data on the local concrete practices. The results of the first field study are promising for both cement and concrete research and sustainable development of the construction. Approximately 30% of the concretes combine good strength and remarkably low CO_2 impact, indeed lower than concrete made in developed countries. This suggests that there is a tremendous potential to make robust and low CO_2 concrete around the world in no standardized conditions. Besides this promising perspective, the analysis of actual concretes showed that the most common wrong practice was the excessive use of water in the concrete mix. Nevertheless, these concretes with extreme (w/c) ratio revealed new insights on the cement hydration mechanisms.

Keywords Sustainable concrete · Field practices · Microstructure · Cement hydration

---

E. Berodier (✉) · L. Aron
Ecole Polytechnique Fédérale de Lausanne, Lausanne, Switzerland
e-mail: elise.berodier@epfl.ch

J. Princeton
Laboratoire National du Bâtiment et des Travaux Publics, Port-au-Prince, Haiti

I. Bartolini
Swiss Agency For Development and Cooperation, Port-au-Prince, Haiti

© RILEM 2020
J. F. Martirena-Hernandez et al. (eds.), *Proceedings of the International Conference of Sustainable Production and Use of Cement and Concrete*, RILEM Bookseries 22,
https://doi.org/10.1007/978-3-030-22034-1_20

1 Introduction

Over the last decades, concrete researchers have put a lot of effort to understand the key parameters to the hardening process of concrete. Today, we know quite well the cement hydration mechanism [1] but little attention has been paid to the actual use of concrete in the world. The large majority of the research activities is done in laboratories with controlled temperature and humidity conditions that are far from the actual field environments. Both lack of knowledge on field conditions, and adapted laboratory research limit our understanding of the actual cement reaction and consequently increases the gap between fundamental research and the real needs in the construction sector.

The objective of this study is to provide further understanding of the use of concrete in the field by characterizing them with mechanical test, microstructural analysis, CO_2 assessment. The study focuses on Haiti as an illustrative example of the situation in emerging and developing countries facing urbanization issues. Today, 90% of the concrete infrastructures and housing are being built in transitional economies [2]. This situation is entirely novel and more challenging than in the past where construction was growing with GDP. In the perspective to develop a sustainable built-environment, it is critical to get urgently a good understanding on the real use of concrete where the future of the construction is shaping. This paper reports the findings from the first randomized control of concrete in informal sector in Haiti.

2 Methodology

2.1 Research Method

We have developed a new method that couples socio-economics to material science to quantitatively measure the current practices with concrete and determine their potential towards a sustainable construction sector.

This research was conducted in Haiti, which offers a fertile ground to examine the challenges faced by emerging and developing economies. Today, a third of the Haitian lives in the capital of the country, Port-au-Prince. This is a major shift compared to the 50 s where 90% of the population lived in the countryside. The urban construction sector is extremely dynamic, and representative of rapid urbanizing countries dominated by the private housing sector.

Twenty-two construction sites were selected randomly in this study to ensure a representative sampling of the local practices.

2.2 Data Collection

On the construction site just after the mixing, a sample of fresh concrete was collected (according to American standard ASTM C172) for measuring the slump (ASTM C143) and making the test specimens for compressive strength (ASTM C31). Although the ASTM procedure does not cover manual mixing for concrete, the sampling was adapted and done after 7 days in different points of the concrete batch to guarantee a representative sample. Three cylindrical ASTM molds were filled with the fresh concrete, labelled, and stored on the job site. Within 48 h after the concrete mixing, the 3 molds were transported to a curing chamber (ASTM C511) in the laboratory, with free water maintained on their surfaces at 23 ± 2 °C. After 7 days of curing, the compressive strength test was carried out according to ASTM C39.

From the broken concrete specimen, a small piece of the residue was kept for microstructural analysis. The small piece was impregnated with epoxy resin and polished to obtain a polished to surface of the microstructure.

3 Results

3.1 Compressive Strength and Slump Test

Figure 1 shows the compressive strength correlated to the slump measurement of the concrete samples. Strength results are dispersed between 2 to 39 MPa with a bimodal distribution indicating that both good and bad quality concrete can be obtained in the field. Concretes in all construction sites of this study were mainly used for similar structural elements such as columns, beams and slabs. The minimal strength requirement for structural element at 7 days would be 10 MPa from international standards. In this study, about 61% of the samples have a strength lower than this value.

Slump measurements are also dispersed from 1.5 cm to approximately 19 cm. Such high slump results on concrete, containing no admixtures, is usually due to an excessive amount of water. This was confirmed with the quantification of the water on sites. The common (water/cement) ratios varies between 0.5 and 2. Consequently, the strength decreases with increasing slumps. Nevertheless, the scatter of the data points suggest that local factors also play a role in concrete failing the strength requirements.

3.2 Microstructural Analysis

The relationship between the water to cement (w/c) ratio in concrete and its capillary porosity is well known. However, this relationship is usually verified for concrete mixes with w/c ratio in a range between 0.35 and 0.7. Here we study the microstructure of three concrete sample with (w/c) ratio varying from 0.5 to extreme (w/c) ratio higher than 1.

Figure 2 compares three microstructures. While in the concrete A with (w/c) of 0.5 the microstructure is filled of hydrates, in the concretes with higher (w/c) there is a substantial remaining porosity (black areas) and uncommon distribution of the C-S-H products. In Concrete B and C, not all former grains are filled by inner C-S-H. Cement grains initially smaller than 20 μm remain empty of inner C-S-H while in (w/c) < 0.8 they are the first to be filled. The remaining empty space from the grain dissolution is usually referred to as 'hollow-shell'. In normal conditions, they are largely present at 1–2 days and then filled by hydrated products [3]. Here, even after 28 days the shells are empty or slightly filled by a light dense product. On the other hand, coarser cement grains larger than 20 μm are filled with inner C-S-H or with a phase identified by edx as Fe-hydrogarnet. It is believed that inner C-S-H would gradually transform into hydrogarnet due the continuous hydration of the clinker phases (notably C_4AF). The hydration can continue as there is no space restriction to limit the reaction [4] neither limited water to react in high (w/c) ratio concretes.

Figure 3 shows a detailed microstructure of that phase. Further analysis on concrete made at high (w/c) ratio in laboratory conditions confirmed that hydrogarnet forms in large hydrated cement grains and normal temperature conditions.

3.3 Environmental Assessment

From an environmental perspective, cement content accounts for 90% of the CO_2 emissions of concrete production [5]. A concrete with less cement is therefore, less impactful on the environment than concrete with a higher cement fraction. One way to calculate this impact is the binder intensity, which corresponds to the amount of binder (cementitious materials) per m^3 of concrete and per strength unit (MPa) [6].

Fig. 1 Compressive strength as a function of slump measurement. For structural elements, standards request a strength higher than 10 MPa at 7 days

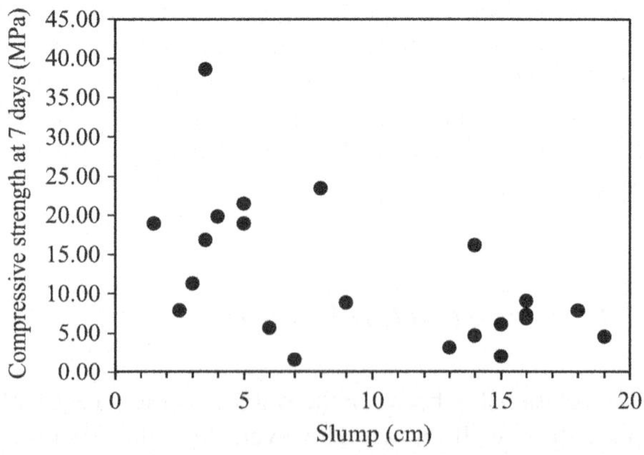

Fig. 2 Comparison of the microstructure at 28 days of three concretes with increasing (w/c) ratio up to 1.5

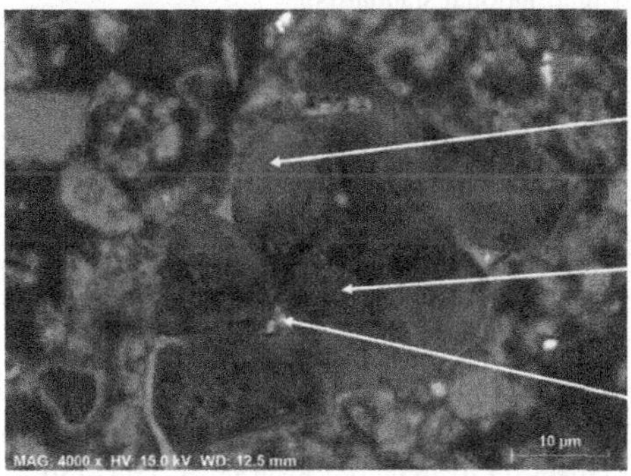

Fig. 3 Details of a hydrated coarse cement grain filled up with inner C-S-H and hydrogarnet

Figure 4 compares the binder intensity of concretes from informal constructions in our study to a database of standardized concretes from Brazil and high income countries published in [6]. The data points from Haiti (represented as diamonds in Fig. 4) show that low strength concrete have high binder intensity whereas higher strength concretes rather have low binder intensity. This suggests that cement is more efficiently used in the high strength concrete. The black dash line represents the minimal binder intensity of concrete complying with the minimal amount of cement from international standards (i.e. 250 kg cement/m^3 concrete). Concrete from the study with a strength higher than 10 MPa, are all below this minimal amount although they provide the same strength than concretes complying international standards. Therefore, in the informal sector in Haiti we can find concrete with both good quality and CO_2 impact lower than in most (or many?) developed countries. This is extremely encouraging and shows that a balance between good quality and low carbon footprint can exist in non-standardized conditions. An additional study coupling material science and socio-economics determined that the educational background

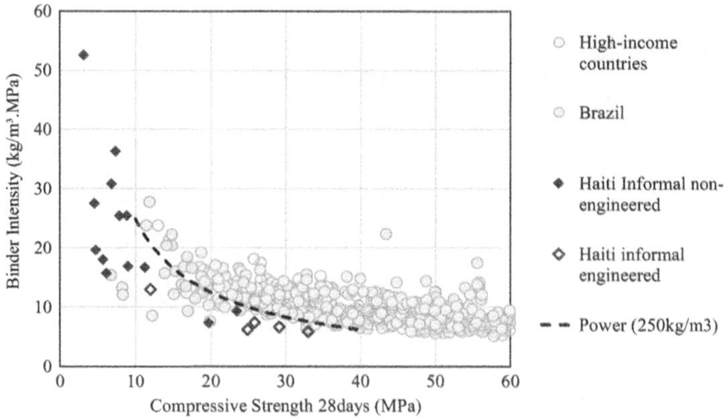

Fig. 4 Binder intensity index as a function of the compressive strength on corresponding concrete sample. Adapted from [7] with additional data from Haiti. Dashed line represents the minimal amount of cement authorized by international standards

of the supervisor of the construction sites was a key factor to achieve this optimized good strength-low carbon concretes.

4 Conclusions

The results of this field study in Haiti provides a better understanding of the actual use of concrete in the field where urbanization issues are faced. The methodology we have developed allowed the analysis of the real-world concrete. Many concretes collected in random construction sites have a strength lower than the international standard requirements, although good strength concrete where also found. These concretes combine good strength and remarkably low CO_2 impact. This suggests that there is a tremendous potential to make robust and low CO_2 concrete around the world in no standardized conditions Indeed, the results show better use of cement in concrete than in developed countries where international standards prevent the reduction of cement fraction below a certain level. Besides this promising perspective, the analysis of real concrete samples showed that the most common wrong practice was the excessive use of water in the concrete mix. These concretes with extreme (w/c) ratio provide new insights on the space filling mechanism of C-S-H.

Acknowledgements The authors would like to thank the Laboratoire National du Bâtiment et des Travaux Publics d'Haiti and the Swiss Agency for Development and Cooperation for funding this research.

References

1. Scrivener, K.L., Juilland, P., Monteiro, P.J.M.: Advances in understanding hydration of Portland cement. Cem. Concr. Res. **78**, 38–56 (2015)
2. Castagnino, S., Rothballer, C., Philipp Gerbert, A.R.: Shaping the future of construction a breakthrough in mindset and technology. World Econ. Forum. Rep. 1–64 (2016)
3. Gallucci, E., Mathur, P., Scrivener, K.: Microstructural development of early age hydration shells around cement grains. Cem. Concr. Res. **40**(1), 4–13 (2010)
4. Berodier, E., Scrivener, K.: Evolution of pore structure in blended systems. Cem. Concr. Res. **73**, 25–35 (2015)
5. Sanchez, Berriel S., et al.: Assessing the environmental and economic potential of limestone calcined clay cement in Cuba. J. Clean. Prod. **124**, 361–369 (2016)
6. Damineli, B.L., Kemeid, F.M., Aguiar, P.S., John, V.M.: Measuring the eco-efficiency of cement use. Cem. Concr. Compos. **32**(8), 555–562 (2010)
7. Scrivener, K.L., John, V.M., Gartner, E.M., Eco-efficient Cements: Potential, Economically Viable Solutions for a Low-CO_2, Cement-Based Materials Industry (2016)

Production of Durable Concrete with a Mineral Addition Blend of Calcined Clay-Limestone-Gypsum (LC2) and Portland Cement

Juan Francisco Garcés Vargas, Flavio Morocho Guamán, Yosvany Diaz Cárdenas, Alina Hereira Diaz and Jose Fernando Martirena-Hernandez

Abstract This article presents the evaluation of different concrete formulations produced with LC3 cement and LC2 mineral addition. The same includes studies of the properties of fresh and hardened concrete and a preliminary evaluation of its durability. LC3-50 2:1 cement-milled by intergrinding was used, Atenas Cements were combined with LC2 mineral addition, 90% Atenas Cement with 10% SikaFume. Atenas Cement, produced industrially at the Cuenca Ecuador plant, was used as a reference and the concrete was formulated for a dosage of certified concrete, for resistance of 35 MPa.

Keywords Calcined clay · Rheology · Mechanical properties · Clinker · Durability

1 Introduction

Cement is one of the essential products for the manufacture of concrete. The contribution of concrete and therefore of this binder to construction makes its production a reflection of a country's economic activity. CO_2 emissions related to concrete cor-

J. F. G. Vargas
Universidad Estatal Península de Santa Elena, Carretero Libertad- Santa Elena, Cantón La Libertad, Provincia de Santa Elena, Ecuador

F. M. Guamán
Hormicreto, Paseo Rio Machangara Bloque 3-18 y Cornelio Vintimilla (Parque Industrial), Cuenca, Ecuador

Y. D. Cárdenas · J. F. Martirena-Hernandez (✉)
Center for Research and Development of Structures and Materials (CIDEM), UCLV, Santa Clara, Cuba
e-mail: martirena@uclv.edu.cu

A. H. Diaz
Universidad Central Marta Abreu de las Villas, CQA, Carretera de Camajuní km 5 1/2, Santa Clara, Villa Clara, Cuba

© RILEM 2020

J. F. Martirena-Hernandez et al. (eds.), *Proceedings of the International Conference of Sustainable Production and Use of Cement and Concrete*, RILEM Bookseries 22,
https://doi.org/10.1007/978-3-030-22034-1_21

respond mainly to emissions associated with the production of cement, its main constituent. CO_2 emissions occur through the decarbonation of raw materials used in the cement manufacturing process and the use of fossil fuels as direct and indirect energy sources [1].

Portland cement clinker, which is the main component of this binder, is formed after calcining limestone and clay at a temperature between 1350 and 1450 °C and some of the materials used as one of its substitutes and which react with calcium hydroxide are commonly called Supplementary Cementitious Materials (SCM) [2].

The amount of clinker that can be replaced by MCS depends on the type of material used. The CSMs traditionally used are wastes from industrial processes, such as fly ash, microsilica and blast furnace slag, as well as natural pozzolans such as clays, zeolitized tuff and volcanic ash. The availability of industrial wastes is limited by the industrial development of each country and its future, conditioned by the efficiency of the processes that generate them, whose tendency points to the reduction of these wastes [3].

Clays, on the other hand, are minerals that are widely scattered throughout the earth's crust and can be activated by thermal processes that require much less energy than that demanded by clinker. However, most of the studies reported in the literature refer to calcination of pure kaolinite clays, commercially called metakaolin. The use of metakaolin as MCS greatly increases clinker replacement levels, but this type of mineral resource is highly demanded by the paper and ceramics industry, which is reflected in its high price.

With calcined clay it is possible to replace a large part of the clinker, almost doubling the amount of cement produced from the same amount of limestone, giving rise to the so-called Low Carbon Cements (LC3). The most common proportion of this binder are 50% Portland cement, 30% calcined clay and 20% limestone [4]. According to studies this new cement makes it possible to reduce CO_2 emissions by 20–30% compared to traditional cement; a great reduction if we consider that cement is responsible for 5–8% of the emissions produced by human activity. It is a low-carbon, low-cost cement that achieves properties similar to or even superior to Portland cement.

One aspect to take into account for concrete and especially produced with LC3 low carbon cement is durability, which is defined as the ability of the concrete to withstand the performance for which it has been designed, as well as taking into account the physical and chemical conditions to which it will be exposed. The durability of concrete depends on different factors including: the water/cement (a/c) ratio, the curing process to which the material is subjected, the type of cement and aggregates used; as well as: the actions of chloride ions, porosity and carbonation [5].

2 Materials and Methods

This research presents the evaluation of seven types of concrete with cement LC3, Atenas Cement/LC2, (50/50%, 55/45%, 70/30% and 85/15%), 90% Atenas Cement with 10% of SikaFume and Atenas Cement as reference. A group of instrumental

Table 1 Physical properties of the cements used

	Dv10%	Dv50%	Dv90%	Clinker%	Blaine cm^2/g
Atenas cement	1.02	13.94	46.49	67	4830.79
15% LC2 + 85% C. ATENAS	0.90	12.56	46.77	57	5836.61
15% LC2) + 85% C. ATENAS-SP	0.90	12.56	46.77	57	5836.61
30% LC2 + 70% C. ATENAS	0.80	9.64	45.13	47	6732.89
30% LC2 + 70% C. ATENAS-SP	0.80	9.64	45.13	47	6732.89
45% LC2 + 55% C. ATENAS	0.74	7.62	42.45	37	8040.20
45% LC2 + 55% C. ATENAS-SP	0.74	7.62	42.45	37	8040.20
50% LC2 + 50% C. ATENAS	0.73	7.21	42.72	34	8557.36
50% LC2 + 50% C. ATENAS-SP	0.73	7.21	42.72	34	8557.36
LC3-50	0.60	4.60	28.98	50	11,546.10

and analytical research techniques were applied such as particle size distribution of the aggregates used. For fresh concrete settlement using the Abrams Cone according to NTE INEN 1578 protocol or ASTM C143, trapped air according to ASTM C231, density of concrete according to ASTM C138. For concrete in hardened state, resistance to 24 h, 3 days, 7 days and 28 days, water absorption test 28 d (effective porosity) and to measure durability, resistivity and resistance to chloride ion penetration by ASTM C 1202.

2.1 *Materials Used*

Cements used: The variables considered in the decision on the use of cements were: Use of cement LC3-50 2:1 (50% clinker and calcined clay/lime ratio 2:1) milled by intergrinding. Combinations of Atenas Cements with the mineral addition LC2 2:1 (combination of calcined clay, limestone and gypsum, with ratio calcined clay/lime 2:1). The combinations used were (Atenas Cement/LC2) 50/50%, 55/45%, 70/30% and 85/15% and a reference cement called Atenas Cement, produced industrially at the Cuenca plant. Additionally, in one of the designs, 10% of Atenas Cement was replaced by the product SikaFume.

Table 1 presents the physical properties of the cements used. In all cases it is verified that the addition of calcined clay significantly increases the specific surface area of Blaine, despite the fact that the particle size distribution is quite similar.

Aggregates: The aggregates used to obtain the concrete test tubes were of two types; the thinner aggregate that is River Sand and the coarse aggregate that is the rubble ¾″, both coming from the Paute River.

Figure 1 shows the particle size distribution of the aggregates used in the manufacture of concrete. Aggregates meet the specifications of the standard.

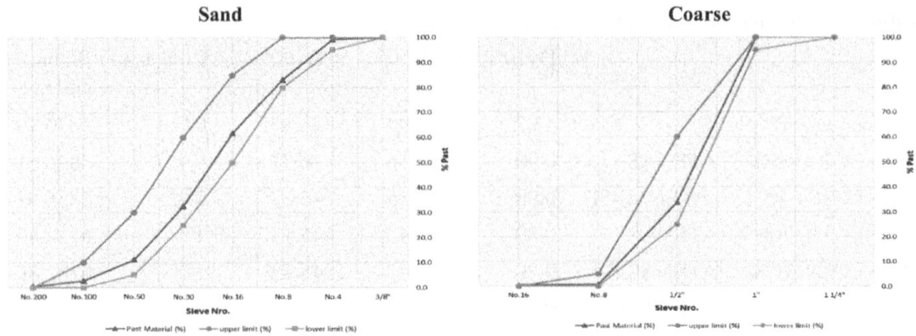

Fig. 1 Particle size distribution of the aggregates used

Silica Fume: Highly reactive pozzolanic material, by-product of the production of siliceous or ferrosilicon metal. It is collected from the gas chimney of electric arc furnaces. Silica fume is an extremely fine powder, with particles about 100 times smaller than the average grain of cement. It is an excellent material as a partial substitute for Portland cement, which provides notable technical and economic advantages.

Superplasticizing Additive: Supplied by BASF, was used. It is a high-range water-reducing additive that complies with ASTM C-494 Type A and F. It is recommended to be used where great fluidity, high resistance and increased durability are required.

2.2 *Experimental Procedure*

For the study we worked with a dosage of certified concrete, for resistance of 35 MPa. Table 2 presents the proportions used, as well as the values of water/binder ratio and slump reached during manufacture.

Given the low water absorption of the aggregates used, the demand for water in concrete produced with Atenas Cement is low and low water/cement ratios are achieved. It is also verified that the presence of calcined clays reduces the settlement and increases the dose of chemical additive for a similar water demand. However, it is necessary to clarify that all the mixtures met the design parameters.

The mix design was modified to include the different cement variants studied, such as the use of Silica Smoke as an alternative for durability, LC3 as the main binder and LC2 in various proportions with Atenas Cement. In all cases it was achieved that the total amount of binder was the same as the original dosage used.

2.2.1 Mechanical Physical Tests Carried Out

In the manufacture of the concrete it was evaluated in fresh and hardened condition. Tests performed for fresh concrete were: Settling using Abrams Cone [6, 7],

Table 2 Dosage used in the manufacture of concrete

Mix types	Portland cement (kg)	LC3-50 co-ground (kg)	Commercial additive (kg)	SF (kg)	LC2 (kg)	Fine arid (kg)	Coarse ari d (kg)	Water	Ratio a/c	Ratio a/agl	% additive	Settlement (cm)	Density (kg/L)	Trapped air %
M1 (reference 1 100% PC)	39.68		337.28			68.71	72.28	12.58	0.32	0.32	0.85	12	2.344	1.7
M2 (reference 2 90% PC + 10% SF)	35.71		357.13	3.97		68.71	72.28	12.88	0.32	0.32	1.00	10	2.345	1.5
M3 (LC3-50 co-ground)		39.68	793.60			68.71	72.28	13.59	0.34	0.34	2.00	7.5	2.345	0.9
M4 (PC (85%) + LC2 (15%))	33.73		396.80		5.95	68.71	72.28	12.44	0.31	0.31	1.00	10.5	2.342	1.6
M5 (PC (70%) + LC2 (30%))	27.78		515.84		11.90	68.71	72.28	12.69	0.32	0.32	1.30	11	2.33	1.4
M6 (PC (55%) + LC2 (45%))	21.82		595.20		17.86	68.71	72.28	13.09	0.33	0.33	1.50	9	2.276	0.9

air occluded according to [8] and concrete density according to (ASTM C138). In the case of hardened concrete, the following tests were performed: resistance to compression at 24 h, 3 days, 7 days and 28 days [9].

This study included tests related to the evaluation of porosity following the protocol of NTE INEN 1855-1 (ASTM C 94) and NTE INEN 1855-2 [10, 11], for the evaluation of the rapid permeability of chlorides the ASTM 1202 standard was used. On the other hand, the resistivity of the concretes studied was evaluated using a Resipod, an instrument designed to measure the electrical resistivity of concrete or rocks, using the Wenner probe principle.

3 Discussion of Results

3.1 Fresh Concrete Properties

Figure 2 presents the results of the measurement of the settlement in the Abrams Cone on the concrete produced. The worst results are those of the concrete produced with the LC3-50 by intergrinding. For increments of up to 30% of LC2, no great differences are observed in relation to the standard series when combinations of Atenas Cement and LC2 are used. The highest consumptions of Superplasticizer are in the mixtures manufactured with the greatest amount of calcined clays in their cement (see Fig. 2). All settlements fall within the range specified by NTE INEN 11578 for plastic consistency concretes.

Figure 3 presents the measured trapped air values in fresh concrete. It is interesting to note that as the calcined clay content increases, the trapped air decreases. This may be related to the ability of clays to retain water in the matrix; this phenomenon will have a very positive influence on durability.

Figure 3 shows the density measurement of fresh concrete. The less dense concretes are those produced with the combination of Atenas Cement and LC2, especially the M6 mixture, which has 45% LC2. The reasons are that the clinker content of the

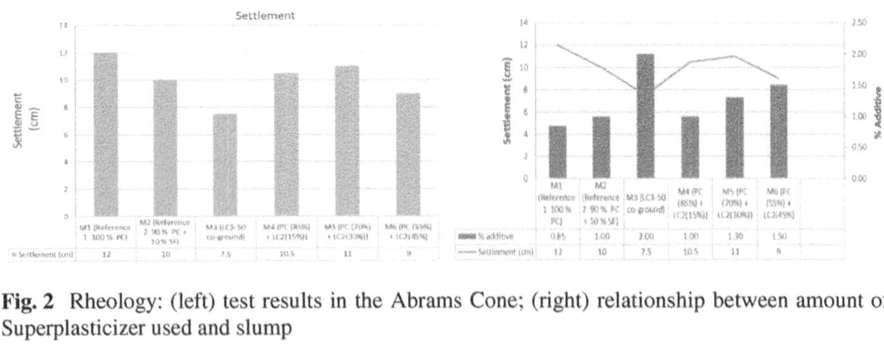

Fig. 2 Rheology: (left) test results in the Abrams Cone; (right) relationship between amount of Superplasticizer used and slump

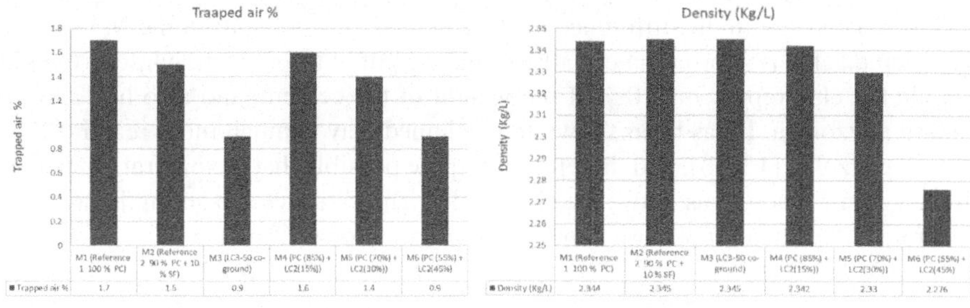

Fig. 3 (Left) Measurement of occluded air in the concrete produced; (right) density of fresh concrete

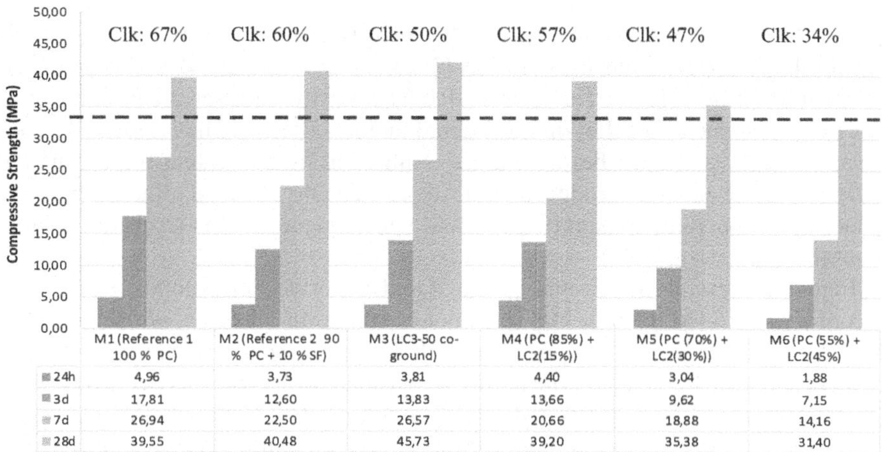

Fig. 4 Resistance values in the studied concretes

resulting binder is very low (around 37%), and the rest of the materials (pozzolans, LC2) are lighter than clinker.

3.2 *Hardened Concrete Properties*

Figure 4 shows the compressive strength values in concrete at 24 h, 3, 7 and 28 days, according to NTE INEN 1573 (ASTM C 39M). The best resistance values are reached in the concrete produced with LC3-50 cement by means of intergrinding. In all cases with the exception of mix 6, with less than 40% clinker in the system; planned design values are achieved. Significant is the difference in values at all ages between mixtures M3 (LC3-50 intergrinding) and M5 (70% Atenas Cement + 30% LC2), which have a similar total amount of clinker.

The reasons are in the difference of the calcined clay content; in the M3 sample the calcined clay is approximately 30% of the weight of the cement, while in the M5 sample the clay represents 20% of the weight of the cement, and also has 20% of natural pozzolana. It has been shown that calcined clay is much more reactive than natural pozzolan [12, 13], and therefore it is quite possible that the natural pozzolana in the system is acting as a low reactivity filler. The compressive strength values of concrete with Silica Smoke are similar to those of Atenas Cement.

3.3 Preliminary Assessment of the Durability of Concrete

In each one of the evaluated concretes the porosity evaluation was carried out following the protocol of the standards (NTE INEN 1855-1 (ASTM C 94) and NTE INEN 1855-2) (see Fig. 5). The lowest effective porosity values are achieved in mixtures M3 (cement LC3-50 by intergrinding) and M4 (85% Atenas Cement + 15% LC2), coinciding with the best resistance results. The highest values of porosity are measured in the concrete produced with Atenas Cement. The total effective porosity values in all concretes are significantly low, which augurs well in terms of concrete durability.

Figure 6 presents the results of the Rapid Chloride Permeability assessment according to ASTM 1202 in the specimens studied. In all cases, with the exception of samples M1 (concrete produced with Atenas Cement) and M4 (85% Atenas Cement and 15% LC2), the permeability is considered very low (between 100–1000 Coulombs), a result that ratifies and corresponds to the low values of effective porosity observed. It is evident that permeability decreases with the content of calcined

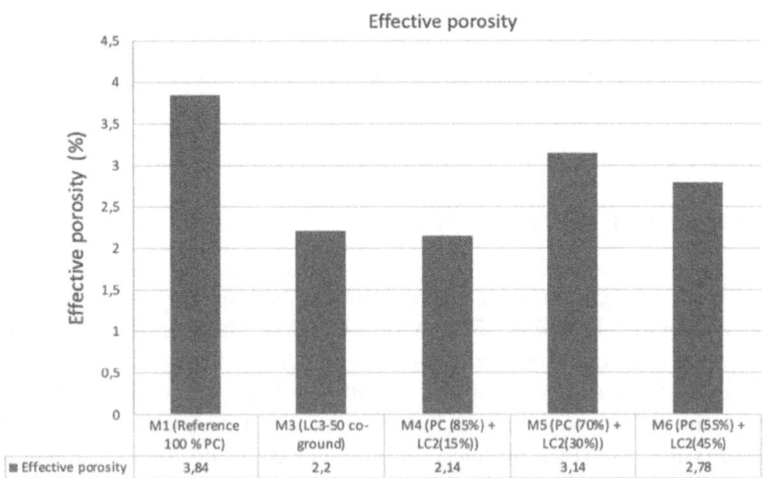

Fig. 5 Effective porosity of the studied concretes

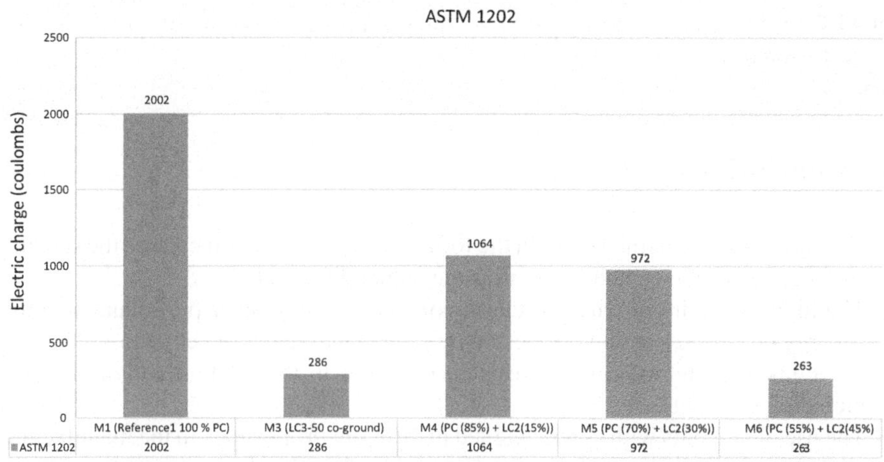

Fig. 6 Rapid chloride permeability values per ASTM 1202

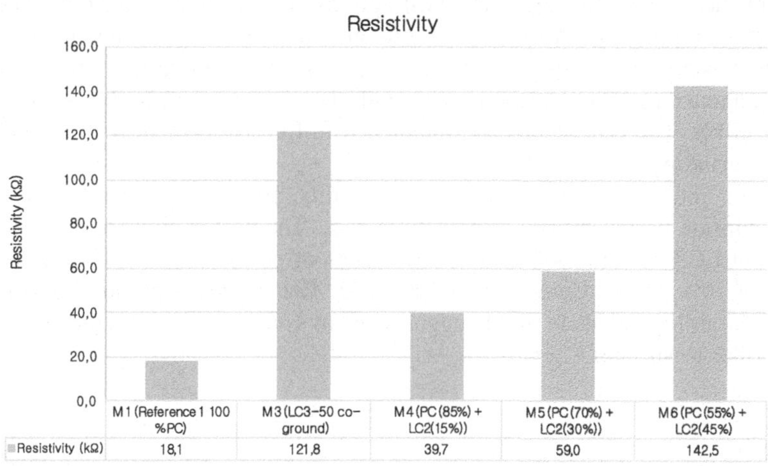

Fig. 7 Resistivity values measured in concrete produced

clays and the presence of less reactive pozzolans in the concrete; the best results are in M3 (LC3-50 by intergrinding, with 30% calcined clay) and M6 (29% calcined clay), with less influence of the total clinker content.

Figure 7 shows the resistivity values measured in cores taken from the concretes produced. Here the highest values correspond equally to the mixture M3 (LC3-50 by intergrinding, with 30% calcined clay) and M6 (29% calcined clay), confirming again the beneficial effect of the clay content on the durability of the concrete. These two series show negligible corrosion risks ($\geq$ 100 kΩ cm).

The worst resistivity values are measured in concrete produced with Atenas Cement, with a moderate corrosion risk rating (10–50 kΩ cm). As the mineral addition LC2 is incorporated the resistivity values improve, and from 30% LC2 the risk of corrosion decreases.

4 Conclusions

- The inclusion of calcined clays in the formulation of cements increases the specific surface area of the cements, and thus the demand for water.
- The high reactivity of calcined clays, compared to any other pozzolana, and the synergy produced with limestones, allows the manufacture of cements with a very low clinker content that reach resistance similar to that of Portland at all ages, including early ages.
- The use of cements with clays calcined in different proportions in the manufacture of concrete ratifies the high demand for water and the need to increase the addition of superplasticisers, but the settlements obtained in all cases meet the requirements established in the standards. The best options are to produce an LC3-50 cement with 50% clinker by inter-grinding, or to mix 70% Atenas Cement with 30% LC2.
- Concretes produced with calcined clay cements achieve high strength values at all ages, and exceed the design strength with the exception of the M6 sample with 34% cliker. The best results are obtained with LC3-50 cement with 50% clinker by co-grinding.
- The concretes produced exhibit high durability, proven through effective porosity, chloride permeability and resistivity tests. The best results are obtained with cement LC3-50 with 50% clinker by means of co-grinding, and from the mixture of 70% Atenas Cement with 30% LC2. The concretes produced must show high durability values in highly aggressive environments, such as the presence of chlorides or sulfates.

References

1. Habert, G., Choupay, N., Escadeillas, G., Guillaume, D., Montel, J.M.: Clay content of argillites: Influence on cement based mortars. Appl. Clay Sci. **43**(3–4), 322–330 (2009)
2. Vizcaíno Andrés, L.M.: Cemento de bajo carbono a partir del sistema cementicio ternario clinquer – arcilla calcinada – caliza. Editorial Universitaria (2015)
3. Sabir, B., Wild, S., Bai, J.: Metakaolin and calcined clays as pozzolans for concrete: a review. Cem. Concr. Compos. **23**(6), 441–454 (2001)
4. Martirena, F., Scrivener, K.: Low carbon cement LC3 in Cuba : ways to achieve a sustainable growth of cement production identification of suitable clay deposits, vol. 2, pp. 3–6 (2018)
5. Tang, S.W., Yao, Y., Andrade, C., Li, Z.J.: Recent durability studies on concrete structure. Cem. Concr. Res. **78**, 143–154 (2015)
6. NTE INEN 1578: Hormigón de cemento hidráulico. Determinación del asentamiento, vol. 1578 (2010)
7. ASTM C143: Método de ensayo Normalizado asentamiento del hormigón de cemento hidráulico

8. ASTM C231: Ensayo para determinar el contenido de aire del hormigón mezclado fresco por el método de presión
9. NTE INEN 1573: Hormigón de cemento hidráulico. Determinación de la resistencia a la compresión de especímenes cilíndricos de hormigón de cemento hidráulico, vol. 1573 (2010)
10. NTE INEN 1855-1: Hormigones. Hormigón premezclado. Requisitos, vol. 1 (2001)
11. NTE INEN 1855-2: Hormigones. Hormigón preparado en obra. Requisitos, vol. 2 (2002)
12. Yanguatin, H., Tobón, J., Ramírez, J.: Pozzolanic reactivity of kaolin clays, a review Pozzolanic reactivity of kaolin clays, a review Reactividad puzolánica de arcillas caoliníticas, una revisión. Ing. Constr. **32**, 13–24 (2017)
13. Alujas, A., Fernández, R., Quintana, R., Scrivener, K.L., Martirena, F.: Pozzolanic reactivity of low grade kaolinitic clays: Influence of calcination temperature and impact of calcination products on OPC hydration. Appl. Clay Sci. **108**, 94–101 (2015)

Behavior of Retraction in Fluid Concretes Produced with Active Mineral Addition LC^2

María B. Díaz García, Reidel Groero Peña, Liset Quesada González and Jose Fernando Martirena-Hernandez

Abstract The Supplementary Cementitious Materials (MCS) are currently, together with the additives, the most sustainable solutions that are worked to achieve high performance concrete and more resistant to aggressive environments. This research focuses on the introduction of additions in the hotel buildings of Cayo Santa María, Cuba. With previous studies showing that some additions can decrease the appearance of cracks due to retraction, a phenomenon very common in the coastal areas of the country, it is decided to begin a research focused on tackling cracking of concrete. The aim is to evaluate the introduction of LC2 (calcined clay, limestone and gypsum) as an active mineral addition concrete. The work is divided into two stages in order to evaluate the introduction of LC2 as an addition. In the first stage, the protocol a of the research is defined, the work is specified with 30% addition, in the second stage the tests are carried out to determine the properties of the fresh concrete as a consistency by the Cone de Abrams, in state hardened, resistance at 7 and 28 days and retraction by ASTM C-157. In this way, a study is made available on the effect of the LC2 addition on the properties of fluid concretes and it is possible to reduce the consumption of cement so harmful to the environment, as well as to make available to the construction a new affordable material in Cuba.

Keywords Retraction · Cement · Structural concrete

M. B. Díaz García (✉) · J. F. Martirena-Hernandez
Center for Research and Development of Structures and Materials (CIDEM), Havana, Cuba
e-mail: mbetania@uclv.cu

R. G. Peña
UCM #4, Matanzas, Cuba

L. Q. González
UCM #3, Santa Clara, Cuba

© RILEM 2020

J. F. Martirena-Hernandez et al. (eds.), *Proceedings of the International Conference of Sustainable Production and Use of Cement and Concrete*, RILEM Bookseries 22,
https://doi.org/10.1007/978-3-030-22034-1_22

1 Introduction

Among the materials used as pozzolans are thermally activated clays, the temperature range for their activation (550–900 °C) is lower than the clinkerization temperature (1450 °C), which implies lower energy consumption with the consequent reduction in CO_2 emission. In addition, one of the reaction products in the activation of clays is H_2O and not CO_2 as in the case of clinker. Therefore, a partial replacement of cement by pozzolans contributes to the reduction of the amount of clinker used and in this way decreases the emission of CO_2, improve also, when used in the appropriate percentage and have the required characteristics, strength and durability [1].

The use of this type of material also favors the protection of constructions since, for example, they develop micropores in their structure that are not intercommunicated and, therefore, make the concrete more impermeable [2].

The addition of pozzolans gives concrete made with Portland cement properties of great practical importance, such as: chemical stability, impermeability, mechanical resistance, low hydration heat, durability, among others [3].

Concrete with pozzolanic additions, having greater chemical stability and impermeability, has greater resistance to weathering and aggressive substances, presenting an excellent resistance to chemical attack against acidic waters. It is resistant to sulfates because the incorporation of pozzolans decreases the proportion of tricalcium aluminate (C3A) in the cement, which is very favorable because this compound is the main responsible for the susceptibility of Portland cement to sulfates. It possesses a more stable behavior to the reaction with the alkalis of the cement, turning it into an inhibitor of the alkali-added reaction, being thus more durable [4].

In the project Development of Ecomaterials for Low Cost Housing, carried out between 2009–2018 by the Federal Polytechnic School of Lausanne and the Marta Abreu Central University of Las Villas, interesting results are obtained when evaluating the pozzolanic properties of a multi-component Cuban clay soil, after being thermally activated. As the main output of the project, it was demonstrated that the most influential clay fraction is kaolinite, which can be activated at relatively low temperatures (800 °C) and that for clay soils with kaolinite contents of 40% a reactive material can be obtained that allows substitutions of up to 30% of the cement, without compromising its mechanical properties and durability [5, 6].

As a continuation of this research, the *Centro de Investigación y Desarrollo de Estructuras y Materiales* (CIDEM) proposes a study on the evaluation of volume changes in concretes made with the pozzolanic addition of LC2, which is nothing more than calcined clay, limestone and gypsum at different proportions. The aim of this research is to achieve a more durable and stable concrete in the face of the existing environmental conditions in Cuba, allowing the useful life of buildings to be extended and reducing maintenance and repair costs.

2 Discussion and Development

The use of pozzolanic additions makes the concretes tend to have a low bleeding speed, increasing relatively the tendency to cracking by plastic retraction, for this reason they require greater control and special and continuous curing to prevent the dangers of cracking and other phenomena.

The changes in volume per drying of the concrete depend mainly on the water content of the mixture. The use of additions requires a greater demand for water than the concretes produced with ordinary portland cement, therefore they tend to suffer a greater contraction and therefore the use of plasticizing additives is recommended for those structural elements susceptible to this type of fissures. For this reason, a study of concretes with the active mineral addition LC^2 is necessary to define the effect that the addition of calcined clay, limestone and gypsum (LC^2) has on the fissures produced by volume changes in concretes, with the action of the additive Superplasticizer.

2.1 Experimental Procedure

The investigation has a procedure that begins with the definition of the dosages to use in the elaboration of concretes for the taking of samples of each mixture. The minimum slump required by Abrams' 16 cm cone test is checked and then cylindrical specimens of 100 mm × 20 mm for resistance are placed in the molds, in the 75 mm × 75 mm × 285 mm prismatic specimen molds for ASTM C-157.

The samples in cylindrical specimens, once hardened, the compressive strength test is carried out in batches of three to 7 and 28 days after the start of the curing and ASTM C 157 test to measure shrinkage in the prismatic specimens for 28 days.

The active mineral addition is formulated in the Faculty of Construction where the thermal activation of the clay is developed by calcination at 800 °C, then proceeds to grinding in a ball mill MB 800, and mixes the clay with limestone in a 2:1 ratio with 7% gypsum total to grind.

The grinding for the manufacture of LC2 is carried out in two stages, first the raw materials separately and then all together until the necessary fineness is achieved for the 90 and 45 μm sieves in compliance with the provisions of the standard (NC:528, 2013). The composition of the addition is presented below. The Superplasticizer Additive is mid-range, high efficiency, chloride-free additive reduces the amount of water and contributes to the permanence of workability.

For the development of the research, a dosage is defined for the tests on cement pastes, where the water/cement ratio is varied, according the characteristics of materials (Tables 1 and 2).

The active mineral addition is formulated in the Faculty of Construction where the thermal activation of the clay is developed by calcination at 800 °C, then proceeds

Table 1 Samples

Samples	Composition
MP1 (0.40)	P35 + SP 0.65%
MP2 (0.40)	P35 + Granite + SP 0.65%
M1 (0.40)	70% P35 + 30% LC2 + SP 1.00%
M2 (0.45)	70% P35 + 30% LC2 + SP 1.00%
M3 (0.50)	70% P35 + 30% LC2 + SP 1.00%

Table 2 Dosage for 20 l

Material	MP1	MP2	M1	M2	M3
Cement (kg)	8.4	18.4	5.88	5.88	5.88
LC^2 (kg)	0	0	2.52	2.52	2.52
Sand of Purio (kg)	18.7	18.7	18.74	18.74	18.81
Coarse of Purio (kg)	17.82	12.46	17.82	17.82	17.82
Water (l)	3.42	3.5	3.41	3.88	4.32
Additive superplasticizer (l)	0.055	0.055	0.084	0.084	0.084

Table 3 Raw materials for the production of LC^2

Raw materials	Origin
Clay	Yaguajay
Limestone	Morejón Quarry, Sancti Spíritus
Gypsum	Punta Alegre

Table 4 Composition of the active mineral addition LC2

Materials	Quantity (kg)	%
Clay	19.8	62
Limestone	9.6	31
Gypsum	0.6	7

to grinding in a ball mill MB 800, and mixes the clay with limestone in a 2:1 ratio with 7% gypsum total to grind (Tables 3 and 4).

The cement to be used is produced at Cienfuegos, according to NC: 95, 2011 Portland Cement—Specifications, classified as Portland P 35 cement.

3 Discussion of Results

The pozzolanic addition LC^2 is a base of calcined clay, limestone and gypsum. This allows a partial replacement of up to 40% of Portland Cement. In previous studies

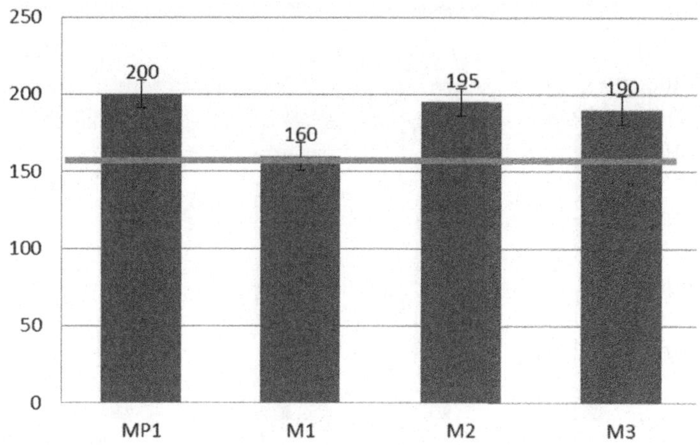

Fig. 1 Slump

the addition is defined only for plastic concretes, but this study tries to evaluate the effect of LC^2 in fluid concretes, which have different characteristics. That is why it is important to analyze the substitutions that are evaluated at an international level (from 10 to 30%).

3.1 Effect of Pozzolanic Addition LC2 on Slump

The comparison is made between the standard sample (MP1) and the samples with 30% LC2 (M1, M2 and M3) in the latter varies the ratio water cement and have 1% additive SP (Fig. 1).

The mixture M1 (LC2 30%, Ra/c = 0.40) is the one with the lowest slump, since it presents a lower water-cement ratio, which is quickly absorbed by the clay. The mixture M2 (LC2 30%, Ra/c = 0.45) with a slump equal to 195 mm is the one that reaches greater slump, because with the relation water cement of 0.45 it is possible to hydrate all the components of the cement. On the other hand, the M3 (LC2 30%, Ra/c = 0.50) although it has a slump of 190 mm, suffers segregation problems due to the excess of water in the mixture.

3.2 Effect of Pozzolanic Addition LC2 on Compressive Strength

That after 7 days all mixtures reach values higher than 35 MPa, except M3 (LC2 30%, Ra/c = 0.50) with a value of 33 MPa. As observed in the samples with LC2, as

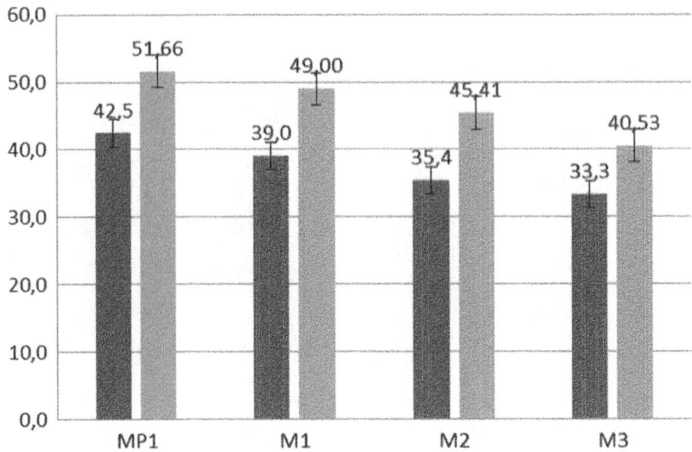

Fig. 2 Compressive strength (7 and 28 days)

the water-cement ratio increases, the resistance decreases by a value of 2 MPa. As referred, when the water-cement ratio is increased, the chemical reactions in the pulp are better, but when aggregates are included, a greater amount of water may remain in the system that does not interact with the cement particles and causes them to become interconnected pores and the resistance in the concrete mass decreases. The evolution of resistance in the manner described is due to the fact that a substitution of the binder by LC2 for practical purposes means a reduction in the actual cement content and for the pozzolanic reaction to occur a certain amount of hydrolysis lime must first be formed (see Fig. 2).

3.3 *Effect of Addition LC2 on Volume Changes*

Figure 3 shows that the M2 sample (LC2 30%, Ra/c = 0.45) at ages 5 to 10 days has a greater retraction, as it shows an increase in 24 h of 50 μm. Mixture M1 (LC2 30%, Ra/c = 0.40) is the best behaviour, does not present variations and decreases in the first 7 days 88 μm. On the other hand, the M3 mixture (LC2 30%, Ra/c = 0.50) is the one that experiences the greatest retraction, reaching 200 μm in the first 24 h, where M1 only reaches 69 μm. It can be defined that M3 is that of values close to the Standard, although after 10 days, they decrease by 30 μm. The M1 and M2 samples reach values lower than the MP1 Standard (P 35, Ra/c = 0.40), but the M1 is the one that has the lowest deformations in time, that is to say that in a period of 28 days it only varies 383 μm, in addition it is the one that presents better behavior in the passage of time. This does not happen with the remaining two samples, since when the water-cement ratio is increased, one percent of the water is free in the system and other phenomena may appear in the hydration process.

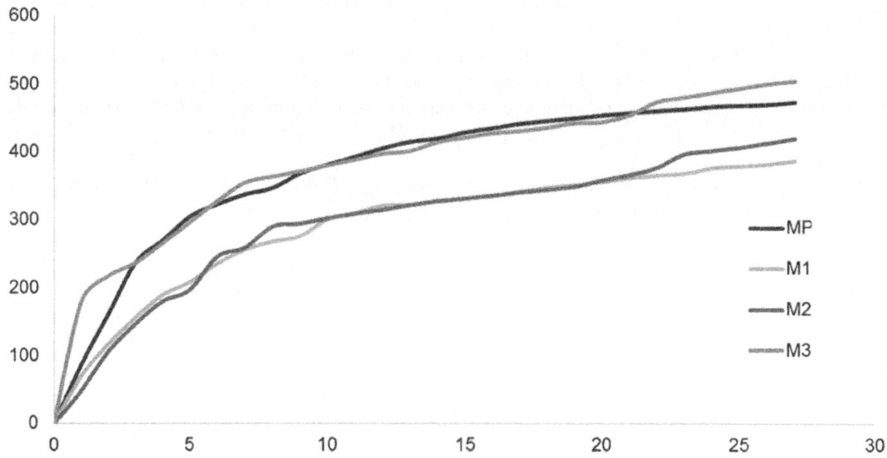

Fig. 3 Volume changes (28 days according ASTM C-157)

4 Conclusions

Pozzolana, both natural and artificial, is capable of partially replacing cement Portland up to 30% and efficiently guarantees the reduction of shrinkage and therefore the cracking of concrete.

The mixtures with the pozzolanic addition LC2 at 30% and water-cement ratio of 0.4, is the one that presents the best results in terms of volume changes decreasing with respect to P35 up to 20%.

The active mineral addition LC2 can be used as a mitigation measure in volume changes for 35 MPa fluid concretes produced in coastal areas of Cuba.

References

1. Habert, G., Choupay, N., Montel, J.M., Guillaume, D., Escadeillas, G.: Effects of the secondary minerals of the natural pozzolans on their pozzolanic activity. Cem. Concr. Res. **38**(7), 963–975 (2008)
2. Famy, C., Brough. A.R., Taylor H.F.W.: The C-S-H gel of Portland cement mortars : Part I. The interpretation of energy-dispersive X-ray microanalyses from scanning electron microscopy, with some observations on C-S-H, AFm and AFt phase compositions. Cem. Concr. Res. **33**, 1389–1398 (2003)
3. Monteagudo S.: Estudio microestructural y de los procesos de hidratación de cementos con adiciones. Thesis Doctoral (2014)

4. Sant, G., Bentz, D., Weiss, J.: Capillary porosity depercolation in cement-based materials: measurement techniques and factors which influence their interpretation. Cem. Concr. Res. [Internet]. **41**(8), 854–864 (2011). http://dx.doi.org/10.1016/j.cemconres.2011.04.006
5. Scrivener, K.L., Nonat, A.: Cement and concrete research hydration of cementitious materials, present and future. Cem. Concr. Res. [Internet]. **41**(7), 651–665 (2011). http://dx.doi.org/10.1016/j.cemconres.2011.03.026
6. Scrivener, K., Snellings, R., Lothenbach, B.: A Practical Guide to Microstructural Analysis of Cementitious Materials. Escuela Politécnica Federal de Lausanne, Suiza

Behavior of Concrete Made with PP-35 Cement in the Province of Cienfuegos

Sergio Betancourt Rodríguez, Betsy Hoyos Rodríguez and Camilo González Díaz

Abstract The problem is derived by the start of the production of PP-35 cement in the Cienfuegos province and the need to know its performance in premixed and precast concrete mixtures, produced with aggregates from this province. Thus, the objective is to characterize the behavior of concrete made with PP-35 cement, to assess its possible application in the prefabricated industry and pre-mixed concrete, with the use of aggregates from the Cienfuegos province. The methodology followed in the work was carried out experimentally in laboratory conditions, evaluating the behavior of the resistance, sorptivity and effective porosity, when the water/cement ratio varies (0.7–0.72). At the same time, this same process was carried out for the cements P-35 and PP-35, achieving a comparison of the behavior models in the three types of cement. The results show that for equal water/cement ratios does not significantly change the resistance of concrete at 28 days when the type of cement used changes, on the other hand, pozzolanic cements improve the sorptivity and porosity compared to cement P-35. The studies allowed to reach the conclusions that it is feasible to use PP-35 cement in the production of premixed concrete and in the prefabricated industry, provided that water-reducing additives, retarders and industrial tests and corrections are used for their application in practice.

Keywords Concrete · Pozzolanic cement · Compressive strength

1 Introduction

The National Direction of the Superior Organization of Business Management (OSDE) of the MICONS Project has oriented to all the Construction Research Units (UIC) of the National Company of Applied Research (ENIA) of the country, to study

S. B. Rodríguez (✉)
Universidad Central "Marta Abreu" de Las Villas, Santa Clara, Cuba
e-mail: sbetancourt@uclv.edu.cu

B. H. Rodríguez · C. G. Díaz
MICONS, Villa Clara, Cuba

© RILEM 2020

J. F. Martirena-Hernandez et al. (eds.), *Proceedings of the International Conference of Sustainable Production and Use of Cement and Concrete*, RILEM Bookseries 22,
https://doi.org/10.1007/978-3-030-22034-1_23

the concretes with the use of cement PP-35, considering that its use has not yet been generalized due to lack of studies that justify its use.

1.1 Problematic Situation

In Cuba, an environmental policy has been established, consisting of reducing energy expenditure, in which the production of cement and concrete is not excluded. The increase in cement production with pozzolanic additions responds to this policy, however, the unjustified tendency to consider Portland cement as the only solution in structural concrete and relegate pozzolanic concrete to secondary uses and masonry has attacked the noble purposes economic and environmental aspects of the policy in question. The reach and potentialities of cement P-35 and to a lesser extent of PP-25 are fully known, however, this is not the case with PP-35, which has not been deepened, at least in the province of Cienfuegos.

The need to demonstrate technically the feasibility of the use of pozzolanic cements in concretes with structural commitments (with their economic advantages) to technicians and executives involved in such decisions, as well as to establish the use and scope of said cements, constitutes the problematic situation of the present investigation.

1.2 Overall Objective

Characterize the behavior of concrete made with cement PP-35, to assess its possible application in the prefabricated industry and premixed concrete.

2 Methodology

The research carried out was based on the experimentation of concretes manufactured at the laboratory level. The experiment design was made with two types of combinations:

A. Cement P-35 of the cement factory "Cienfuegos S.A.", crushed sand from the "Arimao" plant and gravel from the Santiago Ramírez quarry.
B. Cement PP-35 from the cements factory "Cienfuegos S.A.", crushed sand from the "Arimao" plant and gravel from the Santiago Ramírez quarry.

As an initial phase of the research, the characterization of all the raw materials used in the work was carried out: cement P-35, cement PP-35, sand from Arimao, and gravel from the Santiago Ramírez quarry. The results of the characterization are shown in the following section.

Table 1 Dosages for each experimental point with cement P-35

Raw material	Unit	Exp. point I	Exp. point. II			Exp. point III
Cement	kg/m^3	300	375	375	375	450
Water	L	216	216	210	208	210
Sand	kg/m^3	975	943	943	943	912
Gravel	kg/m^3	915	886	886	886	856
Admixture	L	3.0	3.75	3.75	3.75	4.5
W/C rate	–	0.72	0.58	0.56	0.55	0.47
Slump	cm	16	16	15	14	16

Table 2 Dosages for each experimental point with cement PP-35

Raw material	Unit	Exp. point I	Exp. point. II			Exp. point III
Cement	kg/m^3	300	375	375	375	450
Water	L	216	216	210	208	210
Sand	kg/m^3	834	805	813	815	783
Gravel	kg/m^3	945	910	918	923	885
Admixture	L	3.0	3.75	3.75	3.75	4.5
W/C rate	–	0.72	0.58	0.56	0.55	0.47
Slump	cm	10	14	13	12	14

The dosages for the three experimental points were calculated according to the experimental design conceived, they are summarized in Tables 1 and 2.

3 Results and Discussion

3.1 Characterization of Raw Materials

Both, the coarse aggregate and the sand used in the experiments, were in accordance with the specifications established in the Cuban standard NC 251: 2013 [3] (Table 3).

When carrying out an analysis of the results of the tests on the materials used, it is concluded that all the properties studied are in accordance with the standards of current specifications, except the test of percentage of flat and elongated particles made to the gravel of the quarry of Santiago Ramírez, which exceeds the established conformity value (10%).

Table 3 Results of cement tests

Tests		Results		Units	Specifications		Conformity	
		P-35	PP-35		P-35	PP-35	P-35	PP-35
Setting times	Initial	108	124	Min	≥45	≥45	Yes	Yes
	Final	3.0	–	h	≤10	≤10	Yes	–
Standard consistency		23.0	24.3	%	–	–	–	–
Density		3.10	3.08	g/cm^3	–	–	–	–
Fineness		2.8	0.6	%	–	–	–	–
Rc at 7 days		32.6	30.1	MPa	≥25	≥25	Yes	Yes
Rc at 28 days		40.6	40.1	MPa	≥35	≥35	Yes	Yes
Rf at 7 days		4.6	4.3	MPa	–	–	–	–
Rf at 28 days		5.2	5.0	MPa	–	–	–	–

Specifications according to NC 95 [1], and 96/2011 [2]

3.2 *Experimental Design*

The experimental design was carried out according to the following steps:

- Mix design from 300 to 450 kg/m^3 of cement in three experimental points: 300, 375 y 450 kg/m^3.
- The proportion in the mixture of aggregates was carried out according to the dosing method developed by Dr. V. A. O'Reilly, by the experimental procedure of minimum vacuums.
- The admixture was added until the mixture reached a settlement measured in the cone of Abrams 16 cm ± 1.
- In the mixtures with PP-35 the amount of water and admixture determined in the previous point was maintained, and therefore expecting a slight decrease in slump.
- The behavior of the compressive strength was modeled when the water/cement ratio varies.
- The behavior of the porosity was modeled when the water/cement ratio varies.
- The behavior of the sorptivity was modeled when the water/cement ratio varies (Tables 4 and 5).

Table 4 Results of mean compressive strength to 28 days (P-35)

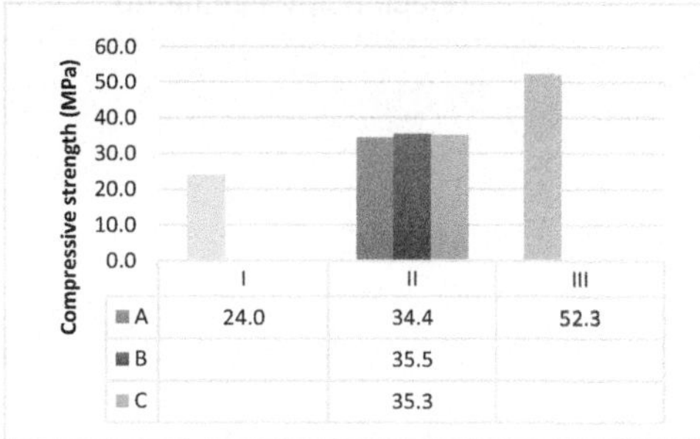

Table 5 Results of mean compressive strength to 28 days (PP-35)

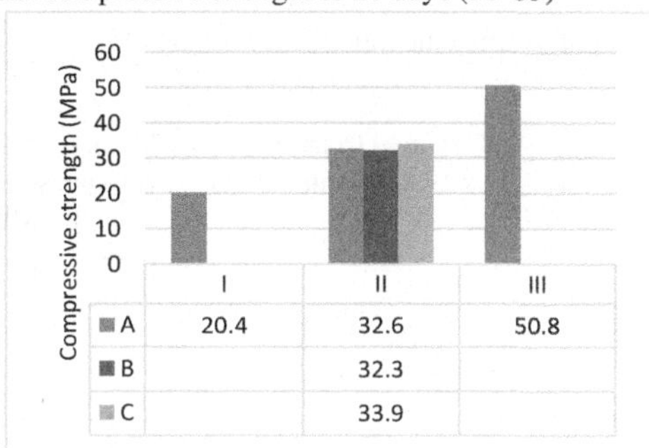

3.3 *Models of Twenty-Eight-Day Behavior of the Average Compression Resistance When the Water/Cement Ratio Varies*

Combination 1

The model that describes the behavior of the strength when the water/cement ratio for cement P-35 varies is the "*Reciprocal Model*", which responds to an equation of the type y = 1/(ax + b). Where "y" is the average strength to compression, "x" relation W/C, a and b model coefficients, $r^2 = 0.98$ (coefficient of determination) and s = 1.21 MPa (standard deviation of the adjustment) (Fig. 1).

The values of the coefficient of determination (0.98) and standard deviation of adjustment (1.21 MPa) evaluate the model as adequate [4]. As can be observed, the resistance increases when the water/cement ratio decreases, there being a notable increase in the extent that the water/cement ratio decreases, note that between the water/cement ratio 0.72 and 0.47 there is a difference of 28.3 MPa in terms of the resistance (Table 6).

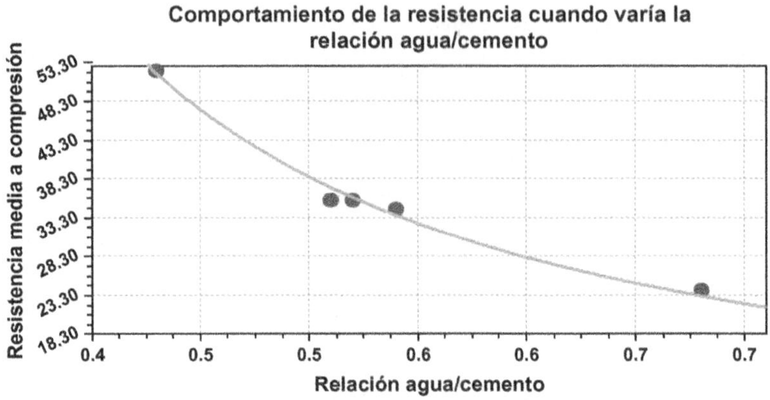

Fig. 1 Behavior of the resistance when the ratio W/C varies (P-35)

Table 6 Cement yield

W/C ratio	Cement content (kg)	Mean compressive strength (MPa)		Yield	
		P-35	PP-35	P-35	PP-35
0.72	300	24.0	20.4	0.80	0.68
0.56	375	35.1	33.0	0.94	0.88
0.47	450	52.3	50.8	1.16	1.13

Combination 2

The model that describes the behavior of the resistance when the water/cement ratio for the PP-35 cement varies is the Reciprocal Model, which responds to an equation of the type $y = 1/(ax + b)$, where "y" is the average resistance to compression, "x" the water/cement ratio, $r^2 = 0.99$ (coefficient of determination) and s = 1.13 MPa (standard deviation of the adjustment) (Fig. 2).

The values of the coefficient of determination (0.99) and standard deviation of adjustment (1.13 MPa) evaluate the model as adequate. As can be seen, the resistance increases when the water/cement ratio decreases, there being a notable increase in the extent that the water/cement ratio decreases, note that between the water/cement ratio 0.72 and 0.47 there is a 30.4 MPa path in terms of resistance.

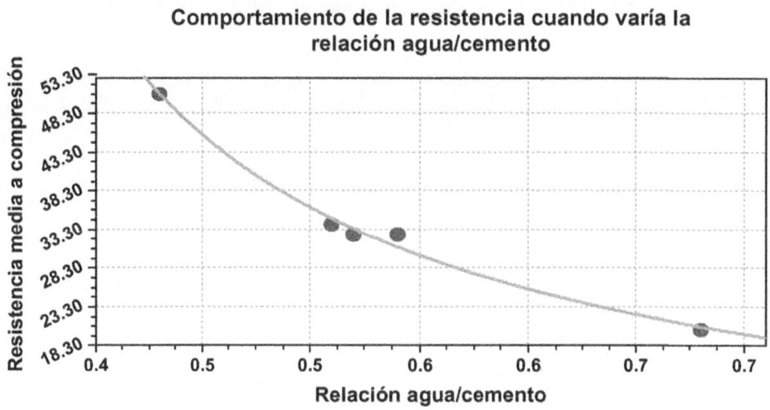

Fig. 2 Behavior of the resistance when the ratio W/C varies (PP-35)

4 Conclusions

1. The studies carried out show that as long as the W/C ratio is maintained, the strength does not vary significantly, regardless of the type of cement used (P-35 and PP-35). In this way, a possible good behavior is predicted in premixed concrete mixtures.
2. The porosity of the concrete decreases with the pozzolan content in the cements studied.
3. Sorptivity improves markedly in cements with pozzolanic additions compared to Portland cement.

References

1. NC 95-2011: Cemento Portland—Especificaciones
2. NC 96-2011: Cemento con adición activa—Especificaciones
3. NC 251 2013: Áridos para hormigón hidráulico. Requisitos
4. Sosa Carrazana, Y.: Modelación matemática de mezclas de hormigón hidráulico aplicando la teoría de los coeficientes separadores en el prefabricado de la Empresa Constructora Militar # 3 de Villa Clara (2011)

Assessment of Addition of Calcinated Clay-Limestone-Plaster to Ordinary Portland Cement in Brickwork Mortars

Dania Betancourt Cura and Jose Fernando Martirena-Hernandez

Abstract The present research work includes the influence of the **LC2** pozzolanic addition in the behavior of type II brickwork mortars from a series of mechanical-physical tests according to specifications established by the 175:2002 Cuban Norm (NC). This addition is made up of 60% calcinated clay, 40% limestone and 10% plaster. The analysis of the LC2 addition behavior in the brickwork mortars was the result of an experimental design in which the definitions of dependent variables according to mortars properties in their different states, as well as independent variables under the analysis of factors, such as: cement volume, lime and addition to be used, were considered. Raw-materials were characterized and forty-eight 40 mm × 40 mm × 160 mm prismatic test tubes for the water-absorption-by-capillarity and flexo-compression tests were elaborated. In the case of adhesion strength by traction, ceramic slabs were used as support. The analysis of the results allowed the satisfactory validation of the use of the LC2 addition in brickwork mortars according to Cuban Norm specifications.

Keywords Pozzolanic addition · Calcined clay · Limestone · Bricklaying mortar

1 Introduction

Cement is the material most produced worldwide. About 150 countries produce cement and in the coming years a steady increase of this trend is highly expected. Its universal use, the industrial production possibility, its great versatility and the great results attained from its use makes its expansion imminent. An increase of 250% in the next 15 years is estimated in Cuba [2] This increase in cement production

D. B. Cura (✉)
Department of Civil Engineering, Constructions College, Universidad Central de Las Villas, Santa Clara, Cuba
e-mail: daniab@uclv.edu.cu

J. F. Martirena-Hernandez
Constructions Faculty, Universidad Central de Las Villas, Santa Clara, Cuba
e-mail: martirena@uclv.edu.cu

© RILEM 2020

J. F. Martirena-Hernandez et al. (eds.), *Proceedings of the International Conference of Sustainable Production and Use of Cement and Concrete*, RILEM Bookseries 22,
https://doi.org/10.1007/978-3-030-22034-1_24

brings about an unfavorable impact upon environment, a reason why a search for new alternatives and strategies to contribute to a sustainable production growth has become necessary [3].

Clinker reduction by means of the use of supplementary cementing materials, being the last one the most viable for the Cuban industry due to minor short-term investments, are among the most extended solutions.

A study on the achievement of an addition made up of calcinated clay + limestone + plaster, this last component in low amount so that it can compensate the increase of aluminates found in the system due to clay use, is being carried out today [1]. This addition is known as LC2 and it promises to be a low-cost, low level of investment, ecological and sustainable product.

The present research work includes the influence and assessment of the **LC2** addition in the brickwork mortars according to specifications established by the 175:2002 Cuban Norm (NC). Brickwork mortars specification [4].

2 Experimental Method

The experiment design was made in order to assess the influence of the LC2 addition (calcinated clay-limestone-plaster) to brickwork mortars with or without the presence of lime, according to NC175, 2002 criteria [4] (Table 1).

Methodology of the experimental design:

(1) Obtaining the constituent materials for mortars elaboration
(2) Raw-materials characterization tests (arid, cement, lime, plaster, clay and limestone)
(3) LC2 production and addition (measurement and ground)
(4) Type II brickwork mortars elaboration by fixing fluency and aware of measurements and parameters established by 175, 2002 CN (NC)
(5) Analysis of elaborated tests on:
    (a) Mechanical strength to 3-day flexo-compression (6 tests a mortar series)
    (b) Mechanical strength to 7-day flexo-compression (6 tests a mortar series)
    (c) Mechanical strength to 28-day flexo-compression (6 tests a mortar series)

Table 1 Measurements tested over the experimental work

No. mixture	Mortar type	Ratios					Quantity of test tubes
		Typo of cement	LC2 addition	Cement	Sand	Lime	
M1 pattern	II	P-35	–	1	5	1	18
M2	II	P-35	0.5	0.5	5	1	18
M3	II	P-35	1.5	0.5	5	–	18
M4 pattern	II	P-35	–	2	5	–	18

(d) Water absorption by capillarity at 28 days (3 tests a mortar series)
(e) Adhesion strength by traction at 28 days (5 tests a mortar series).

3 Results

Results Analysis. Mortars Mechanical Strength

Compression and flexion strength of brickwork mortars were determined at ages 7 and 28 days. Six 40 mm × 40 mm × 160 mm tests tubes were used for the analysis.

Next, compression strength results attained by test tubes at different ages are shown.

Figure 1 shows that mortars corresponding to samples M2 and M4, though with lower strength regarding M1 and M3 respectively, accomplish the 175:2002 CN (NC) specifications which for type II mortars establish 3.5 MPa at 28 days [4].

Results analysis of water-absorption by capillarity in brickwork mortars

According to values in Fig. 2, capillary absorption in mortars elaborated with LC2 addition is higher than those from mortars elaborated with the P-35 reference cement, mainly due to clay content of the aforementioned addition.

Results analysis of Adhesion strength by traction

The adhesion strength by traction was carried out 28 days after curing; in this case each mortar series was placed on four slabs which worked as a support. Five samples

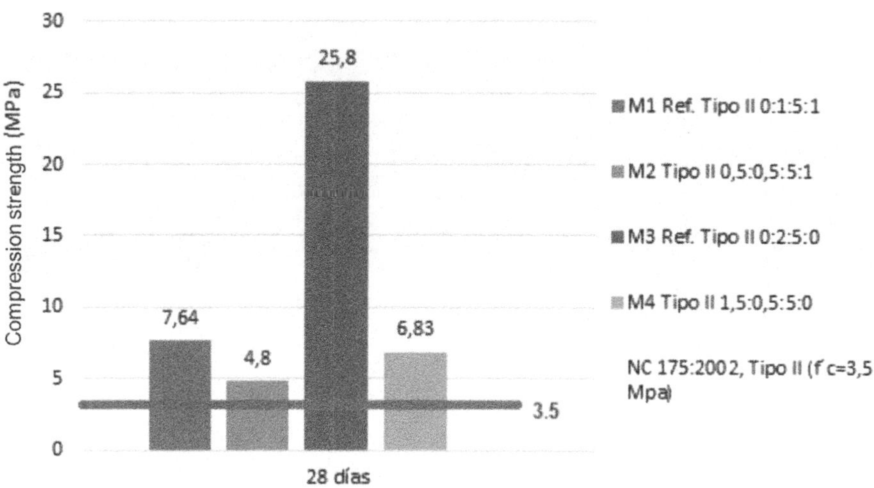

Fig. 1 Compression strength for mortars at 28 days

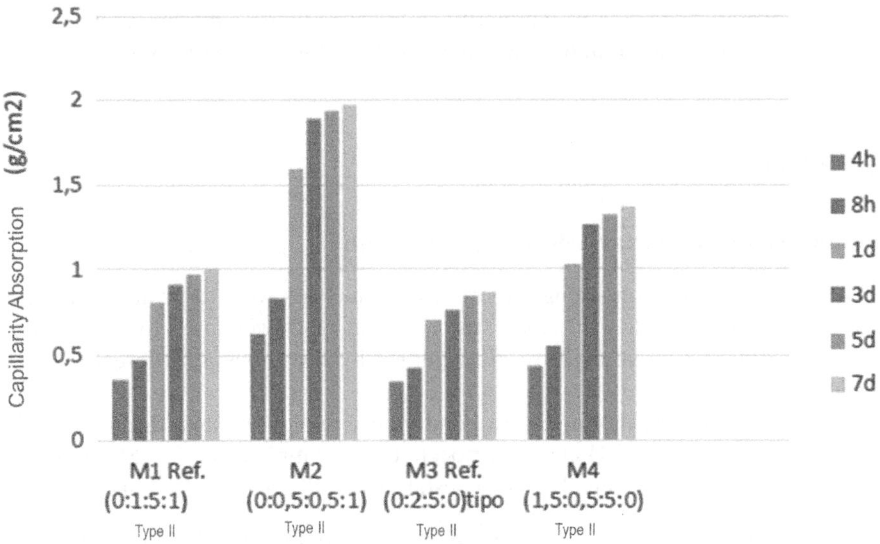

Fig. 2 Water absorption by capillarity graphic

Table 2 Results of the adhesion strength by traction test

Results of the adhesion strength by traction test. Mortar series	Mean Adhesion Strength (MPa)
M1 ref. (0:1:5:1) type II	0.39
M2 (0:0.5:0,5:1) type II	0.31
M3 ref. (0:2:5:0) type II	0.33
M4 (1.5:0.5:5:0) type II	0.32

Fig. 3 Samples elaborated for the adhesion test

per test were analyzed 70 mm separate one from the other. Table 2 shows results of test for each type of mortar (Fig. 3).

As seen in Table 2, the adhesion strength by traction test results produced no significant differences concerning results attained by mortars elaborated with Portland cement (reference sample) and those elaborated with LC2 addition and all samples tested accomplished the 175:2002 CN (NC) in effect today [4].

4 Conclusions

1. Compression strength at age 28 days of mortars elaborated with LC2 pozzolanic addition complies with specifications established by the 175:2002 CN (NC) [4].
2. Capillary absorption in mortars elaborated with LC2 pozzolanic addition gets higher values than reference mortars elaborated with P-35 cement.
3. The values of adhesion strength by traction attained by mortars elaborated with LC2 addition are similar to those obtained by mortars elaborated with Portland cement and comply with the regulations in effect.

References

1. Alujas, A.: RE: Obtención de un material puzolánico de alta reactvidad a partir de la activación térmica de una fracción arcillosa multicomponente (2010)
2. Martirena, F.: Reporte sobre Estudio de factibilidad para la producción de metakaolín en Cuba. Cuba (2011)
3. Martirena, J.F.: Una alternativa ambientalmente compatible para disminuir el consumo de aglomerante de clínker de cemento Pórtland: el aglomerante cal-puzolana como adición mineral activa. Tesis de Doctorado, Universidad Central de Las Villas (UCLV) (2003)
4. NC175: Morteros de albañilería. Especificaciones. Ciudad de La Habana: Oficina Nacional de Normalización (2002)

Concrete Manufactured with LC3 Following the Cuban Standard NC 120: 2014

Eilys Valdés Alemán, Viviana Rodriguez Rodriguez and Jose Fernando Martirena-Hernandez

Abstract As part of the development strategy of the LC3 in Cuba, a procedure is developed for the preparation of concretes with different doses (H1, H2, H3, H4) according to the levels of atmospheric aggressiveness established in NC 120: 2014 "Hydraulic concrete—Specifications Concretes will be made that respond to three designs of mixtures where parameters such as cement content and water/cement ratio vary according to what is established in the aforementioned standard. The rheological and physico-mechanical behavior of concrete will be evaluated in a fresh and hardened state, for which the following tests will be carried out: Determination of the slump by the cone of Abrams and the strength of the hardened concrete at 3, 7 and 28 days of curing. The concretes will be made with low carbon cement (LC3) up to 50% replacement of the clinker, calcined clays and limestone, and the patterns, made with Portland cement 35 (P35). Curing influence for the H1 series will be evaluated in both standard and LC3 manufactured concretes.

Keywords Levels of atmospheric · Rheological · Physical-mechanical aggressiveness · Slump · Resistance · Concrete

1 Introduction

Since the appearance of cement its production has been considered a key indicator in the development of a country for various reasons, the main one is that it constitutes an essential product for the manufacture of concrete, considered as the most used material in the construction industry currently [6]. However, its high global production has resulted in negative consequences such as the depletion of natural resources,

E. V. Alemán (✉)
Faculty of Constructions of the UCLV, CIDEM, Havana, Cuba
e-mail: evaleman@uclv.cu

V. R. Rodriguez
ECOA #37-Havana, Havana, Cuba

J. F. Martirena-Hernandez
CIDEM, Havana, Cuba

© RILEM 2020

J. F. Martirena-Hernandez et al. (eds.), *Proceedings of the International Conference of Sustainable Production and Use of Cement and Concrete*, RILEM Bookseries 22,
https://doi.org/10.1007/978-3-030-22034-1_25

such as raw materials and fossil fuels used in its production, and the industry is not sustainable in environmental terms [1]. One of the alternatives to these problems is the implementation of a new kind of cements that allows the substitution of 50% of the clinker content in the cement production known as LC3 low carbon cement [5]. The implementation of the production of said cement and its commercialization requires the existence of a Cuban regulation that establishes the specifications for its production and use in concretes [7].

Currently, the research carried out to reach the elaboration of this regulation is limited, and the demand of the market is increasing, so it is necessary to check if the concretes made with these cements comply with the technical parameters established in the NC 120: 2014 Hydraulic Concrete—specifications, norm that defines these parameters for the concretes made with the different cements in Cuba [2].

The present investigation shows the procedure developed for the elaboration of concretes with different dosages (H1, H2, H4) according to the levels of atmospheric aggressiveness established in NC 120: 2014 "Hydraulic Concrete—Specifications" as well as the results obtained in the tests carried out [2]. The work began with the realization of test pieces that respond to three mix designs where parameters such as cement content and water/cement ratio are varied according to the aforementioned standard. The rheological and physico-mechanical behavior of concretes in fresh and hardened conditions is evaluated, for which the following tests were carried out: Determination of the slump by the Abrams cone and strength of the hardened concrete at 3, 7 and 28 days cured. The concretes were made with low carbon cement (LC3) with up to 50% substitution of clinker, calcined clays and limestone, and the standards, made with Portland cement 35 (P35). In addition, the influence of curing for the H1 series is evaluated both in the standard concretes and those made with LC3.

2 Materials and Methods

For the development of the corresponding tests, three concrete series (H1, H2, H4) will be developed, each of them with specific dosages depending on the level of aggressiveness for which the concrete is manufactured according to NC 120: 2014 "Hydraulic Concrete. Specifications" [2]. The mix design to be used for the series H1, H2 and H4 are shown in Table 1.

2.1 Test Protocol

- Slump aided by the Abrams cone: The test consists of filling a conical metal mold of standard dimensions, in three layers tamped with 25 strokes of rod and then remove the mold, measure the slump that the mass of concrete placed inside

Table 1 Dosages for the H1, H2 and H4 series for PC and LC3

Type mix	Aggressiveness NC 120:[2]	f'c (MPa)	W/C	Cement (kg)	SP (kg) PC	SP (kg) LC3	A. fine (kg)	A. medium (kg)	A. coarse (kg)	Water (kg)
H_1	Very high	35.0	0.40	430	2.79	4.3	742	410	622	172
H_2	High	30	0.45	400	2.6	4	740	409	620	182
H_4	Low	20	0.55	345	2.24	3.45	753	416	631	190

with respect to the mold removed as established in the NC—ISO 1920-2: 2010 "Concrete Testing-Part 2: Properties of fresh concrete".

- Compressive strength: For this, three test pieces of (10 cm × 20 cm) were taken for each curing time 3, 7 and 28 days and for each mixture for both PC and LC3, for a total of 9 mixespecimens. The requirements of the compression test are found in NC 244-2005: Hormigón endurecido [4]. Resistance of the concrete in hardened state. To put the test into practice, the cylindrical specimens were placed centrally on the plate of the machine and a force was applied with a continuous load at a rate of 1 mm × minute until its break.
- Influence of the curing: The H1 series will be replicated in order to analyze the influence of the curing for this series, so that one of the replicated concretes of this series will not be cured and the other if and they will be practiced both the same tests both physical-mechanical and rheological, which will allow reaching conclusions regarding the curing for this series.

3 Results on Fresh Concrete. Slump

Tables 2 and 3 show the results of the test carried out on fresh concrete (consistency with the use of the Abrams cone) where slumps were obtained between 16 and 22 cm for concrete with PC and between 17 and 23 cm for concrete with LC3, it was achieved a fluid consistency in accordance with the stipulated in the NC 120: 2014 "Hydraulic Concrete. Specifications", facilitating the workability of the mixtures [2]. This was possible due to the use of SikaPlast 9100CU chemical additive in the proper proportions for each mixing design.

Table 2 Slump in the mixture with PC

PC			
Level of aggressiveness	Slump	SP (%)	Consistency
H1 (cured)	16	0.65	Fluid
H1 (not cured)	16	0.65	Fluid
H2	17	0.65	Fluid
H4	22	0.65	Fluid

Table 3 Slump in the mixture with LC3

LC3			
Level of aggressiveness	Slump	SP (%)	Consistency
H1 (cured)	17	1	Fluid
H1 (not cured)	17	1	Fluid
H2	23	1	Fluid
H4	19	1	Fluid

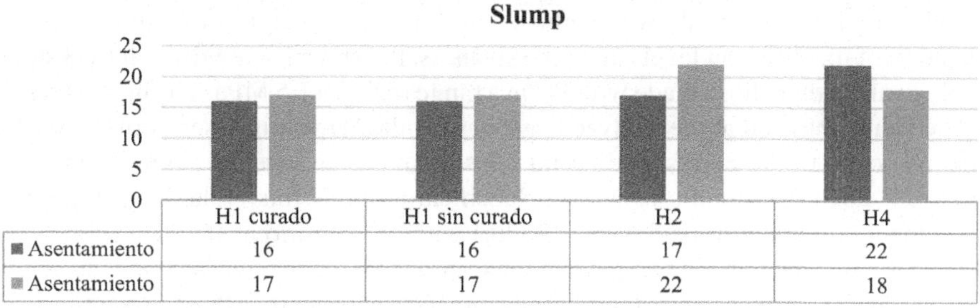

Fig. 1 Slump for concrete with LC3 and PC (Prepared by the author)

The slump of the concretes made with PC and those made with LC3, was similar in all the manufactured mixtures, as shown in Fig. 1.

4 Hardened Concrete

To analyze the results of the resistance tests, 3 samples were taken per mixture for each of the cements used to be tested at 3, 7 and 28 days. The influence of curing on compression resistance at each of these ages in the H1 series was evaluated.

4.1 Result of the Compression Resistance Trial After 3 Days

Figure 2 presents the compressive strength of concrete made with LC3 and PC for each of the manufactured concrete series (H1, H2, H4) at the age of 3 days.

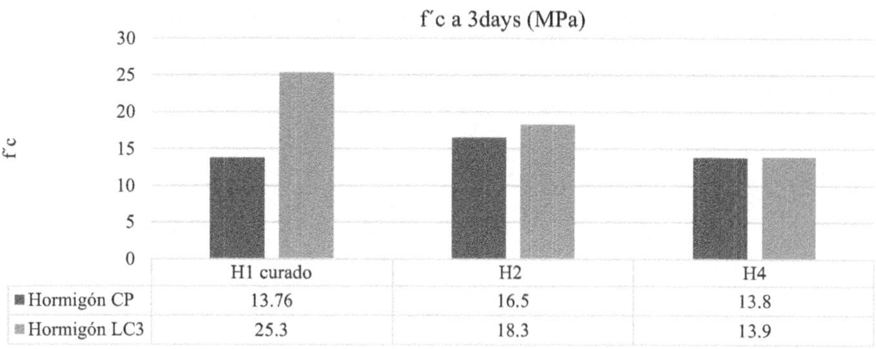

Fig. 2 3d strength on concrete

The previous results allow to deduce that, for all the designs of mixtures used in function of the different levels of aggressiveness, the concretes made with LC3 show higher values than those made with PC in a range of 0.1–11.5 MPa. The highest resistance value obtained at the analyzed age was for the H1 (curing) series of 25.3 MPa that responds to the corresponding mix design at a high level of aggressiveness in NC 120: 2014 "Hydraulic Concrete. Specifications", as this is the largest amount of cement contained in its design (430 kg) with an a/c ratio of 0.4 [2]. The lowest value of resistance at 3 days was obtained in the H4 series that responds to a low level of aggressiveness with a minimum cement content in its 345 kg mix design for an a/c ratio of 0.55. On the other hand, the results of the statistical analysis show that both for reference concretes and for those made with LC3, we worked with a level of reliability of 95% since the typical error does not exceed 5% in any of the cases studied. The standard deviation allows to deduce that there is precision in the results, by limiting the mean of uncertainty to values between 1.2 and 1.9 for the tests performed on concretes with LC3 and 1.1 and 2.1 for those made with PC.

4.2 Result of the Compression Resistance Trials After 7 Days

Figure 3 shows the comparison between the strength of the concretes made with LC3 and PC for each of the manufactured concrete series (H1, H2, H4) at the age of 7 days.

In an analysis of the results plotted above, it can be concluded that concretes made with LC3 show higher values than those made with PC in a range of 1.1 to 4.75 MPa in the healed H1 and H2 series, while in the series H4 those made with PC exceed in 3.4 those made with LC3. The highest resistance value obtained at the analyzed age was for the H1 (cured) series of 34.3 MPa that responds to the corresponding mix design at high aggressiveness level in NC 120: 2014 "Hydraulic Concrete. Specifications", as this is the largest amount of cement contained in its design (430 kg) with an a/c ratio of 0.4 [2]. The lowest resistance value at 7 days was obtained in the H4 series, which responds to a low level of aggressiveness with a

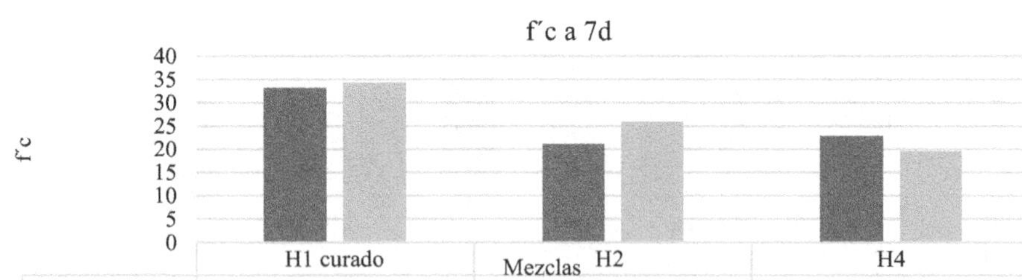

	H1 curado	H2	H4
Hormigón CP	33.2	21.15	22.9
Hormigón LC3	34.3	25.9	19.5

Fig. 3 7d strength results

minimum cement content in its 345 kg mix design for an a/c ratio of 0.55. The results of the statistical analysis show that, in the standard concretes and those made with LC3, a level of reliability of 95% was used in all the curing series of the concrete with PC and LC3 since the typical error is less than 5% given to the correct elaboration of the manufactured specimens. The standard deviation allows to deduce that there is precision in the results, when limiting the uncertainty average to values between 0.8 and 1.4 for the tests performed on concretes with LC3, in the ones elaborated with PC the standard deviation is of 0.5 and 2.1 for all the series offering precision of the results.

4.3 Result of the Compression Resistance Tests at 28 Days

After analyzing the results plotted in the previous figure, it can be concluded that the concretes made with LC3, as at the age of 3 and 7 days, show values higher than those made with PC in a range of 11.8 to 4.1 MPa in all curing series. The compressive strength achieved by these concretes was always greater than the design strength of each mixture, which was set according to the specifications of NC 120: 2014, taking into account the atmospheric aggressiveness [2] (Fig. 4).

The results of the statistical analysis show us that we worked for a level of reliability of 95% for all the curing series in the standard concretes and those made with LC3, since the typical error is less than 5%. The standard deviation allows to analyze that there is precision in the results, when limiting the uncertainty average to values between 0.9 and 2.3 for the tests performed on concretes with LC3, in the ones elaborated with PC the standard deviation varies between 1.1 and 1.9 for all the series offering precision of the results.

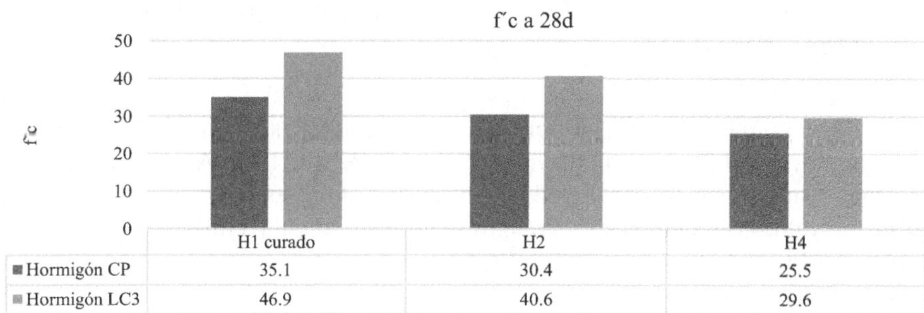

Fig. 4 Resistance of the concretes at 28 days (Prepared by the author)

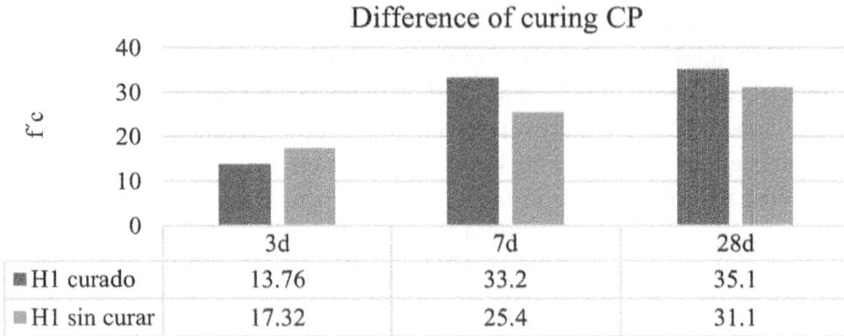

Fig. 5 Influence of curing on mixture H1 for PC

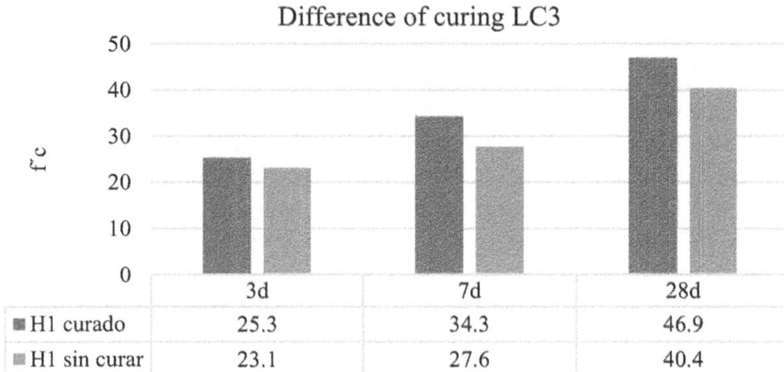

Fig. 6 Influence of curing on mixture H1 for LC3

4.4 *Influence of Curing: The Influence of Curing on the H1 Series for the PC and for the LC3 Is Presented Below (Figs. 5 and 6)*

In the above graphs, the curing difference in the H1 series is shown for the three ages of curing (3, 7 and 28 days) in the concretes with PC and LC3. After analyzing the previous results, it can be concluded that the curing of the specimens positively influences the values of compressive strength in all ages studied, either in the concretes made with PC as in those made with LC3.

5 Conclusions

It is necessary to use more additives in the mixtures made with LC3, because the clay has a greater surface area than clinker, to achieve the consistency fixed in the concrete, maintaining the same water cement ratio. With the use of LC3 and superplasticizer additive fluid consistency is achieved in concretes that meet the specifications according to the level of aggressiveness of NC 120: 2014. The concretes made with LC3 show higher resistance values than those made with PC at all ages studied [2]. The curing of the specimens to be tested has a positive influence on the compression resistance results, since the cured specimens of the H1 series offer higher values in relation to these results, compared to those that were not cured.

References

1. Mehta, P.: La reducción del impacto ambiental del hormigón. El hormigón (2001)
2. NC 120: Hormigon hidráulico—especificaciones. El Vedado (La Habana): s.n (2014)
3. NC 174: Hormigón Fresco-Medición asentamiento por el cono. El Vedado (Ciudad de La Habana): s.n (2002)
4. NC 244-2005: Hormigón endurecido—determinación de la resistencia a la compresión en probetas cilíndricas
5. Martirena, J. F., et al.: LC3 Manual de Calidad (2015)
6. Serpell, R.: (2013). Seminario Arquitectura en Hormigón. Posibilidades y desafios. Chile (2013)
7. Vizcaíno, L.: Cemento de bajo carbono a partir del sistema cementicio ternario clínquer-arcilla calcinada-caliza, Tesis Doctoral, Universidad Central «Marta Abreu» de Las Villas (2014)

Evaluation of the LC^3 to Be Used for the Local Production of Materials in Cienfuegos

Ariens Laritza Yrigoyen Acosta, Gercier Amaniel Gradaille and Ivan Machado-Lopez

Abstract The high demand for binders in Cuba as a result of the constructive needs of a developing country and the unavailability of pozzolans determined that specialists from the Center for Research and Development of Structure and Materials (CIDEM) of the Central University "Marta Abreu" of Las Villas, together with a technical team from the Federal Institute of Technology of Lausanne, Switzerland, will elaborate a proposal of ternary cement based on the substitution of elevated quantities of clinker by a combination of calcined clay and calcium carbonate, a formulation known internationally as LC^2 and/or LC^3. In this work the use of the pozzolanic addition is evaluated in a workshop of local production of construction materials in the Cienfuegos province, the suitability of the aggregates usually used in relation to the national standards and its relation with the resistance for manufactured concretes is determined with the new type LC^3 binder. Results are exposed in the manufacture of small format elements such as hollow blocks with the use of concretes designed and tested in a specialized laboratory of the Ministry of construction, resistance levels higher than 5 MPa are obtained in correspondence with the national standards.

Keywords Portland cement · Mechanical resistance · Calcined clays

1 Introduction

The international experience in construction using concrete blocks has demonstrated the excellent behavior of this construction system and its adaptability to any type of building, especially to low-rise series constructions such as housing, the use of this

A. L. Y. Acosta · G. A. Gradaille
ENIA–Cienfuegos, Cienfuegos, Cuba
e-mail: laritza@eniacfg.co.cu

I. Machado-Lopez (✉)
UCLV-CIDEM, Santa Clara, Cuba
e-mail: ivanm@uclv.edu.cu

© RILEM 2020

J. F. Martirena-Hernandez et al. (eds.), *Proceedings of the International Conference of Sustainable Production and Use of Cement and Concrete*, RILEM Bookseries 22,
https://doi.org/10.1007/978-3-030-22034-1_26

system can achieve important advantages in terms of saving resources compared to other systems.

Hence, currently in Cuba, local producers of building materials have become major manufacturers of these elements, however, in many cases the shortage of Portland cement (CPO) limits the development of these productions, resulting in research aimed at the search for a binder that can replace and/or increase the CPO offering advantages from all points of view to local producers of materials.

In this sense, an important way to improve sustainability in the production of cement is to reduce the clinker factor by adding CPO to supplementary cementitious materials (SCM). Recent work developed in the CIDEM have demonstrated the feasibility of using a binder resulting from the tertiary mixture of thermally activated clinker + clay (TAC) + calcium carbonate, this product has been called Low Carbon Cement or LC^3 [1–3].

An important variant that can be used advantageously is to obtain the addition of TAC + calcium carbonate (LC^2) in industrial form and mix it properly with Portland cement P-35 in the local production workshops. This product is proposed for use in the production of mortars and concretes, as well as in small format elements such as hollow concrete blocks and others [4].

However, the growth of CPO production does not fully satisfy the current demand, hence the production of LC^3 and/or LC^2 is emerging as a solution that allows multiplying the production volumes with an important environmental contribution. For the local industry the application of the LC^3 is an important solution that allows to multiply the availability of binder, in Cuba the program for the production and sales of local materials is developed, the local production enterprises of construction materials maintain a sustained growth that can be related to the application of this new binder.

An industrial test was carried out in the Siguaney cement factory to obtain LC^2 (TAC + calcium carbonate + gypsum), it is intended to use this material in the local production of elements for construction, the objective of this work is to obtain different formulations LC^2 + P 35 and elaborate different dosages for its use in several applications.

2 Analysis and Results

In the experimental design to locally produce an LC^3 cement, which maintains the physic-mechanical properties required by current regulations for cements with pozzolanic addition, as well as hollow concrete blocks, the following methodology was followed:

3 Characteristics of the Raw Materials Used for the Manufacture of the LC^3-50 and LC^3-65 Cements

To obtain the LC^3, the P35 from the Carlos Mark factory in the province of Cienfuegos and the LC^2 from the industrial test carried out in Siguaney are used. For the LC^3-50, 50% of the P-35 cement is substituted by 50% of LC^2 and in a second variant, LC^3-65 is obtained from 70% of P-35 plus 30% of LC^2. Tables 1 and 2 show the characteristics of P-35 and in Table 3 those of LC^2.

4 Aggregates

As fine aggregate, the sand from the ¨Arimao¨ quarry and the coarse aggregate (fraction 10–5) from the quarry of "Los 500" from the province of Cienfuegos are used, which were characterized in the laboratory of the National Enterprise of Applied Research. (ENIA) of said province obtaining the following results (see Tables 4, 5 and 6).

Table 1 Characteristics of cement P-35

Blaine NC 980: 2013	3166
CN NC 524: 2015	24.6
NC volume stability 524: 2015	0.39
Retained sieve 90 microns NC 980: 201	2.2
Setting time NC 524: 2015	
Initial (min)	105
Final (horas)	2.8

Table 2 Análisis Químico del P-35

SiO_2	22.37
Al_2O_3	5.19
Fe_2O_3	3.13
CaO	62.11
MgO	1.52
SO_3	2.96
CaOL	0.75
RI	3.04
PF	2.18
Álcalis equiv.	0.65

Table 3 Chemical analysis of LC^2 cement

Clay	SiO_2	Al_2O_3	Fe_2O_3	CaO	MgO	SO_3	Na_2O	K_2O	TiO_2	P_2O_5	Mn_2O_3	Otros	PPI
Yaguajay	46.58	20.06	14.41	2.94	0.74	0.04	0.11	0.06	1.12	0.13	0.73	0.14	12.74

Table 4 Physical characteristics of the aggregates under study

Test	Sand	Granite 10–5	Units	NC 251:2013
T-200	0.7	0.88	%	≤5% para árido fino
Specific weight current	0	0	0	≤1% para árido grueso
Specific weight current	0	0	0	0
Specific weight current	2.59	2.69	g/cm^3	0
Adsorption	0	0	0	≥2.50
Weight Volumetric	2.63	2.7	g/cm^3	0
Weight Volumetric Compacted	2.71	2.75	g/cm^3	0
% Vacuum	1.7	0.86	%	<3%
Module of finer	1228	1362	kg/m3	0

Sand: ¨Arimao¨; Granite: 10–5 ¨Los 500¨

Table 5 Granulometry, Sand Arimao

Sieve (mm)	**9.52**	**4.76**	**2.38**	**1.19**	**0.59**	**0.297**	**0.149**
Specifications	**100**	**90–100**	**70–100**	**45–80**	**25–60**	**10–30**	**2–10**
%Past	100	98	87	74	53	26	7

Table 6 Granulometry Granite 10–5 ¨Los 500¨

Sieve (mm)	**12.7**	**9.52**	**4.76**	**2.38**	**1.19**
Specifications	100	**85–100**	**15–35**	**0–10**	**0–5**
%Past	100	100	91	23	1

5 Manufacturing Process of Hollow Concrete Blocks

The production of the blocks was done with the Vibrocom compaction machine [5]. 100 and 150 mm blocks were manufactured, using different dosages, one for LC^3-50 and another for LC^3-65. This process was executed by a staff of experience in the development of blocks.

The design of the concrete mix was made according to the Technical Instructions for the use of Low Carbon Cement. Then, in Tables 7 and 8, the dosages used are shown.

The results obtained show that the use of the LC^3 in its application for hollow blocks of concrete meets in both variants with the values of resistance to compression required by Cuban standards.

Table 7 LC^3-50, Sand "Arimao" y Granite Los 500 (10–5 mm)

Dosif.	Quality (mm)	Type	LC^3-50 (kg)	Sand (kg)	Granite (10–5) (kg)	W (l)	W/C
1	100	Gravimetric	260	823	1227	130	0.50
		Volumetric	1	2.4	3.2	0.48	
	150	Gravimetric	320	784	1159	153	0.48
		Volumetric	1	1.9	2.5	0.46	

Table 8 LC^3-65, Sand Arimao y Granite Los 500 (10–5 mm)

Dosif.	Quality (mm)	Type	LC^3-65 (kg)	Sand (kg)	Granite (10–5) (kg)	W (l)	W/C
2	100	Gravimetric	260	823	1227	130	0.50
		Volumetric	1	2.5	3.3	0.5	
	150	Gravimetric	320	784	1159	153	0.48
		Volumetric	1	1.9	2.6	0.48	

6 Test of Compression Resistance to Hollow Concrete Blocks

The compression resistance test was carried out at 7 and 28 days. A Russian Class 2 press, Type P-125, brand ZIL, with a resolution of 2.5 kN and an upper measurement limit of 1250 kN was used.

First, a layer of paste consisting of cement mortar P-35 is placed on the load and support surface of the blocks in order to level them as shown in Fig. 1. The batch is considered to be compliant when the average value of the Compressive strength is greater than or equal to that established in the NC 247-2010, provided that the coefficient of variation is less than or equal to 0.20, 3 hollow concrete blocks were tested for each of the variants described above (see Table 9).

7 Absorption Test for Hollow Concrete Blocks

The test of absorption of the hollow blocks of concrete was made to the blocks of 150 mm that are required by the standard for it was taken a sample of 3 blocks of each of the variants manufactured. For this, a technical balance of 20 kg with a precision of 20 g and an automatic regulation stove was used.

The applied procedure comprises placing the blocks in the oven at a temperature of 105~110 °C to dry them and obtain the dry mass. Once this process is completed, the blocks are placed inside a pond full of water for a period of 24 h, then they are dried with a damp cloth and weighed to determine the wet mass. The lot is considered to be compliant when the average of the units in the sample complies with the absorption

Fig. 1 Blocks made with LC^3 ready to be tested in compression

Table 9 Compressive strength of the blocks manufactured with LC^3

Blocks	Type	R'm (MPa)	SR' (MPa)	V (%)
Compressive strength 7 days				
100	LC^3-50	3.0		
150	LC^3-50	4.3	0.17	0.05
100	LC^3-65	3.3	0.45	0.15
150	LC^3-65	4.9		
Compressive strength 28 days				
100	LC3-50	4.2		
150	LC3-50	5.2	0.13	0.025
100	LC3-65	4.8	0.19	0.036
150	LC3-65	5.9		

established in the Cuban standard NC 247-2010, the results obtained are shown in Table 10.

The results of the absorption test show in both cases that they comply with the value specified in NC 247: 2010, which must not exceed 10%, so both variants are shown to guarantee the levels of absorption required by Cuban manufacturing standards. of blocks.

Table 10 Absorption test results

Type	Msi (kg)	Mhi (kg)	Am (%)	Sa (%)
150 LC^3-50	12.58	13.10	4.2	0.1
150 LC^3-65	13.46	14.14	5.0	0.06

8 Conclusions

The application of formulations for binders based on the CPO where the clinker factor decreases, either industrially or in local production, is an appropriate contribution to the reduction of energy consumption and thereby support the reduction of the effects of change climate.

LC^3 cement, when produced locally with appropriate technologies and different sources of ceramic waste, reaches values of fineness, strength and other parameters similar to the CPO, which allows its application in the manufacture of prefabricated elements of small format such as hollow concrete blocks, masonry mortars and other uses.

The hollow blocks of concrete produced from the LC^3 in the local production in Cienfuegos show adequate results according to the national norms so the use of the LC3 in the two studied variants is feasible.

References

1. Alujas, A.A.A., Betancourt, S., Leyva, C.: Pozzolanic reactivity of low grade kaolinitic clays: influence of mineralogical composition. In: Proceedings of the 1st International Conference on Calcined Clays for Sustainable Concrete RILEM BOOKSERIES Volume 10 ISSN 2211-0844 ISSN 2211-0852 Springer Dordrecht Heidelberg New York London ©RILEM 2015 (2015)
2. Avet, F., Alujas, R.S.A., Scrivener K.: Development of a new rapid, relevant and reliable (R3) testing method to evaluate the pozzolanic reactivity of calcined clays. In: Proceedings of the 1st International Conference on Calcined Clays for Sustainable Concrete RILEM BOOKSERIES Volume 10 ISSN 2211-0844 ISSN 2211-0852 Springer Dordrecht Heidelberg New York London ©RILEM 2015 (2015)
3. Fernandez, R., Martirena, F., Scrivener, K.L.: The origin of the pozzolanic activity of calcined clay minerals: a comparison between kaolinite, illite and montmorillonite. Cem. Concr. Res. **41**(1), 113–122 (2011)
4. Machado, I.L., Moya, H.I, Sánchez, S.B, Martirena, F.: Improves of the environmental energy sustainability in the production of cement Portland with addition of clays activated thermally. In: 2nd International Conference on Calcined Clays for Sustainable Concrete, La Havana. CUBA (2017)
5. Martirena, F., M. I., Martinez, L., Betancourt, S., Gonzalez, R. (Ed.): Manual Instructivo para la producción de bloques huecos de hormigón con la maquina Vibracom. Santa Clara. Cuba, Samuel Feijoo (2012)

Using a Physical Model Based on Particle Mobility for Mix Design of Commercial Concretes in Order to Increasing Eco-Efficiency

Mariana Menezes, Rafael G. Pileggi, Markus Rebmann and Carlos Massucato

Abstract Concrete greenhouse gas emissions are mostly from cement. A relevant strategy for the concrete greenhouse gas emissions abatement is the increase of the efficiency in the use of its binder. For this, it is important to understand the packing and the mobility of the concrete particles and their effects on its rheological behavior in the fresh state and the mechanical properties in the hardened state. Currently, the mix design methods are empirical, as opposed to this experimental optimization, the computational optimization emerges, based on predictive models. Thus, the objective of the study is to use the physical model based on particle mobility to predict the rheological behavior of commercial concretes. The method consisted in the following steps: characterization of raw materials; eco-efficiency analysis of 60 formulations of a concrete plant using the binder intensity concept; and a parameter estimation for a descriptive model of slump in function of mobility variables. These concrete formulation binder intensities were between 7.5 and 10.5 kg/m^3/MPa, in which the rise of specified fck and maximum aggregate size have a positive impact on this index, while the slump growth has a negative effect. For the base formulations, the mobility variable MPT (Maximum Paste Thickness) showed good correlations with the specified slump (R^2 0.99), as well as the model presented a good adjustment to the data (R^2 0.94). These results allow an improvement in the mix design methodology using computational optimization, which can lead to an increase on the eco-efficiency of the commercial concretes.

Keywords Particle packing · Mobility · Interparticle separation distance · Slump · Rheological behavior · Model · Physical model · Mix design · Mix proportioning

M. Menezes (✉) · R. G. Pileggi · M. Rebmann · C. Massucato
Universidade de São Paulo, Cidade Universitária, São Paulo, Brazil
e-mail: mariana.menezes@lme.pcc.usp.br

M. Menezes
InterCement, Av Das Nações Unidas 12.495, São Paulo, Brazil

© RILEM 2020

J. F. Martirena-Hernandez et al. (eds.), *Proceedings of the International Conference of Sustainable Production and Use of Cement and Concrete*, RILEM Bookseries 22,
https://doi.org/10.1007/978-3-030-22034-1_27

1 Introduction

Greenhouse gas emissions from concrete are mostly from cement, with the cement industry accounting for approximately 7% of global CO_2 emissions [3]. In the concrete field, a relevant strategy is to increase the efficiency in the binder use, that is, to achieve better properties with lower binder consumption.

To increase the efficiency of concrete binders, it is necessary to understand the structure of this material and its consequences on the properties of the concrete, both in the fresh and hardened state. For this, it is important to understand the packing and mobility of the concrete particles and their effects on the rheological behavior of the material in the fresh state and on the mechanical properties in the hardened state, a theme that has already been widely discussed by several concrete researches [4–6].

The present work was developed with the objective of advancing the use the physical model based on particle mobility to predict the rheological behavior of commercial concretes in order to improve the current mix design methodology.

2 Methodology

In order to understand the current logic of mixture design, data were obtained from conventional concrete formulations of a concrete operation in São Paulo/Brazil. This concrete plant has 60 base formulations. These concretes have specified compressive strength (f_{ck}) ranging from 20 to 45 MPa, slump between 90 and 200 mm and maximum aggregate size 19 mm or 9.5 mm.

With these data, the concrete binder intensities (Eq. 1) were evaluated to understand their eco-efficiency. In order to elucidate the observed trends, the physical models of packing and mobility were used to evaluate the formulations. In this study, the packing porosity was calculated by Westman and Hugill [7] model and the distances between particles (Eq. 2) were used: IPS (Interparticle Separation Distance, for the particles finer than 100 μm) and MPT (Maximum Paste Thickness, for the particles greater than 100 μm) [2, 5]. These two variables were correlated with the fresh concrete property, slump, and a model was adjusted.

$$BI = b/pr \tag{1}$$

BI = Binder Intensity (kg/m^3/MPa)
b = binder content (kg/m^3)
Pr = performance requirement, compressive strenght at 28 days (MPa)

$$\mathrm{D} = \frac{2}{VSA}\left[\frac{1}{Vsol} - \left(\frac{1}{1-\varepsilon}\right)\right] \tag{2}$$

D = Distance between particles—IPS for fine or MPT for coarse (μm)
VSA = Volumetric Superficial rea (m^2/cm^3)
V_{sol} = Solids volumetric fraction
ε = Packing porosity

To use these models, it was necessary to characterize the raw materials: particle size distribution by laser granulometry and dynamic image analysis and density by helium pycnometry.

3 Results and Discussion

3.1 *Material Characterization*

In this section, the material characterization is presented in Table 1 and Fig. 1.

(1) Granulometric volumetric area was calculated from particle size distribution.

Table 1 Material set characteristics

Characteristic	19 mm Gravel	9.5 mm Gravel	Artificial Sand	Natural Sand	Cement
Density (kg/m^3)	2.71	2.70	2.70	2.65	3.06
Granulometric volumetric area (m^2/cm^3)[1]	7.8×10^{-3}	4.0×10^{-3}	8.2×10^{-2}	4.2×10^{-2}	1.17

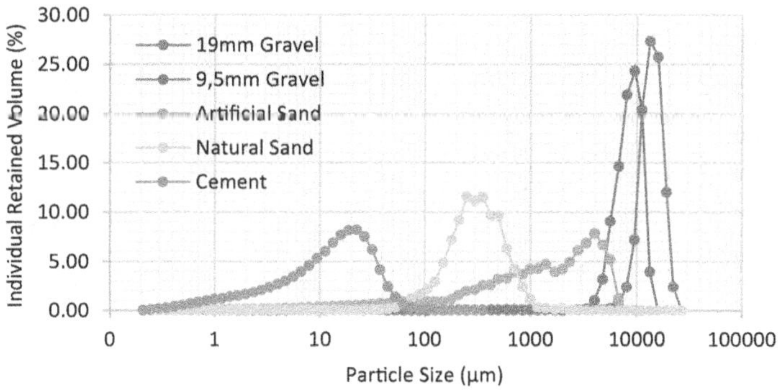

Fig. 1 Material set particle size distribution

3.2 Eco-Efficiency Analysis Using the Binder Intensity

A concrete eco-efficiency can be first analyzed by the binder intensity, since clinker is the most intense source of CO_2 emissions in a concrete.

Looking at the change of the binder intensity in function of f_{ck}, as shown in Fig. 2, it is possible to notice that there is a decrease of this index between 20 and 30 MPa, but, from 30 to 35 MPa the intensity stagnates and grows slightly when reaching 40 MPa. Comparing the results with Daminelli [1], the intensities of the evaluated concretes are in the same experimental region of the concretes mapped by him, however with binder intensities well above the best-found intensities, below 5 kg/m^3/MPa. The name of each series was given following the pattern: it starts with "form1", followed by which was the maximum aggregate size ("b1" for 19 mm and "b0" for 9.5 mm), and then which was the slump (ex. 140).

3.3 Mobility Analysis and Descriptive Model

In order to understand the concrete mobility, coarse packing porosity, the MPT and the IPS, as defined by Eq. 2, of the base formulations were calculated. First, the formulations for different f_{ck}—consequently different paste compositions—were analyzed plotting the slump as a function of the coarse packing porosity (Fig. 3).

With these results, it is not possible to find a correlation between slump and coarse packing porosity. Different formulations with the same slump present different porosity.

Thus, the slump was plotted as a function of MPT (Fig. 4). With these results, it was noticed that there is a high correlation between the MPT and the slump, with the coefficient of correlation (R^2) of the adjusted linear curve 0.99, given a paste

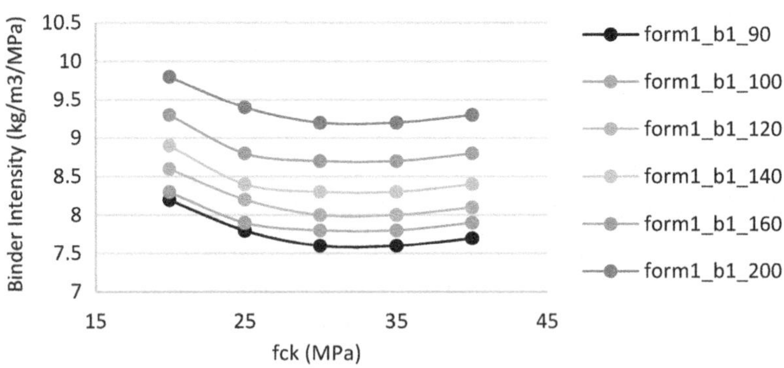

Fig. 2 Binder intensity per f_{ck} for different slumps

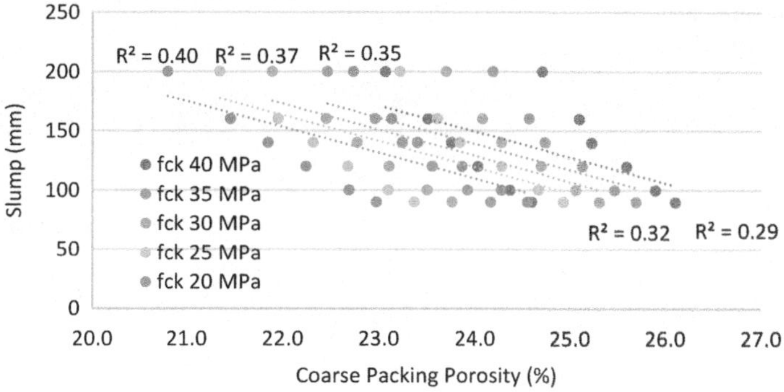

Fig. 3 Slump in function of coarse packing porosity per f_{ck}

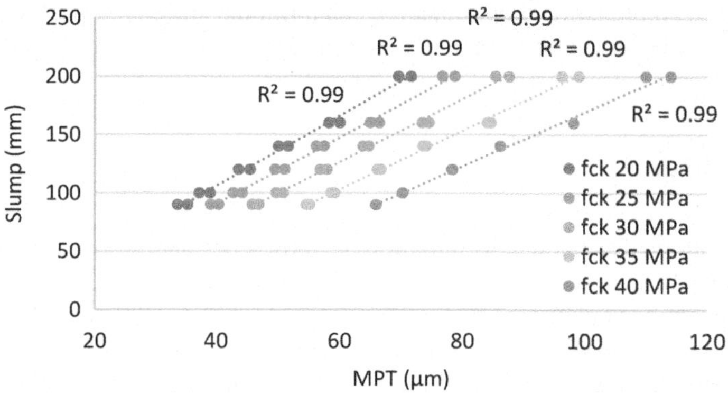

Fig. 4 Slump in function of MPT per f_{ck}

composition. That is, formulations that use different types of paste (in the graph represented by fck) do not present the same MPT for a slump.

The paste mobility is described by the IPS, which in this set of concretes has a high correlation (R^2 0,99) with the f_{ck}. Understanding the effect of the paste—reflected by the IPS—as well as the MPT, as discussed in the both previous paragraphs, a model was adjusted. This model (presented in Eq. 3) has the lower limit zero and is asymtoptic in 300, so covering the slump test range. Figure 5 shows the high correlation between real slump and model slump (R^2 0.94).

$$s = 300\frac{(MPTIPS/210)^2}{\left(1 + (MPTIPS/210)^2\right)} \tag{3}$$

This realization allows the creation of a mobility-based concrete design strategy. For a specified concrete (f_{ck}, slump and maximum aggregate size), it is desired to increase the coarse solid fraction keeping a fixed MPT by reducing the coarse packing

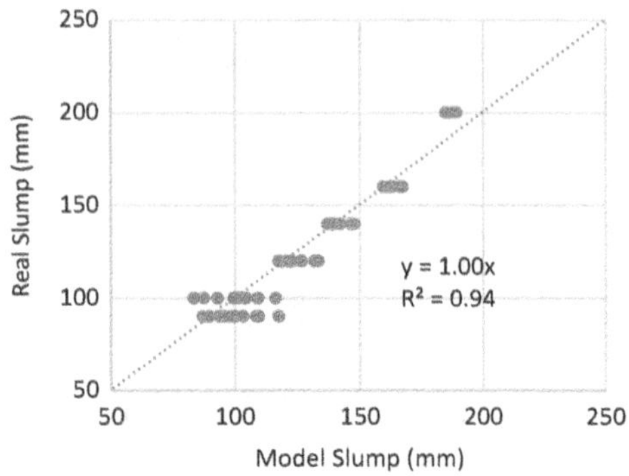

Fig. 5 Correlation between real slump and model slump

porosity. The computational optimization can minimize the coarse packing porosity, strategy that will reduce the concrete binder intensity and increase eco-efficiency.

4 Conclusion

This study allowed a concrete formulation eco-efficiency analysis of a concrete plant in São Paulo/Brazil. For a given strength, the formulations with the lowest binder intensity are the formulations with maximum aggregate size 19 mm and lower slump: about 7.5 to 8 kg/m3/MPa. The formulations with the lowest eco-efficiency are those with a maximum aggregate size 9.5 mm and higher slump, reaching values between 9 and 10 kg/m3/MPa.

For the base formulation mobility, there is a high correlation between the MPT and the slump, with the correlation coefficient (R^2) of the adjusted linear curve 0.99 given a paste composition. Based on that, a descriptive model of slump in function of MPT and IPS was adjusted and has a R^2 0.94 between the model slump and the real one. This model can be used by computational optimization of mixture design, which can lead to an increase on the eco-efficiency of the commercial concretes.

References

1. Damineli, B.L.: Conceitos para formulação de concretos com baixo consumo de ligantes: controle reológico, empacotamento e dispersão de partículas. São Paulo: Universidade de São Paulo (2013)
2. Funk, J.E., Dinger, D.: Predictive process control of crowded particulate suspensions applied to ceramic manufacturing. 1. ed. Springer Verlag, Clemson (1994)
3. IEA; WBCSD.: Technology Roadmap: low-carbon transition in the cement industry. [s. l.], (2018)

4. Larrard, F., Sedran, T.: Mixture-proportioning of high-performance concrete. Cement and Concrete Research, [s. l.], v. 32, n. 11, p. 1699–1704, (2002). Disponível em: https://www.sciencedirect.com/science/article/pii/S000888460200861X. Acesso em: 2 dez. 2017
5. Pileggi, R.G.: Ferramentas para o estudo e desenvolvimento de concretos refratários. São Carlos: UFSCar (2001)
6. Powers, T.C.: The properties of fresh concrete. 1. ed. Michigan
7. Westman, A.E.R., Hugill, H.R.: The packing of particles. J. Am. Ceram. Soc., [s. l.], **13**(10):767–779 (1930)

Influence of Temperature and Co_2 Partial Pressure on Carbonation Curing for Cement-Free Steel Slag-Based Materials

Pedro S. Humbert and João P. Castro-Gomes

Abstract Sustainable development requires alternative possibilities for industrial sector and green binders are feasible solution for construction industry due their environmental impact reduction by valorising waste, reducing energy consumption and the use of natural resources in production process. Steel slag carbon dioxide activated binder is a cement-free construction material which consumes low amount of water and stores CO_2 through its activation. The effect of temperature and CO_2 partial pressure on the carbonation reaction was studied aiming to reach optimal conditions to achieve more desirable mechanical properties for the carbon dioxide activated binder. A Portland cement paste under the achieved optimal carbonation conditions were designed to compare compressive strength and reaction products which were determined by thermogravimetric analysis and scanning electron microscope. An electric arc furnace slag from the national steel industry in Portugal and a Portland cement CP-I were used as binders. This work shows the compressive strength development of a steel slag-based binders under 40, 50, 60 and 70 °C of temperature and 0.5, 1.5 and 2.5 bars of CO_2 partial pressure on the carbonation curing. Portland cement and steel slag binders had similar compressive strength development and reaction products which were mainly calcium carbonates and calcium silicate hydrates. The optimal temperature on the carbonation reaction was 60 °C achieving an average of 97.4 MPa. The partial pressure also showed a strongly influence on the compressive strength development achieve up to 128.1 MPa under 2.5 bars however, the difference between 1.5 and 2.5 bars was not substantial increasing less than 2 MPa in the compressive strength result.

Keywords Carbon dioxide activation · Steel slag-based binders · Industrial waste utilization · Carbon dioxide utilization

P. S. Humbert (✉) · J. P. Castro-Gomes
Centre of Materials and Building Technologies, University of Beira Interior, Covilha, Portugal
e-mail: pedro.humbert@ubi.pt

© RILEM 2020

J. F. Martirena-Hernandez et al. (eds.), *Proceedings of the International Conference of Sustainable Production and Use of Cement and Concrete*, RILEM Bookseries 22,
https://doi.org/10.1007/978-3-030-22034-1_28

1 Introduction

Global warming limits has exceeded in 2015 and temperature keep increasing, new solutions to reduce greenhouse gases emissions are needed [1]. CO_2 emissions and industrial wastes pose negative impacts on human health and on the environment [2]. The steel industry by-products are mainly dumped in landfills resulting in possible environmental pollution [3]. Moreover, these landfills also affect the quality of life for local population once cause serious landscape impacts and environmental pollution [4]. Some studies evidence that mining, quarrying and other industrial wastes can be reused in different construction applications [5–7]. This study was done aiming to find new circular economy solutions in terms of waste valorisation and utilisation, together with carbon dioxide emissions reduction.

The iron carbonation forms complex iron carbonates which have binding properties that can be interesting for construction applications. CO_2 captured from Ferrybridge Power Station's carbon in the UK was used as a feedstock to synthesize polymers and high value chemicals for composite manufacturing [8]. Other products such as textiles, coatings, adhesives, polycarbonate plastics used in bulletproof glass, spectacle lenses and electronic parts are being synthesised from captured CO_2 [9].

Few studies have reported the feasibility of producing construction materials combining industrial waste with CO_2. The calcium carbonates are responsible for the binding properties and the construction material valorise and give a proper destination to the industrial waste moreover, it works as a CO_2 storage [10]. Bonenfant et al. demonstrated that 24.7 g of CO_2 can be stored for each 100 g of slag used which evidence the great potential [11]. Controlling carbonation conditions such as fineness and water to solid ratio, the binder can reach high compressive strength [12]. The mechanical properties increase within the carbonation duration and the after 4 days the samples evidence compressive strength which are similar to ordinary Portland cement binders [13]. The carbon dioxide activated binder is a potential construction material for precast applications, once it should be manufactured in closed and controlled environment. Moreover, further research is needed on this field once different wastes should be tested, the optimal carbonation conditions should be defined, the industrial production design should be drawn and the cost analysis must be done [14].

2 Experimental Program

2.1 *Materials and Procedures*

An electric arc furnace slag was received from the national steel industry at Maia and Aldeia de Paio Pires, Seixal, Portugal. This slag is currently being used in pavement constructions in Portugal [15], however the slag has a CO_2 storage potential through which this cement-free binder is one of the alternative application being a carbonated mixture of slag powder and water. The as-received slag is put in the oven at 60 °C for 24 h to dry, afterwards crushed into a small size at a crusher mill and then further pulverized into powder at a ball mill. This slag powder was sieved and the powder which passed through a sieve of 45 μm was used and subjected to a density test which showed a result of 3.7712 g/cm^3, then to a Blaine test that found the Blaine number of 529 m^2/kg. Portland cement CP-I was subjected to a density and Blain test evidencing density of 2.977 g/cm^3 and fineness of 382 m^2/kg.

The chemical composition of the electric arc furnace (EAF) slag and Portland cement CP-I were determined by the energy dispersive spectroscopy (EDS) analysis. Through the analysis was possible to see that the slag is rich in calcium (CaO), iron (Fe_2O_3) and silica (SiO_2) with some other minor oxides on its composition while the ordinary Portland cement was rich mainly in calcium (CaO) and silica (SiO_2). Chemical composition, density and Blaine number is shown at Table 1.

Each carbonation condition had three cylindrical samples compacted by 30 MPa in order to achieve less porosity and higher compressive strength into a mould of 20 mm of diameter and 40 mm of height with a water to slag ratio of 0.10. The samples were subjected to carbonation activation right after casting with a concentration of 100% of CO_2 inside the chamber with 0.5, 1.5 and 2.5 bar pressure. The CO_2 gas used had 99.9% of purity and was injected to the chamber for 24 h, being replenished as it was being consumed by the samples. The chamber was inside an oven at 40, 50, 60 and 70 °C of temperature and 100% of relative humidity. After 24 h of carbonation the samples were taken from the chamber to dry in an oven for 20 h at 40 °C of temperature and then tested on compressive strength in a rate of 0.5 kN/s. The samples that varied the CO_2 partial pressure had the oven at 40 °C and the samples that varied the temperature had the CO_2 partial pressure fixed at 0.5 bar. All seven different carbonation conditions are evidenced at Table 2 plus optimal condition which was used for comparison with a Portland cement binder in same conditions (Table 3).

2.2 *Microstructural Characterization*

Microstructural analysis was done on the carbonated binder samples. Optimal conditions samples were submitted to thermogravimetric analysis (TGA) analysis aiming to identify the formed products. The carbonated samples were smashed into powder

Table 1 Slag chemical composition, density and Blaine number

Material	Oxides percentage by mass						Density (g/cm^3)	Blaine number (m^2/kg)
	CaO	SiO_2	Al_2O_3	MgO	SO_3	Fe_2O_3		
Slag 45	30.21	14.18	12.00	5.58		29.51	3.771	529
OPC	62.20	12.31	3.08	1.27	4.33	2.42	2.977	382

Table 2 Carbonation conditions

Changed parameters	Fineness (m2/kg)	Water to solid (%)	Compacting pressure (MPa)	Temperature (oC)	Partial pressure (bar)	Duration (h)
p1	529	10.0	30	40	0.5	24.0
p2					1.5	
p3					2.5	
T1				40	0.5	
T2				50		
T3				60		
T4				70		
Optimal	529–382[a]			60	2.5	

[a]529 and 382 were respectively for steel slag and Portland cement both under optimal conditions

Table 3 Carbon dioxide activated binder compressive strength

Samples	Compressive strength (MPa)
p1	72.1
p2	126.6
p3	128.1
T1	72.1
T2	74.6
T3	97.4
T4	58.0
Optimal conditions—steel slag	151.5
Optimal conditions—Portland cement	86.0

and sieved under 125 μm to be tested at TGA. The carbonated samples were crushed into small pieces to be submitted to preparation for scanning electron microscopy (SEM) analysis to ratify TGA results. A small piece of the fractured binder was coated with gold to improve the imaging of the samples at the SEM analysis which showed the reaction products in a fractured piece the products and its distribution was seen through the analysis.

3 Results and Discussion

3.1 Compressive Strength

The temperature showed that compressive strength increase up to 60 °C and after that has a significant reduction. The carbon dioxide activated binder at 60 °C reached compressive strength of 97.4 MPa. Compressive strength at 40 and 50 °C were similar around 25% lower than the 60 °C while at 70 °C the compressive strength were 40% lower reaching only 58.0 MPa. The CO_2 partial pressure showed higher influence reaching 126.6 and 128.1 MPa at 1.5 and 2.5 bar respectively which were more 75% higher than the 0.5 bar. The steel slag binder under optimal conditions reached a compressive strength of 151.5 MPa while the Portland cement under the same conditions reached 86.0 MPa. The carbon dioxide activated binder can be characterised as a building material with potential properties to be a Portland cement replacer for structural and non-structural elements such as beams, panels or blocks for ordinary and high performance applications. Table 2 shows the compressive strength of each carbonation conditions.

Due to the production process and procedure the application might be for precast elements, once the compaction, carbonation time, partial pressure, carbon dioxide replenishment, temperature and humidity might be controlled. Moreover, due size effect further studies should be done in order to analyse the feasibility to manufacture structural elements.

3.2 Thermogravimetric Analysis

The thermogravimetric and differential thermogravimetric curves showed that there is no weight loss at the non-carbonated steel slag sample which was expected once its manufacturing process occurs over 1000 °C making the steel slag with no mineralogical phases which could changing by heating up to 1000 °C. The carbonated steel slag samples showed two different peaks of weight loss representing the product formation when the steel slag reacts with CO_2. The peaks were between the intervals of 555–748 °C on the steel slag binder and 650–790 °C on the Portland cement binder which correspond to the calcium carbonate product (Figs. 1 and 2).

3.3 Scanning Electron Microscopy

Through the SEM analysis was possible to confirm the formed products identified by TGA. At the steel slag carbonated sample mainly calcium carbonates were formed. Figure 3 shows few calcium silicate hydrates which have been mainly carbonated as well that formed more calcium carbonates and silica gel. The Portland cement

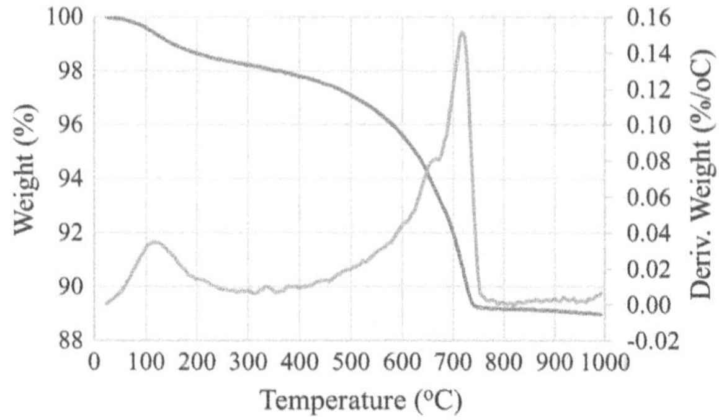

Fig. 1 Thermogravimetric and differential thermogravimetric curves of the steel slag carbonated sample with temperature and weight loss indicated

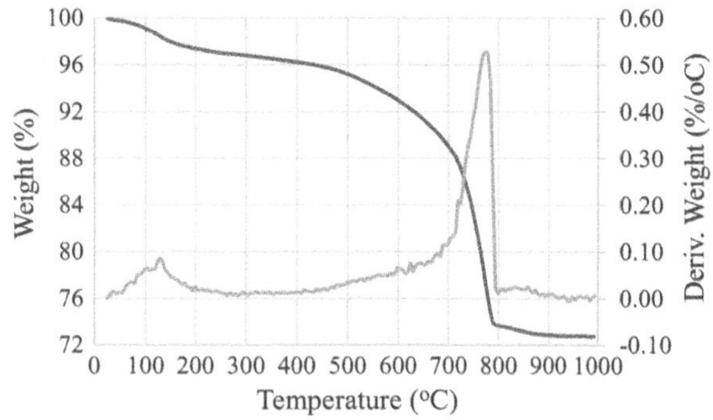

Fig. 2 Thermogravimetric and differential thermogravimetric curves of the Portland cement carbonated sample with temperature and weight loss indicated

Fig. 3 SEM image of a steel slag carbonated binder fracture

carbonated sample is evidenced also calcium carbonates, carbonated calcium silicate hydrates and ettringite which is the main difference on the formed products. Figure 3 shows a fractured piece of steel slag carbonated sample while Fig. 4 shows a fractured piece of Portland cement carbonated sample.

Fig. 4 SEM image of a Portland cement carbonated binder fracture

4 Conclusions

This experimental study has analysed and discussed the influence of the temperature and carbon dioxide partial pressure on the carbonation curing for cement-free steel slag based binders. The following conclusions can be drawn from this research:

- The studied binder has good potential as a Portland cement replacer for precast applications of structural and non-structural building material with ordinary or high performance strength due its mechanical properties.
- Further research is needed to studied the size effect for these binders and the properties related with the potential applications.
- Calcium carbonates is the main formed products of the carbonation reaction. Calcium silicate hydrate is also seen as a product in less percentage. Ettringite is evidenced only for the Portland cement binders.
- The compressive strength development increase with the increase of CO_2 partial pressure however, after 1.5 bar the compressive strength growth is not significant. The compressive strength development also increase with the increase of the oven's temperature up to 60 °C. At 70 °C there is a strong loss of compressive strength.

References

1. Szulejko, J.E., Kumar, P., Deep, A., Kim, K.H.: Global warming projections to 2100 using simple CO_2 greenhouse gas modeling and comments on CO_2 climate sensitivity factor. Atmos. Pollut. Res., 1–5 (2016)
2. Eurostat : Greenhouse gas emission statistics—statistics explained. Eurostat (2016) [Online]. Available: http://ec.europa.eu/eurostat/statistics-explained/index.php/Greenhouse_gas_emission_statistics#Further_Eurostat_information. Accessed 02 Feb 2017
3. Eurostat. Waste statistics main statistical findings-total waste generation. no. September 2015, p. 3 (2016)

4. European Commission (EC).: Management of tailings and waste-rock in mining activities. Ref. Doc. Best Available Tech., no. January, pp. 1–414 (2009)
5. Sedira, N., Castro-Gomes, J., Kastiukas, G., Zhou, X., Vargas, A.: A review on mineral waste for chemical-activated binders: mineralogical and chemical characteristics. Min. Sci. **24**, 29–58 (2017)
6. Humbert, P.S., Castro-Gomes, J.P., Bernardo, L.F.A., Pinto, C.M., Paszek, N.: Elastic modulus and stress-strain curve analysis of a tungsten mine waste alkali-activated concrete. In: REMINE International Conference, (2017) p. in press
7. Pacheco-Torgal, F., Jalali, S., Castro Gomes, J.P.: Geopolymers structure, processing, properties and industrial applications. In: Geopolymers Structure, Processing, Properties and Industrial Applications, pp. 267–293 (2009)
8. Chapman, A.M., Keyworth, C., Kember, M.R., Lennox, A.J.J., Williams, C.K.: Adding value to power station captured CO_2: Tolerant Zn and Mg Homogeneous Catalysts for Polycarbonate Polyol Production. ACS Catal. **5**(3), 1581–1588 (2015)
9. Lim, X.: How to make the most of carbon dioxide. Nature **526**(7575), 628 (2015)
10. Humbert, P.S., Castro-Gomes, J.P.: CO_2 activated steel slag-based materials: A review. J. Clean. Prod., vol. accepted (2018)
11. Bonenfant, D., Kharoune, L., Hausler, R., Niquette, P.: CO_2 Sequestration potential of steel slags at ambient pressure and temperature. Ind. Eng. Chem. Res. **47**, 7610–7616 (2008)
12. Humbert, P.S., Castro-gomes, J.P., Saafi, M.B.S.: Effect of water-binder ratio and Blaine fineness on the carbon dioxide activation of a cement-free binder. In: 5th International Conference on Civil and Urban Engineering, (2018), p. in press
13. Das, S., Souliman, B., Stone, D., Neithalath, N.: Synthesis and properties of a novel structural binder utilizing the chemistry of iron carbonation. ACS Appl. Mater. Interfaces. **6**(11), 8295–8304 (2014)
14. Humbert, P.S., Castro-Gomes, J.P.: Cement-Free binder activated by carbon dioxide: potential and challenges aglomerante livre de cimento ativado por dioxido de carbono: potencial e Desafios. In: *30 Congresso Luso Brasileiro Materiais de Construcao Sustentaveis* (2018)
15. Larguesa, A.: Waste from the Steel Industry Gives 'Ground' to Cars, Shops and Footballers," p. (in Portuguese), Nov-2015

Use of Residues of Crushed Pet Bottles in the Form of a Scale as an Addition to the Manufacture of Concrete

Joaquín Raúl Cuetara Ricardo

Abstract With the increase in the consumption of plastic, human influence has become a problem, since most of these new upstarts are not biodegradable. When there is inefficiency in the activities of recycling or reuse, the effects produced by human activity on the environment are aggravated. The present work proposes a realistic solution in order to avoid that all that plastic that is not recycled or reused correctly reaches the seas and landfills. For the analysis of the results, a completely random factorial design was used. In the design of the experimental program, the determination of the resistance to compression at 7 and 28 days was designed, and the total porosity test at 28 days. The non-destructive tests, ultrasonic pulse speed and electrical resistivity, were conceived at the age of 3; 7; 14 and 28 days. The study showed that the addition of up to 1% of PET during the manufacture of concrete does not influence the quality of the same, in terms of its resistance to compression and porosity.

Keywords Recycling · Addition of PET · Compressive strength · Total porosity

1 Introduction

The impact of the waste to the environment has been so much that the inhabitants of the planet call contamination to any manifestation of objects thrown into landfills, roads and streets. They are the "plastics" the upstarts, moldable objects. Hence the term and, that was the word with which this great invention was popularized worldwide in the first quarter of the 20th century.

According to Guillermo Perdomo [1], the idea of the macromolecule was not accepted by most of those who worked on the subject at the beginning of the 20th century. Early in the 1920s, a great scientist: Hermann Staudinger, later awarded the Nobel Prize in 1953 for his work, managed to get many colleagues to understand the

J. R. C. Ricardo (✉)
Construction Department, Technological University of Havana (CUJAE), Havana, Cuba
e-mail: jrcuetara@ceinpet.cupet.cu

© RILEM 2020

J. F. Martirena-Hernandez et al. (eds.), *Proceedings of the International Conference of Sustainable Production and Use of Cement and Concrete*, RILEM Bookseries 22,
https://doi.org/10.1007/978-3-030-22034-1_29

idea that there were macromolecular compounds that were not molecular aggregates or colloids [1].

The situation presented by plastics, especially terephthalate polyethylene (PET), is that on a rating scale it is the least sustainable packaging and, once it has been used properly, it is thrown into landfills or oceans where they will last between 500 and 1000 years [2]. Bacteria do not know these materials and therefore do not digest them. They are very young and have not been built on, that is why the accumulation of this waste in landfills and oceans.

To have an idea, in the Spanish magazine "Muy interesante" [3] it was commented that to date 8300 million tons of plastics had been generated, equivalent to 822 thousand Eiffel towers or 80 million blue whales. Every year, approximately 8 million tons of plastic are finished in the oceans and it is estimated that by 2050, the weight of these in the oceans will exceed that of fish. In addition, researchers from the University of Alicante found that there can be between 60 and 280 micro plastic particles per kilogram of salt, with PET being the most frequent compound for an 83.3% presence [4].

1.1 Current Status of Recycling and Its Environmental and Economic Benefits

Projects with excellent results on an international scale have been proposed and developed. In the more developed countries, what is known as "Energy Recovery" is practiced, where thermal technology processes (gasification, pyrolysis and thermal depolymerization) and non-thermal processes (anaerobic digestion, fermentation and mechanical biological treatment) are obtained gas, methanol, lactic acid, among others [5]. In this aspect the main exponents are the European countries [6] and, cities like San Francisco with their project "Zero Waste" where 80% of garbage has been prevented from being taken to landfills [7] and the sale of PET water bottles has been prohibited [8].

At the national level, measures have been taken and projects have been proposed to reduce the source, reuse and recycle PET. In 1975, Law 1288 was enacted, which obliges all those legal entities that generate waste in their production processes or service provision, which will not be reused by them, to deliver them to recycling. Lack of culture of recycling, both in the population and in the state sector [9] but, we have directed our interests towards an activity that promises and that substitutes imports. So much so that the recycling activity reported a saving of 212 million dollars in 2014 due to the combined effect of the substitution of imports and exports [10].

About the recycling of PET and its reuse, its main destination once crushed are local industries. According to Granma newspaper [11] the engineer Isabel Alfonso director of non-metallic products of the Union of Raw Material Recovery Companies (URMRP), the import of each recycled ton of high density polyethylene (discarded shampoo bottles, deodorant, plastic boxes and others) costs Cuba more than 840 euros, while the sale on the international market of a ton of PET containers (water bottles, soft drinks and others) brings between 300 and 400 dollars.

1.2 Use of Plastics (PET) in the Manufacture of Concrete

All the measures taken to reduce the source, reuse and recycle plastics have been insufficient. There are some companies in our countries that demand high levels of PET, a total of 1368.30 tons, if of this considerable weight, only 8% is recycled, ¿what happen with the rest of the plastics?

The use of PET waste, crushed in the form of flake and fibers, in the manufacture of concrete has been considered in many investigations [6]. Recently, papers have been published that demonstrate the potential of using PET waste to improve some concrete properties, even using it as a substitute for aggregates [2].

Dora Foti [12] studied the concretes reinforced with recycled PET fiber waste and concluded with important results:

- Greater development of concrete ductility
- High adhesion between PET-concrete.

As a result of the study of the results of this work, the reinforcement of concrete with recycled PET fibers was considered in substitution of steel in the case of elements not subject to great stress, although it is recommended to deepen the studies for its possible use in the future. On the other hand, D. Foti concluded that the use of recycled PET fibers and strips as reinforcement in concrete represents important advances in the reduction of PET waste, the cost of production and degradation due to the corrosion of steel.

In other investigations, the properties of concrete made with PET waste in the form of flakes as a substitute for fine aggregate have been studied and the results have been surprising [2]. Slightly decreases the workability of the concrete with the increase in PET up to 5%, henceforth the values are significant. It did not happen in the same way with the rest of the properties that were measured, in the case of the tensile strength and the ultrasonic pulse velocity measurements decreased slightly starting with the substitution of PET for fine aggregate.

Similar results were obtained by AM Azhdarpour and others [13] when they studied the properties of the concretes made with fragments of PET in substitution of fine aggregate, although, they achieved replacements of up to 10% without being will affect the properties under study. A work that was very interesting was the one developed by J. Thorneycroft and others [14] where the influence of the use of different plastics, its shape and fineness for its use in the manufacture of concrete. Based on the study, they concluded that flake PET, and polypropylene strips and fibers are among the plastics studied that most influence the tensile and compressive strength of concrete. Likewise, in the manufacture of concrete with PET plastics as a replacement for fine aggregates, which are well graduated influences more than the fineness they present if the purpose is to obtain more resistant concrete.

This versatile way of reusing PET waste has gained attention in recent years and this can be seen in the work developed by Lei Gu and Togay Ozbakkaloglu [6], which makes a critical review of all the works that have been directed to use any type of plastic in the manufacture of concrete, from mechanical properties to durability. In the present work, the influence of the addition of fragments of PET plastic cut manually on some properties of concrete is studied.

2 Materials and Methods

For the manufacture of the concretes, a OPC was used that corresponds to the P-35 denomination (type I) of the Cementos Curasao factory, of the Mariel municipality, in the province of Artemisa. Table 1 shows its main characteristics and the values are compared with those established by the Cuban norm [15].

The aggregates used in the study are of natural origin, from the La Molina quarry, located in the Mariel municipality, in the province of Artemisa. The tests were carried out according to the Cuban norms NC 186; 181; 177 and 182, all of the year 2002. Table 2 shows its main physical properties. The maximum size of the coarse aggregate is 19.1 mm and the fine aggregate fineness module is 3.58. The fragments of PET were cut manually with scissors until their size was between 2 and 4 mm. The fragments of PET were cut manually with scissors until their size was between 2 and 4 mm. The shape of the particles that were obtained were rhombic, square and triangular.

2.1 Experimental Method

The interest of the study is to know how much the addition of fragments of PET influences the mechanical and durability properties. In addition, the obtained values

Table 1 Physical properties of cement (P-35)

Properties	Values	NC 95: 2011 [15]
Specific surface area (cm^2/g)	3317	≥2800
Fineness (%)	1.7	≤10
Initial setting time (h)	102:00:00	≥0.75
Final setting time (h)	3:02:00	≤10
Volumetric weight (kg/m^3)	1168	
Bulk Density (g/cm^3)	3.15	
Normal consistency (%)	24.5	

Table 2 Physical properties of aggregates

Properties	Sand	Gravell
Current specific weight (g/m^3)	2.61	2.63
Specific saturated weight (g/m^3)	2.65	2.66
Apparent specific weight (g/m^3)	2.72	2.70
Absorption percent (%)	1.8	1.0
Loose volumetric mass (kg/m^3)	1477	1371
Compact volumetric mass (kg/m^3)	1663	1510
Percent of holes (%)	37.0	45.0
Finest material sieve 200 (%)	9.9	1.2

of ultrasonic pulse velocity (UPV) and electrical resistivity (ER) of the concretes with the addition of PET will be compared with the control concrete, that is, without the addition of PET.

To study in detail the compressive strength and porosity of the concretes made with the addition of PET, it was decided to use a completely random design. In the Addition PET factor three levels were included, low, intermediate and high. Each treatment was assigned three blocks to be tested, that is, an experiment and two replications were performed for each treatment and for each block to be tested three concrete specimens were made.

The analysis of the data was made from a fundamental statistical tool for this type of design, the analysis of variance [16]. This is a statistical procedure, and was created by RA Fisher in 1925 to decompose the variability of an experiment into independent components that can be assigned to different causes, it is the basis of statistical methods used mostly in the field of design of experiments [17]. In the case in question, the use of this procedure implies whether the expected value changes or not when the response variable is subjected to different levels of a given factor. In the present work, the levels of the factor that can lead to an optimal response with the help of statistical software StatGraphics Centurion XV [18] were determined experimentally.

Porosity was jointly measured by the method established in the ASTM standard [19]. For the study in question the witnesses were extracted from the cubic samples of 15 × 15 × 15 cm in order to avoid the wall effect that occurs in the areas close to the surface. To determine the porosity by the proposed method, three witnesses were analyzed for each type of concrete representative of each batch.

2.2 Procedure Description

A standard concrete with a characteristic compressive strength of 30 MPa at 28 days was considered, and depending on the degree of control in the laboratory, its average resistance to compression will be 32 MPa. The mixtures considered in the

Table 3 Dosage and quantity of materials

Materials	PP	PR_0.5%	PR_1.0%
Relationship a/c	0.45	0.45	0.45
Cement P-35 (kg)	37.52	37.52	37.52
Water (L)	15.44	15.44	15.44
Coarse (kg)	85.76	85.76	85.76
Fine (kg)	72.2	72.2	72.2
PET (kg)	–	0.1876	0.3752
Additive (%-ml)	0.70–128	0.70–128	0.70–128

study have the same type of aggregate and dosed in the same proportion in order to obtain a constant granulometric curve. A proportion of fine and coarse aggregate of 45/55% respectively was used. The water-cement ratio (0.45) was considered for an environment where aggressiveness is high, as established by the Cuban norm [20].

To improve the workability of fresh concrete, the Dynamon SRC-20 chemical admixture belonging to a new MAPEI system was used. It is an acrylic-based water-reducing admixtures (second generation advanced) modified for premixed concretes characterized by a low water-cement ratio, high mechanical resistance and long maintenance of the workability. The desired settlement measured by the Abrams cone will be between approximately 14 and 16 cm. The dose used did not vary and the amount of materials to be used for each type of concrete to be elaborated is shown in Table 3.

3 Results and Discussion

The test of porosity and compressive strength of concrete was applied according to the methodology established by the standards [19] and [21] respectively, using cores with dimensions (diameter x thickness) of 10 × 4.5 cm and cylindrical test pieces of 10 × 20 cm respectively. The analysis in detail is summarized in the following paragraphs.

3.1 Analysis of the Compressive Strength

In the study in question, the Addition_PET effect has a P-value greater than 0.05, which indicates that there are no statistically significant differences between the means of the compression resistance values at 7 and 28 days of the PET addition levels with a level of 95% confidence. However, although the analysis indicates there is no difference between the levels under study, the average values for the 1.0% addition is slightly higher than the standard concrete, but not enough for create

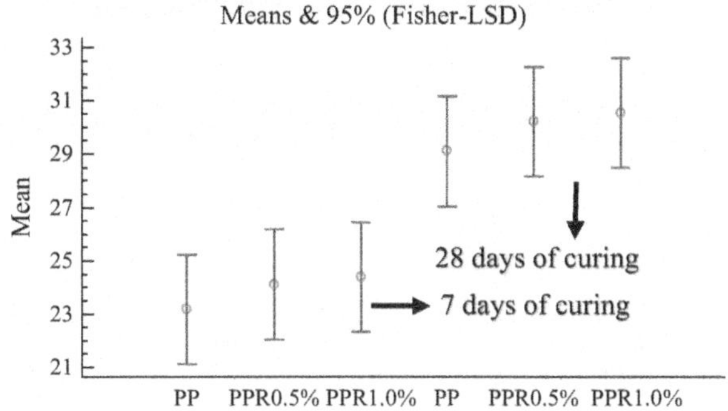

Fig. 1 Diagram of sample means of CS values at 7 and 28 days

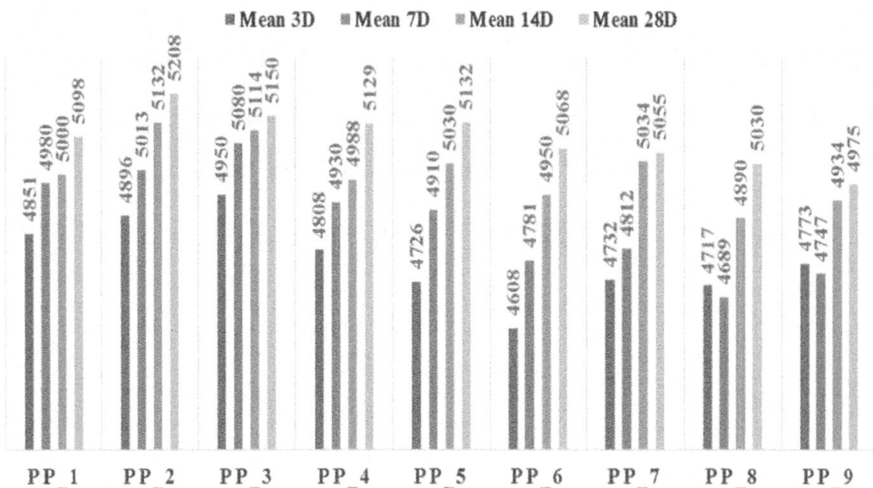

Fig. 2 UPV values (m/s) in the standard concrete

significant differences For this reason it can be stated that the Addition_PET factor does not influence the compression resistance in any of the ages studied. In Fig. 1 the graphs of the sample means of the CS values are detailed (Figs. 2, 3 and 4).

3.2 *Analysis of the Ultrasonic Pulse Velocity*

The measurements were made in cured concretes until 3, 7, 14 and 28 days with 0.5 and 1.0% of PET addition, in order to compare the VPU values with the standard concrete. Taking into account that these values were obtained in saturated specimens, the values measured in each specimen are shown in Figs. 5, 6 and 7.

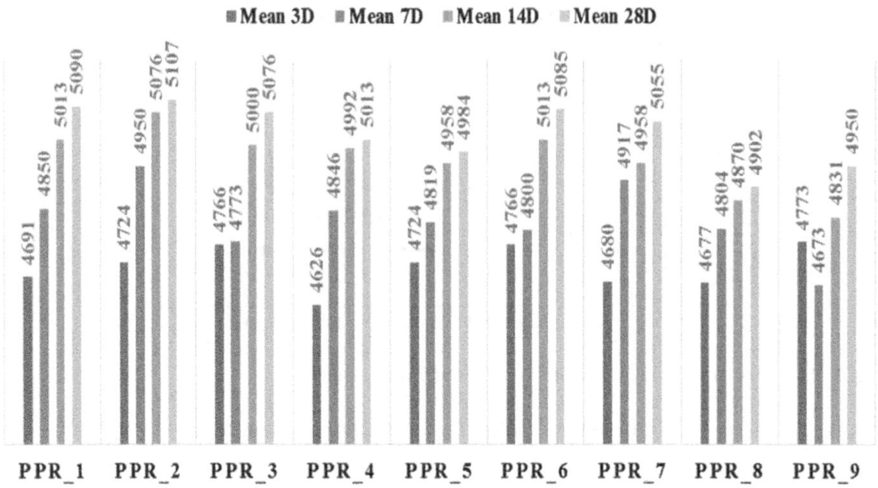

Fig. 3 UPV values (m/s) in the standard concrete with 0.5% PET addition

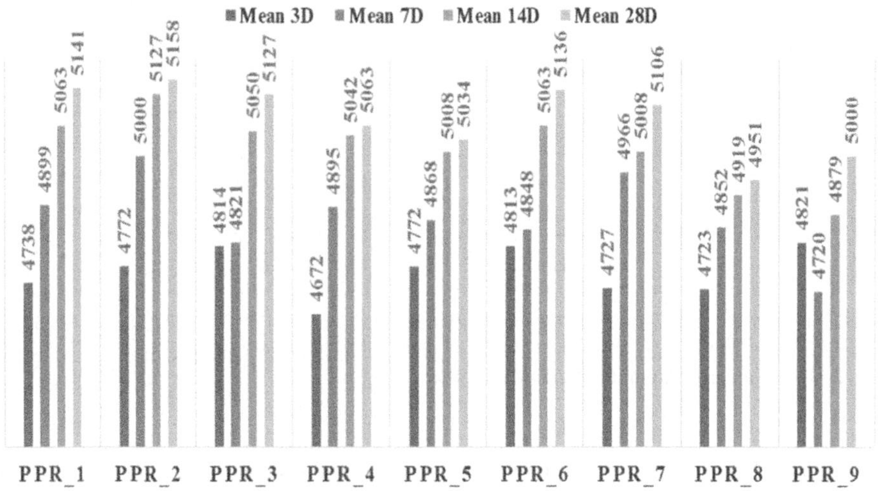

Fig. 4 UPV values (m/s) in the standard concrete with 1.0% PET addition

The graphs of columns shown above indicate a progressive increase in the VPU values as the hydration of the cement particles evolves, therefore, the addition of PET to the concrete does not affect the quality of the same. Likewise, according to the RED DURAR [22] and based on a hypothesis test, where a value of 4000 m/s was taken as a null hypothesis, it was determined that in all cases the quality of the concrete is durable. In addition, a simple comparison was made from the values obtained and it was found that, although there are no statistically significant differences between

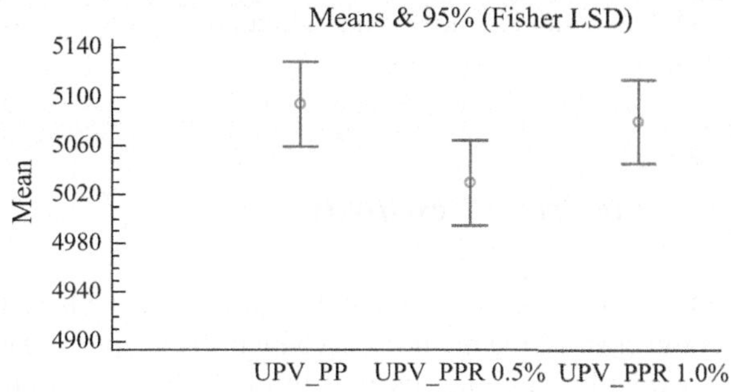

Fig. 5 Diagram of sample means of UPV values (m/s)

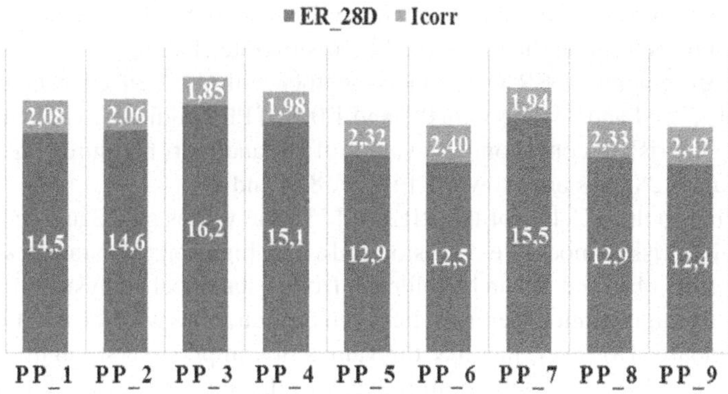

Fig. 6 Values of ER (k·ohm·cm) in the standard concrete

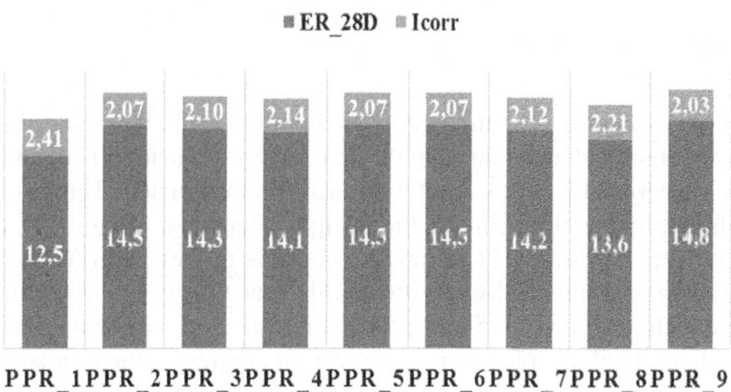

Fig. 7 ER values (k·ohm·cm) in the concrete with 0.5% PET addition

the means of the analyzed values, there is a slight tendency to grow as PET is added; this is shown in Fig. 5.

3.3 Analysis of Electrical Resistivity

In order to estimate the risk of corrosion that the reinforcing steels that are embedded in this concrete could present, the method based on the principle of the Wenner probe was used, and it works by applying current through the two external probes and measuring the resulting potential between the interior probes. The moisture content (water) present in the pores of the concrete is able to transport the current between the probes, which makes it possible to calculate the resistivity of the material between them, and thus determine the theoretical corrosion rate (Icorr).

The measurements of ER were made together with those of UPV in cured concretes until 3, 7, 14 and 28 days with 0.5 and 1.0% of PET addition, with the objective of knowing the risk of corrosion. The values of ER and Icorr measured only in cured concretes up to 28 days are shown in Figs. 7, 8, 9 and 10.

According to the TC-154 of the RILEM [23], the values of RE and Icorr show a concrete with a risk of moderate corrosion and a very high corrosion rate, respectively. With the results obtained, it can be inferred, from a statistical analysis, that there are no statistically significant differences between the concretes with the addition of PET and the standard. However, just like the values of compressive strength, the means of the measured ER values tend to improve with the increase of the PET addition.

3.4 Analysis of Porosity

Based on the dry weights, after immersion, after boiling and submerging, the densities in the dry, apparent and absolute state of the controls are determined, and with this the volume of permeable voids and total porosity is estimated. The test is used to establish durability requirements in the design of concrete. Figure 10 shows the total accumulated porosity (PT) values. An analysis of variance was carried out to compare if there is a statistically significant difference between the results obtained for concrete with addition of 0.5 and 1.0% with respect to the standard concrete. In Fig. 11 the diagram of sample means is shown. In contrast to the standard concrete, those with 0.5 and 1.0% replacement have a reduction in total porosity of 6.63 and 5.56% respectively.

Although the means of the obtained values indicate a reduction in the total porosity of the concretes studied, the data set is insufficient to infer that there are statistically significant differences between the standard concrete and those with PET addition with 95% of trust.

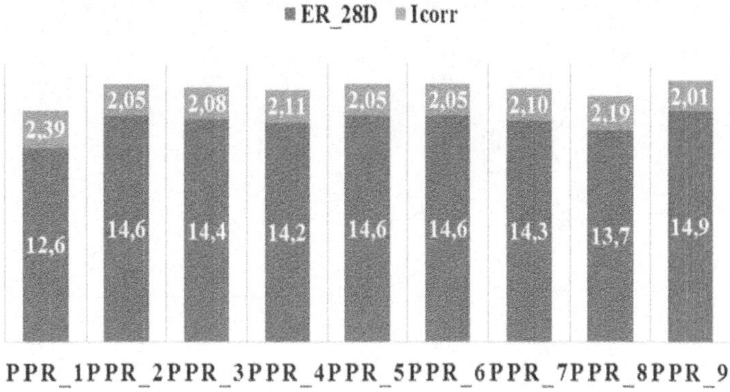

Fig. 8 ER values (k·ohm·cm) in the concrete with 1.0% PET addition

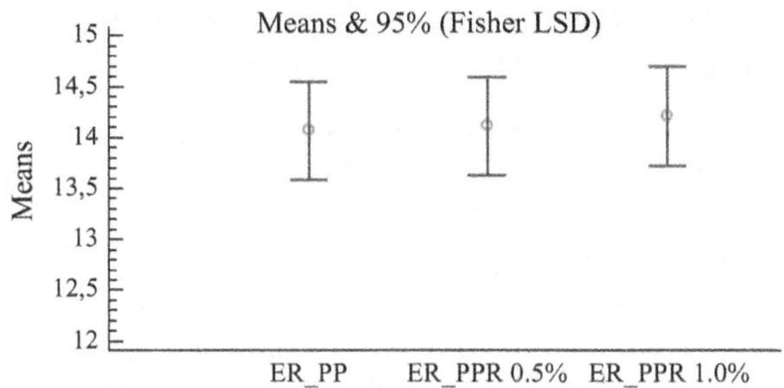

Fig. 9 Diagram of sample means of RE values (k·ohm·cm)

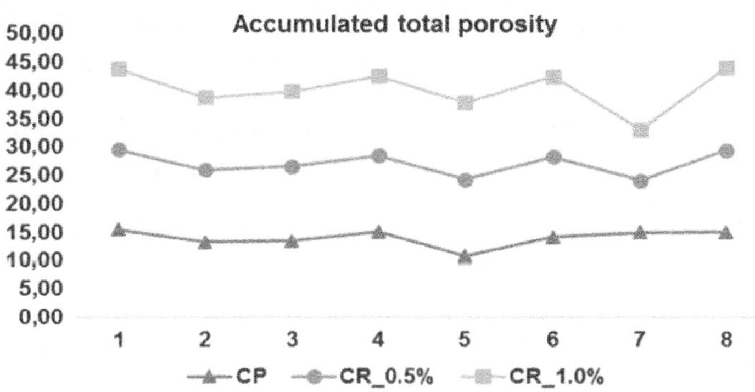

Fig. 10 Graph of cumulative total porosity values

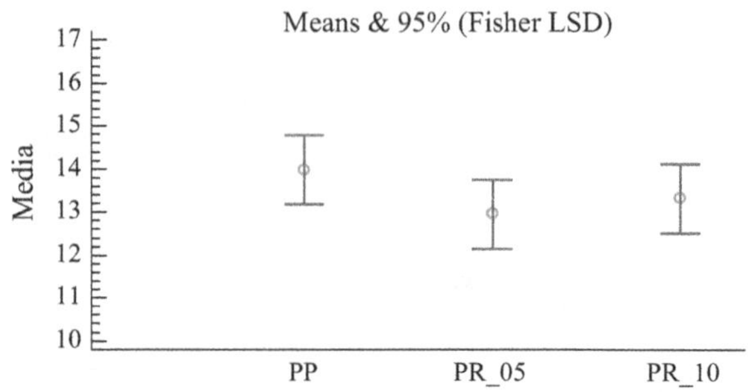

Fig. 11 Diagram of sample means of PT values (%)

4 Conclusions

From the analysis of the results obtained during the study that was performed on the concretes with the addition of PET, it can be concluded that:

- There are no statistically significant differences between concretes with addition of PET and conventional ones. With the current data set it can be inferred that the addition of fragments of PET waste does not influence the values of compressive strength and total porosity of the same.
- The values of VPU indicate a concrete of durable quality while the values of RE a risk of moderate corrosion, both in the standard concrete and in the concretes with the addition of PET. The values of Icorr, which were determined from the RE, indicate a very high corrosion rate in both types of concrete.
- Although the results reveal that the addition of fragments of PET in the manufacture of concrete does not affect the total porosity of these, it can be deduced that there are slight reductions in the volume of the permeable voids and the total porosity of the same.
- The study carried out proposes a practical solution with a view to the future of all PET plastics that are not reused or recycled efficiently and, therefore, are thrown into the seas and landfills.

References

1. Perdomo, G.A.: Plásticos y medio ambiente. Revista Iberoamericana Polímeros **3**(2), 13 (2002)
2. Rahmania, E., et al.: On the mechanical properties of concrete containing waste PET particles. Constr. Build. Mater. **47**, 7 (2013)
3. Aguilera, C., Jodar, F.: Muy data, in Muy interesante. España, p. 124 (2017)
4. Miller, A.: Sazón plástico. In: muy interesante. México, p. 106 (2017)
5. Book, W.: Waste to energy in Austria, in waste to energy in Austria. Austria (2010)
6. Gu, L., Ozbakkaloglu, T.: Use of recycled plastics in concrete: a critical review. Waste Management, p. 24 (2016)
7. Korosec, K.: Top 10 cities leading in urban sustainability. Ecointeligencia (2013)
8. Muñoz, M.: Una ciudad sin botellas de plástico. Eldiario.es (2016)

9. Rey, L.: Reciclar el reciclaje. Periódico Trabajadores (2014)
10. Morales, Y.C.: Eslabones cuadrados de una cadena redonda. Granma (2015)
11. Morales, Y.C.: Reciclar es ahorrar, in Periódico Granma. Combinado Poligráfico Granma, La Habana, p. 16 (2011)
12. Foti, D.: Use of recycled waste pet bottles fibers for the reinforcement of concrete. Compos. Struct. **96**, 9 (2013)
13. Azhdarpour, A.M., Nikoudel, M.R., Taheri, M.: The effect of using polyethylene terephthalate particles on physical and strength-related properties of concrete; a laboratory evaluation. Constr. Build. Mater. **109**, 8 (2016)
14. Thorneycroft, J., et al.: Performance of structural concrete with recycled plastic waste as a partial replacement for sand. Constr. Build. Mater. **161**, 7 (2018)
15. NC-95, Cemento Portland. Especificaciones. Pficina Nacional de Normalización, La Habana (2001)
16. Montgomery, D.C.: Diseño y análisis de experimentos. 3 ed, ed. G.E.I. S.A. México. 589 (1991)
17. Serpa, A.D.C., Guerra, E.F., Vallina, M.L.B.: Diseño y análisis estadístico de experimentos, CUJAE, (ed.). La Habana (2011)
18. Polhemus, N.W.: Statistical analysis using StatGraphics Plus. In: Corporation, S.G. (ed.) Quality control and experimental design, Englewood Cliffs, NJ (1999)
19. ASTM-C-642: Standard test method for density, absorption, and voids in hardened concrete. American Society for Testing Materials, West Conshohocken (2006)
20. NC-120: Hormigón hidraúlico. Especificaciones. Oficina Nacional de Normalización, La Habana, p. 74 (2014)
21. NC-244: Hormigón endurecido. Determinación de la resistencia a la compresión en probetas cilíndricas. Oficina Nacional de Normalización, La Habana (2005)
22. RED-DURAR, Manual de inspección, evaluación y diagnóstico de corrosión en estructuras de hormigón armado. 2000. 208
23. Polder, R., et al.: Test methods for on site measurement of resistivity of concrete. Materials and Structuresj **33**, 603–611 (2000)

Evaluation of the Aggregates of the Quarry "El Purio" in the Province of Villa Clara

Daniel Martínez León and María Betania Díaz Garcia

Abstract In Cuba the aggregates are the raw material of construction of lower cost, they occupy around 75% of the total volume of the concretes that are produced and the characteristics they present in the future will have a direct repercussion in the performance and obtainment of good quality concretes. This work includes a study of the aggregates that are produced in one of the four quarries in the province of Villa Clara, specifically the gravel, granite and sand that are produced in the quarry "El Purio" in the municipality of Encrucijada. For the characterization of the aggregates, it has realized a group of tests according to NC 251: 2013 [11], and subsequently in the results was proceded in the elaboration of concrete test pipe using cement P-35 and the SikaPlast additive. All this with the purpose of evaluating the real performance that present these aggregates and its repercussion in the obtaining of high resistance concretes. Samples were taken in the storage area and on the conveyor belt, in order to evaluate the technological process and its effect on the final product. This study revealed the influence of the production processes in the studied quarry on the physical and chemical characteristics on fluid concretes of more than 35 MPa.

Keywords Aggregates · Concrete · High resistance

1 Introduction

Concrete considers at least three basic elements within its traditional composition; these elements are cement, water and aggregates. It is well known that aggregates present a great importance in the behavior of concrete, where they occupy 60–80% of the volume of concrete. The aggregates were originally seen as an inert and inexpensive material that was dispersed over the cement paste to produce a greater

D. M. León
Empresa de Materiales de Construcción de Villa Clara, Santa Clara, Cuba
e-mail: daniel@matvcons.co.cu

M. B. D. Garcia (✉)
Universidad Marta Abreu de las Villas, CIDEM, Santa Clara, Cuba
e-mail: mbetania@uclv.cu

© RILEM 2020

J. F. Martirena-Hernandez et al. (eds.), *Proceedings of the International Conference of Sustainable Production and Use of Cement and Concrete*, RILEM Bookseries 22,
https://doi.org/10.1007/978-3-030-22034-1_30

volume of concrete. Aggregates are not truly inert because their physical, thermal and sometimes chemical properties influence the behavior of concrete [12]. In order to predict the behavior of concrete, an understanding of the effects of the type of aggregate, its size and its content is required [2]. An increase in the amount of aggregate in the concrete corresponds to a lower use of cement, which can bring a whole series of beneficial effects. Examples of these benefits are the reduction of production cost and reduction of the use of cement which leads to decrease the pollution caused by the production of clinker. It also causes a decrease in the problems of durability, a reduction in the effects of shrinkage and cracking, among others [1].

2 Materials and Methods

2.1 Materials

In carrying out this research work, P-35 cement from the Karl Marx factory in Cienfuegos was used as raw material. The chemical additive SikaPlast 9100CU was used as an additive, while the aggregates taken came from the quarry "El Purio" in the municipality of Encrucijada, specifically the gravel, granite and crushed rock sand.

Cement

To determine the quality of the P-35 cement, the physical and chemical properties of the cement were studied, in accordance with the procedures established by Cuban standards.

The chemical analysis and the physical tests revealed that this cement complies with the standards NC 95: 2017 [3] and NC 96: 2017 [4].

Additive

The SikaPlast 9100CU additive is a mid-range water-reducing additive with high efficiency and permanence of workability, free of chlorides. It complies with the ASTM-C-494 standard as Type A (depending on the dosage and the materials used) with a density of 1.13 kg/ L. It is advisable to use 0.4–2% of the weight of the cement, depending on the degree of fluidity that is sought in the mixture or the reduction of water.

Aggregates

The samples of aggregates were extracted according to the provisions of the standards NC 54-29: 1984 and NC 671: 2008. The material was stored in bags perfectly identified with all the necessary data concerning the type of aggregate and transported without it suffered any loss until the arrival to the laboratories of the Faculty of Constructions of the Central University "Marta Abreu" of Las Villas.

2.2 Methods

The manufacture of the samples was carried out in the Materials Laboratory of the CIDEM of the Faculty of Constructions in the Central University "Marta Abreu" of Las Villas.

Aggregates

A whole group of tests was carried out on the samples of aggregates, which are mentioned below.

- Aggregates. Granulometric analysis. NC 178: 2002 [6].
- Aggregates. Determination of % of holes. Testing method. NC 177: 2002 [5].
- Aggregates. Determination of the volumetric weight. Testing method. NC 181: 2002 [7].
- Sand. Specific weight and water absorption. Testing method. NC 186: 2002 [9].
- Coarse aggregate. Specific weight and water absorption. Testing method. NC 187: 2002 [10].
- Aggregates. Determination of the material finer than the 0.074 mm sieve (No. 200). Testing method. NC 182: 2002 [8].
- *Concrete*

The manufacture of the concrete on a laboratory scale is produced in a Torgar concrete mixer with a capacity of 40 L, according to the procedure established by the NC 412: 2005 "Guide for the preparation, mixing, transport and pouring of concrete".

The dosages used in this investigation are shown in Table 1. From the data shown in this table can be seen the quantities of materials used for each dosage, as well as the variations in the water cement ratio.

It is important to note that in this table, D1 corresponds to the first dosage that was made with the sample M1 corresponding to the aggregates collected in the first sampling. The same procedure is applied for D2 where the aggregates corresponding to M2 are taken, D3 with M3 and finally D4 with M4.

Table 1. Dosifications of each sample

Materials	D1	D2	D3	D4
Cement (kg)	420	420	420	420
Fine aggregate (kg)	938	945	937	937
Gravel (kg)	623	891	623	891
Granite (kg)	259	0	259	0
Water (L)	172	166	172	167
Sika Plast (L)	2.8	2.8	2.8	2.8

3 Results and Discussion

The aggregates of the quarry "El Purio" come from the crushing of limestone rocks, a process that is carried out by means of an impact mill. Its lithology is pure organogenic limestones, hard, compact, white to cream, massive or slightly crooked. There are three types of limestones, aphanitic, fine-grained, and coarse-grained fragmentary limestone. The granulometric fraction corresponding to gravel is (5–20) mm, for granite (5–10) mm and for crushed rock sand (0.15–5) mm. For its characterization, samples of gravel, granite and sand were worked, each with sieve 200, granulometry and physical properties, respectively.

Gravel

- Granulometry (Fig. 1).
- Physical properties (Table 2).

Granite

- Granulometry (Fig. 2).
  - Physical properties (Table 3).

Sand

- Granulometry (Fig. 3)
- Physical properties (Table 4).

Sieve 200 (Fig. 4)

Influence of the properties of aggregates on concrete

Slump Influence (Fig. 5)

The M4 sample undergoes the lowest slump due mainly to having the most variable characteristics, greater absorption, does not comply with the granulometry and has

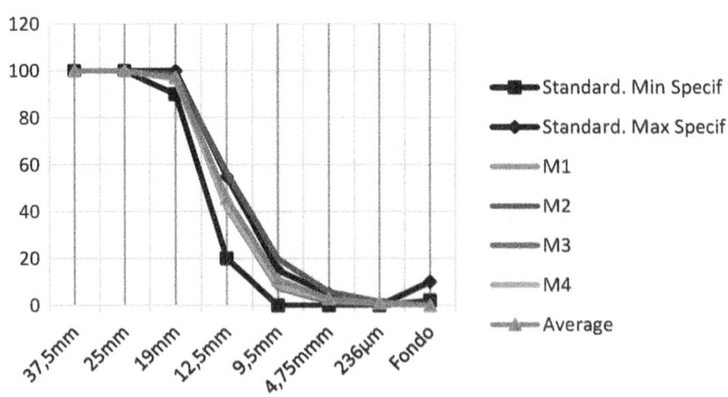

Fig. 1. Granulometry of the gravel from the Purio

Table 2. Physical properties of the gravel of El Purio Quarry

Tests	M1	M2	M3	M4	According to NC 251: 2013
Current specific weights	2.39	2.52	2.50	2.44	2.50
Saturated specific weights	2.52	2.62	2.55	2.51	
Apparent specific weights	2.55	2.63	2.60	2.6	
Absorption (%)	2.17	2.73	2.17	2.53	≤3.0
Loose volumetric mass (kg/m^3)	1446.7	1486.7	1420.65	1452.801	
Compact volumetric mass (kg/m^3)	1587.8	1598.5	1567.86	1579.927	
Percent of gaps (%)	26.7	27.3	32.70	35.25	

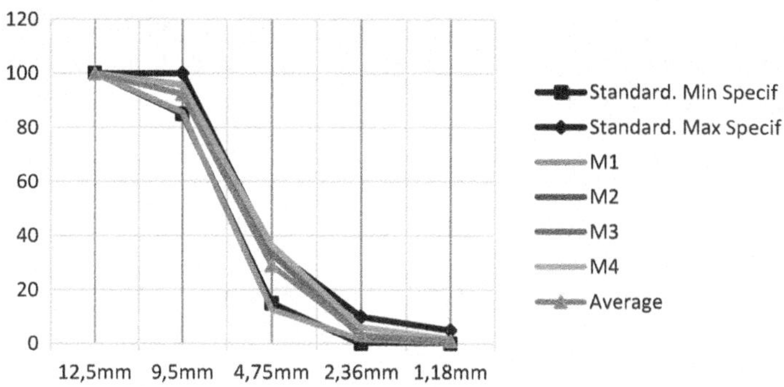

Fig. 2. Granulometry of granite from El Purio quarry

Table 3. Physical properties of granite from El Purio quarry

Tests	M1	M2	According to NC 251: 2013
Current specific weights	2.39	2.40	2.50
Saturated specific weights	2.46	2.47	
Apparent specific weights	2.6	2.6	
Absorption (%)	3.03	2.76	≤3.0
Loose volumetric mass (kg/m^3)	1401.9	1388.31	
Compact volumetric mass (kg/m^3)	1577.251	1497.0345	
Percent of gaps (%)	34.01	37.50	

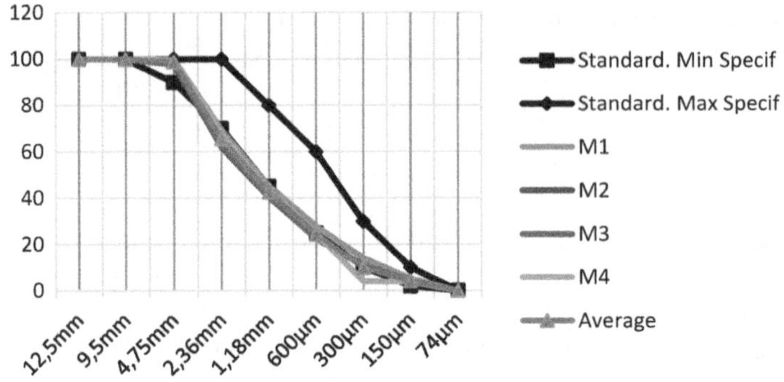

Fig. 3. Granulometry of the sand from El Purio quarry

Table 4. Physical properties of the sand from El Purio quarry

Tests	M1	M2	M3	M4	According to NC 251: 2013
Current specific weights	2.45	2.56		2.40	2.50
Saturated specific weights	2.67	2.61		2.48	
Apparent specific weights	2.7	2.7		2.61	
Absorption (%)	2.23	2.27		3.33	≤3.0
Loose volumetric mass (kg/m^3)	1489.4	1495.53		1430.10	
Compact volumetric mass (kg/m^3)	1634.5	1626.72		1614.12	
Percent of gaps (%)	35.3	36.33		32.71	
Fineness module	3.24	3.37		3.42	2.20–3.58

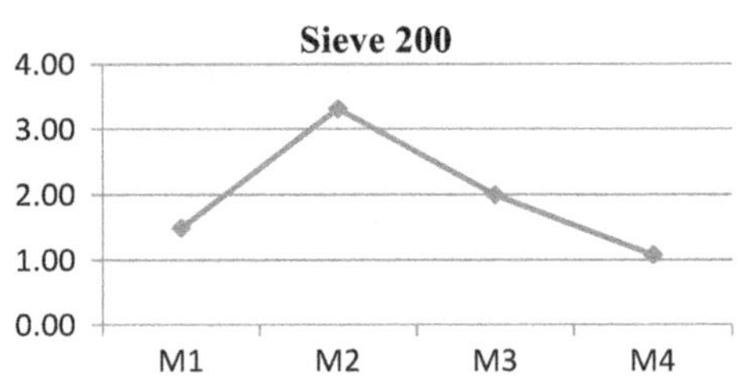

Fig. 4. Sieve 200 test of the sand from the El Purio quarry

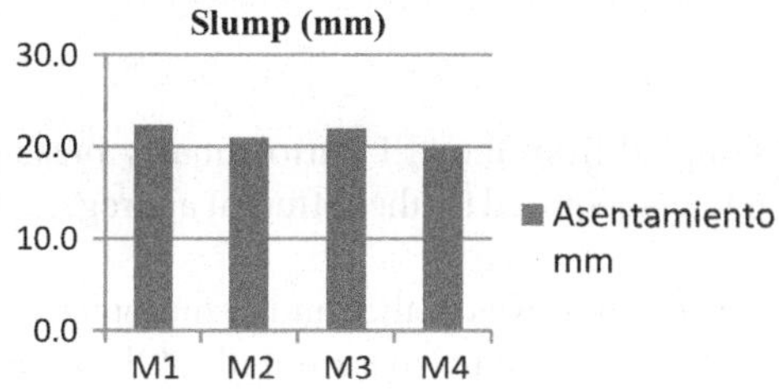

Fig. 5. Slump through the Cone de Abrams

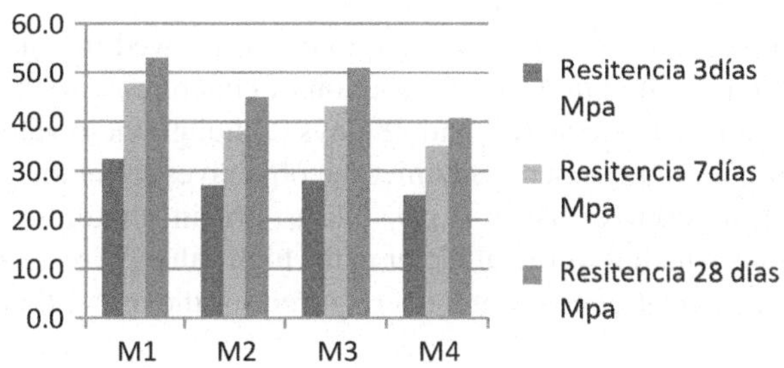

Fig. 6. Resistance to 3, 7 and 28 days

a modulus of fineness that qualifies it as coarse sand (3.10–6.00) mm. On the other hand, the sample that has the most slump is the M1 where a slump of 22 cm is obtained, which is classified as fluid. It is important to mention that the use of the SikaPlast additive influences the slump of the mixtures, as happens in the concretes produced in the ECOT Cayo Santa María.

Influence on compression resistance

The concrete mixes are designed for 35.0 MPa (Fig. 6). All the mixtures comply with respect to seven days, where the highest result is obtained in sample M1 and the lowest value in M4. In these tests the influence of the properties of the aggregates in the manufacture of concrete is evidenced. The influence is not given so much by the properties of the coarse aggregates, but by the fine aggregate. As previously seen, the samples of coarse aggregate are very similar and all comply according to the granulometry and physical properties. In the case of fine aggregate, there are differences between samples 1 and 4, which are the highest and lowest values of slump and resistance respectively. The fineness of grinding has an influence on the resistance, it is observed that the finer samples M4 decreases the resistance in the 28 days, only increases in proportion in 4 MPa, while the sample M1 and M3 that are in the range of the standard are those with the highest resistance values, with 1.5 and 2 of the 200 sieve.

4 Conclusions

(1) The aggregates sampled from the "El Purio" quarry present values that correspond to the parameters required by the different aggregate standards, especially NC 251: 2013 [11].
(2) The lack of fines in certain sieves in the sand granulometric test was not a factor that affected the values of compressive strength of the concretes tested.
(3) The concretes made with the aggregates studied reach strengths above 35 MPa after 7 days of age, which shows a positive influence of the aggregates as a fundamental component of the concretes.
(4) Of the samples analyzed, M4 was the one that showed the most unfavorable results, which resulted in the conformation of concretes with the lowest values of slump and resistance at 3, 7 and 28 days, although these values were above of the minimum requirements and met the objectives set in the investigation.
(5) The SikaPlast additive makes it possible to obtain mixtures with low water cement ratios, which is a crucial factor in the high values of resistances obtained, although an adequate fines content is required for the use of this additive.

References

1. Ghasemi, Y.: Aggregates in concrete mix design. Academic thesis. Principal supervisor Prof. Mats Emborg. Luleå University of Technology, Sweden (2017)
2. Kozul, R., Darwin, D.: Effects of aggregate type, syze and content on concrete strength and fracture energy. University of Kansas Center for Research, June 1967 (1997)
3. NC 95: 2017. Portland cement. Specifications (2017)
4. NC 96: 2017. Cement with active addition. Specifications (2017)
5. NC 177: 2002. Arid. Determination of the percentage of holes. Testing method (2002)
6. NC 178: 2002. Aggregates. Granulometric analysis (2002)
7. NC 181: 2002. Aggregates. Determination of the volumetric weight. Testing method (2002)
8. NC 182: 2002. Aggregate. Determination of the material finer than the 0.074 mm sieve (No. 200). Testing method (2002)
9. NC 186: 2002. Sand. Specific weight and water absorption. Testing method (2002)
10. NC 187: 2002. Coarse Aggregate. Specific weight and water absorption. Testing method (2002)
11. NC 251: 2013. ggregate for hydraulic concrete. Requirements (2013)
12. Neville, A.M., Brooks J.J.: Concrete technology. British library. Second edition (2010)

Characterization of Ferrocement Designed with GRP Reinforcement. Deformation and Displacement

Dayana de la Caridad Trelles Barroso

Abstract Cuba, because it is a Caribbean country, has a very aggressive environment for structures that have reinforcing steel in its structural composition. Engineers have sought new technologies to solve these problems that greatly limit the useful life of the structures projected in alkaline environments. Reinforced concrete is one of the main materials in the construction, however its biggest problem is the corrosion of steel. Nowadays, with the advancement of technology, several studies have been carried out with the use of ferrocement since it is a material that works with small thicknesses, however the use of steel as main reinforcement remains an obstacle to be avoided due to the inevitable danger of corrosion that it has. A new variant has emerged in recent years to reinforce the structures, the use of composite materials of Polymers reinforced with glass fibers (GRP) that is the ideal solution for a tropical island with a highly aggressive environment. In this work, a study of the bending behavior of ferrocement reinforced with GRP meshes using slab-type specimens with different mesh layers will be made, to reach conclusions to be taken into account in the use of this material in the construction sector and in our company (ETECSA) that could sometimes replace reinforced concrete. This could be the structural solution in buildings in aggressive environments, building lighter structures, resistant fronts to the effects of corrosion and much more durable.

Keywords Reinforced concrete · Composites · Glass fiber · Ferrocement

1 Introduction

Nowadays, in the construction sector, the material that has the greatest demand is reinforced concrete because it is resistant, durable, of constructive ease; But the main drawback of the concrete is to protect the steel bars against corrosion, because they oxidize slowly in contact with the environment. Studies carried out to eliminate or avoid corrosion have shown that the process can be delayed but not eliminated.

D. de la Caridad Trelles Barroso (✉)
Empresa de Telecomunicaciones de Cuba SA (ETECSA), Havana, Cuba
e-mail: dayana.trelles@etecsa.cu

© RILEM 2020

J. F. Martirena-Hernandez et al. (eds.), *Proceedings of the International Conference of Sustainable Production and Use of Cement and Concrete*, RILEM Bookseries 22,
https://doi.org/10.1007/978-3-030-22034-1_31

With the aim of eliminating corrosion, the use of polymers has been put into practice, a durable and resistant material that fulfills the function of steel in concrete.

In this work we propose to use reinforced ferrocement with fiberglass reinforced polymer mesh fabrics that compete from an economic, aesthetic and durability point of view with structures built reinforced with steel and other materials.

The ferrocement is a sheet of mortar that has a thickness of 2–5 cm, and is reinforced with wire mesh, glass fibers or other synthetic material, of small diameter; This material must be of high quality, have strengths greater than 25 MPa and superior impermeability to conventional concrete.

For the aforementioned, a study of the bending behavior of ferrocement reinforced with GRP mesh using slab-type specimens with different layers of meshes will be made.

1.1 Analysis of the Tests Carried Out on Aggregate and Cement

1.1.1 Fine Aggregate

The sand to be used is that of Arimao, to which a series of tests was carried out to determine its properties and determine if it was suitable for the specimens.

Tests:

- Passed through the sieve 200:

This test is carried out to know the percentage of impurities that the aggregate presents, since these impurities tend to adhere to the larger particles, preventing a good adhesion of the cementitious material to the aggregate, damaging the resistance to the mechanical stresses of the mortars. made with said aggregates. The finest material that the #200 sieve is expressed in percent through (Eq. 1)

$$\text{Tamiz}\,\#200 = \frac{\text{A} - \text{B}}{A} * 100 \tag{1}$$

A Weight of the material before the test
B Dry weight of the material after washing

The test is based on separating through the washing and sieving the finer particles than the #200 sieve, it was carried out three times [1].

- Grain size distribution:

This test consists of determining the distribution of sizes that the fine aggregate has for its use since this is very important in the properties of the mortar. The objective of this essay is to know if it complies with what is established in the standards.

Fig. 1 Sieving machine

The test was carried out by taking several samples from the different sands, which were placed in the stove at a temperature of 1100 °C approximately for 24 h, these were sifted from the #8 screen, making use of the CECAT electric sieve machine (Fig. 1) [2].

From this test, the maximum aggregate size and the fineness modulus can be determined.

1.2 Reinforcement Material

A GRP mesh with spacing of 25 mm (Fig. 2) will be used as reinforcement material, which is imported from the Russian Armotex company, which is part of the Yaroslavl Composite Materials Factory. The characteristics of the mesh are shown in (Fig. 2).

Fig. 2 GRP mesh

2 Results and Analysis of the Tests

2.1 Bending Test Carried Out on Reinforced Ferrocement Specimens with GRP

Upon completion of the 28 days of curing of the specimens, they were removed from the curing tank and the surface was left to dry for a few minutes. Once dry, a layer of calcareous paint was applied to the bottom of the specimen for when if it were to be tested, cracks could be seen at a glance. After finishing this activity it is left to dry and then transported to the laboratory where they will be tested later in this case the tests were carried out in the CECAT.

The test is carried out in compliance with and established by the provisions of CN 724, the interval of time elapsed between the extraction of the test pieces from the curing tank and their breaking less than three hours [3]. Due to the low thickness of the specimens, it is advisable to use a press with a low load capacity so that more accurate measurements can be made, in this case the test was carried out manually by placing loads on the specimen 10 kg bone 0.1 kN (Fig. 3) until ruptured. In addition to that you must have the necessary accessories to make measurements in such a way that loads can be applied to the specimen to one third of the light between supports, thus generating a section in the constant moment specimen, and thus allow measurements with greater degree of precision (Fig. 4). Besides that for the precision of the measurements, it should be counted on equipment of measurements of deformations so much vertical as horizontals automated, as they are the case of extensómetros and defórmetros.

During the tests, the following instruments and accessories were available:

- Weights (measured with load cell) which simulate the loads on the test piece, these vary between 2, 3, 5, 10 and 50 kg. The load was increased by 10 in 10 kg.

Fig. 3 Attachment for bending test a specimen with load in two points [3]

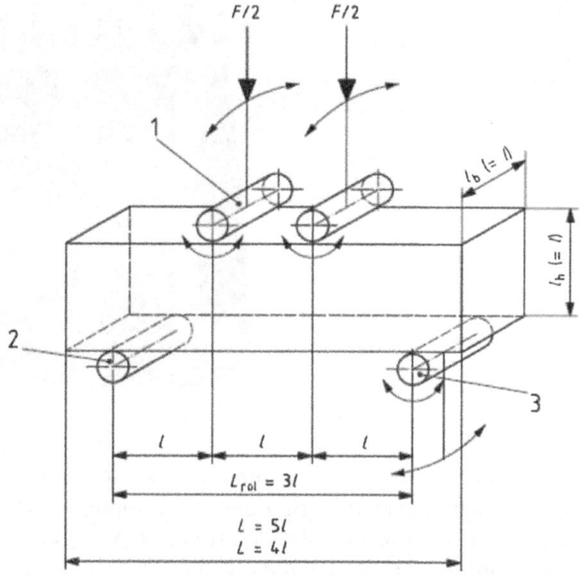

Fig. 4 Placement of loads (weights) and breakage of the specimen due to failure of the mortar

Fig. 5 Measuring attachment of the modulus of elasticity (STDM-1)

- Millimeter sheet placed behind the test piece at the time of the test as a reference to measure the arrow using the Level (topographic equipment).
- Camera of professional video (brand CANON) for obtaining photos and videos that later help to estimate the values of the deformations.
- Horizontal and vertical parameters with an accuracy of 0.0001 mm.
- Adjustments of measurement to obtain the modulus of elasticity (STDM-1).

Operating principle:

(1) The upper and lower blades are fixed with a gap according to the test space and these exert some pressure on the test element, then when applying the loads to the test element will lead to an axial displacement which will turn the blade toward down in the center, in the deformed area, causing the defometer to start measuring the horizontal deformation.

(2) At the end, the modulus of elasticity according to the spacing between blades will be obtained (Fig. 5).

Regulation of measurement up to half a millimeter of precision (60 cm). Used to measure the arrow with the help of the millimeter sheet and the Level.

Accessory to generate the load in two points of the slab. This was manufactured in the workshop of the CECAT consists of two angles of section 40 × 40 × 4 mm and a length of 250 mm separated from each other by 116 mm. Which are welded to two bars of 16 mm in diameter and 170 mm in length separated to 150 mm, which simulate the rollers to distribute the load in two points of the slab (Fig. 6).

Level with its tripod (Equipment used mainly in topographic studies). In this case it was used to measure the arrow, placing it at the same height and facing the test specimen and with the help of the millimeter sheet and the ruler placed on the other side of the specimen the arrow is measured as the specimen it was flexed by the applied loads.

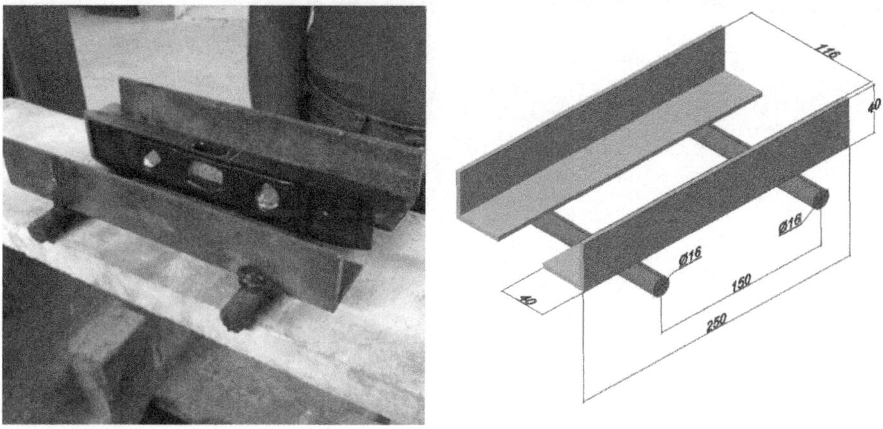

Fig. 6 Attachment to distribute the load in two points

Laser level and liquid to level. It is important that both the test area and the attachment and the specimen itself were level to avoid that the loads were displaced from the center of it and are evenly distributed obtaining a better result.

References

1. NC-182, Árido: Determinación del Material más fino que el Tamiz de 0.074 mm (no. 200) (2002)
2. NC-178, Áridos: Análisis Granulométrico (2002)
3. NC-724: Ensayo de resistencia del Hormigón endurecido (2015)

Part III
Durability

Atmospheric Corrosion Behaviors of Reinforcement Steel in Reinforced Concrete in a Coastal City as Havana, Cuba

Abel Castañeda, Francisco Corvo, Juan J. Howland, Rigoberto Marrero and Alejandro Fernández

Abstract Atmospheric corrosion behaviors of reinforcement steel in reinforced concrete in a coastal city as Havana was carried out. Concrete quality was characterized before to expose the reinforced concrete specimens in outdoor exposure condition. An outdoor exposure site at a distance of 10 m from the sea was selected in coastal city of Havana, Cuba. Three reinforced concrete specimens (RCS), one for each water/cement ratio 0.4, 0.5 and 0.6 with concrete covering thicknesses 20 and 40 mm were exposed to outdoor exposure condition. Corrosivity categories of the atmosphere in the outdoor exposure site were Very high (C5) and Extreme (CX). Atmospheric corrosion behaviors of reinforcement steel in reinforced concrete specimens served to obtain service life from the sum of Time-to-corrosion-initiation and Time to corrosion with induced cracking in reinforced concrete structures in a coastal city as Havana, Cuba. Compressive strength and ultrasonic pulse velocity are necessary test but not enough in the concrete quality characterization. Percentage of effective capillary porosity should be determined as a deeper factor for concrete quality characterization. Designs conditions for the concrete elaboration with a quality high and more durable were established, before to carry out the building works in a coastal city as Havana City, as well as in all shoreline north and south of Cuba with the purpose to increase the service life in reinforced concrete structures.

Keywords Concrete · Atmospheric corrosion · Reinforcement steel

A. Castañeda (✉)
Environment Division, Material Protection Group, National Center for Scientific Research, Havana, Cuba
e-mail: abel.castaneda@cnic.edu.cu

F. Corvo
Corrosion Research Center, Autonomous University of Campeche, San Francisco de Campeche, Campeche, Mexico

J. J. Howland · R. Marrero · A. Fernández
Technological University of Havana, Havana, Cuba

© RILEM 2020

J. F. Martirena-Hernandez et al. (eds.), *Proceedings of the International Conference of Sustainable Production and Use of Cement and Concrete*, RILEM Bookseries 22,
https://doi.org/10.1007/978-3-030-22034-1_32

1 Introduction

Coastal city of Havana is located at western of Cuba (2° 58′, 23° 10 NL and 82°30′, 82° 06′ WL). Many reinforced concrete structures built to short distance from the sea are affected by atmospheric corrosion of reinforcement steel. Study the behavior of atmospheric corrosion of reinforcement steel in reinforced concrete from experimental measurement of electrochemical corrosion velocity was the main objective of this research work. Reinforced concrete specimens of water/cement ratios 0.4, 0.5 and 0.6 with concrete cover thickness of 20 and 40 mm were elaborated. Specimens were exposed during three years in outdoor exposure site at a distance of 10 m from the sea in a coastal city zone of Havana. Corrosivity categories of the atmosphere were estimated in the outdoor exposure site. Measurements of compressive strength, ultrasonic pulse velocity, percentage of effective capillary porosity were used for concretes quality characterization for the three water/cement ratios. Time-to-corrosion-initiation and Time to corrosion with induced cracking were obtained. Service life real that can present the reinforced concrete structure from the sum of both times was shown.

2 Results and Discussion

2.1 *Monthly Behavior of Deposition and Meteorological Parameters*

Monthly behavior of Chloride deposition rate and sulphur compounds deposition rate, as well as Relative humidity, air temperature and wind speed in outdoor exposure site are shown during first year of study (Fig. 1a and b). A high deposition in both aggressiveness agents is noted during the winter months (October to April) (Fig. 1a). Wind speed reached the highest values (Fig. 1b). Relative humidity, temperature and wind speed were 77%, 25.3 °C and 3.4 m s^{-1} respectively. Therefore, Chloride deposition rate occurs in saline dissolution form in the reinforced concrete structure due to high annual average value of Relative humidity [1]. Atmospheric corrosion rate values for carbon steel, zinc and copper were determined [2]. Regarding to classification ranges, atmospheric corrosion rate values confirm that reinforced concrete structures built a short distance from the sea are exposed to corrosivity categories of the atmosphere Very high (C5) and Extreme (CX) in coastal city of Havana (Table 1). Carbon steel used as reinforcement in the reinforced concrete is exposed to Corrosivity categories of the atmosphere Extreme (CX).

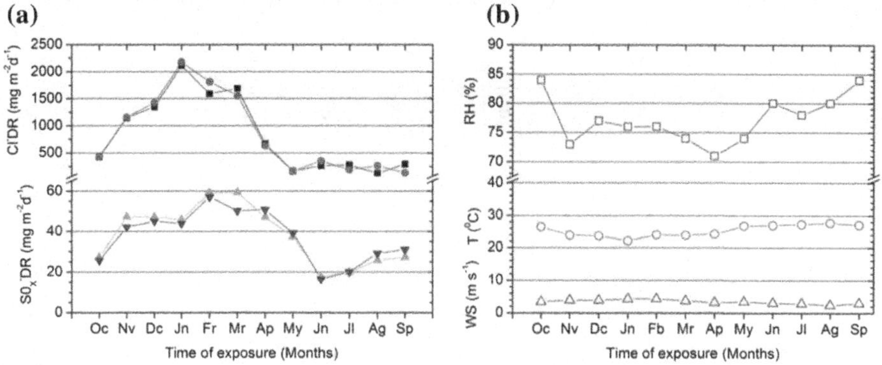

Fig. 1 Monthly behavior of Chloride deposition rate (**a**) and sulphur compounds deposition rate (**b**)

Table 1 Corrosivity categories of the atmosphere in the outdoor exposure site

Corrosion rate (μm year^{-1})			Corrosivity category (C)		
Carbonsel	C	Zinc	C	Copper	C
425.40	CX 200 > r_{corr} > 700	22.55	CX 8.4 > r_{corr} > 25	3.59	C5 2.8 > r_{corr} > 5.6

2.2 *Measurements in Concretes*

Standard deviation values and Coefficient of variation values are very low. Therefore, average values of compressive strength (f_{ck}), ultrasonic pulse velocity (UPV) and percentage of effective capillary porosity can be used in the concrete quality characterization. Eight values for each parameter were determined (Table 2).

According to DURAR network of CyTED program, concretes with compressive strength values higher than 20 MPa is indicative of its good quality and adequate durability [3]. Average values of compressive strength (f_{ck}) were higher than 20 MPa for the three concretes of water/cement ratios 0.4, 0.5 and 0.6. Cuban normative established that compressive strength value of concrete should be 30 MPa to guarantee a high durability in reinforced concrete [4]. Average values of Ultrasonic pulse velocities were higher than 4000 m s^{-1}. Concrete of water/cement ratio 0.6, quality is high. Average values were between 3001 and 4000 m s^{-1}. Results about percentage of effective capillary porosity show that only concrete with water/cement ratio 0.4 indicates a concrete of good quality and compactness. Average value was lower than 10%. Not being so in the concrete of water/cement ratio 0.5. Concrete quality is moderate due to average value was between 10 and 15%. Average value higher than 15%, durability is inadequate for the concrete of water/cement ratio 0.6.

Table 2 Measurements in concrete for each water/cement ratio

w/c	ε_e (%)	f_{ck} (MPa)	*UPV* (m s^{-1})	w/c	ε_e (%)	f_{ck} (MPa)	*UPV* (m s^{-1})	w/c	ε_e (%)	f_{ck} (MPa)	*UPV* (m s^{-1})
0.4	7.0	37	4160	0.5	11.6	30	4109	0.6	19.7	25	3900
	7.7	37	4280		12.8	30	4110		19.9	26	3910
	7.4	36	4200		12.3	30	4170		19.5	26	3910
	7.2	36	4290		14.9	30	4100		21.3	26	3930
	6.3	35	4270		15.7	30	4050		20.6	26	3930
	6.7	36	4270		13.5	31	4070		18.3	26	3960
	7.6	36	4240		14.9	30	4090		18.2	26	3980
	6.6	36	4260		14.0	29	4150		20.2	26	3970
Av.	*7.06*	*36*	*4246*	*Av.*	*13.7*	*30*	*4106*	*Av.*	*19.7*	*26*	*3936*
S.	*0.50*	*0.64*	*44.70*	*S.D*	*1.42*	*0.53*	*39.24*	*S.D*	*1.06*	*0.35*	*30.20*
C.V (%)	*7.08*	*1.77*	*1.05*	*C.V (%)*	*10.3*	*1.76*	*0.95*	*C.V (%)*	*5.38*	*1.34*	*0.76*

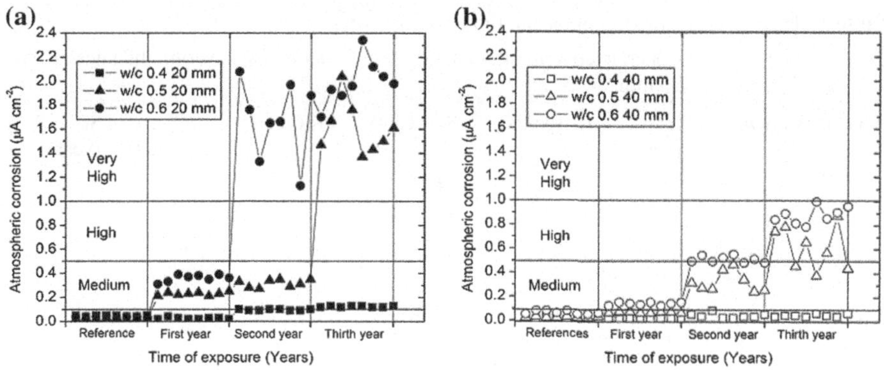

Fig. 2 Atmospheric corrosion behaviors of reinforcement steel in RCS

2.3 *Electrochemical Corrosion Velocity Behaviors in RCS*

Electrochemical corrosion velocity as an indicator of the atmospheric corrosion of reinforcement steel ($\mu A\ cm^{-2}$) was increased in the time of exposure (2008-first year, 2009-second year and 2010-third year) for the RCS of water/cement ratio 0.5 and 0.6 in both concrete covering thicknesses in the outdoor exposure site (Fig. 2a and b). Criteria of electrochemical corrosion velocity in reinforcement steel in service life terms for the reinforced concrete structures existing in the DURAR network of CyTED program are shown for both concrete covering thicknesses (Fig. 2a and b).

2.4 *Time-to-Corrosion-Initiation and Time to Corrosion with Induced Cracking*

Time-to-corrosion-initiation (t_i) was considered when electrochemical corrosion velocity values reached a medium level in service life terms for the reinforced concrete structures ($r_{corr} > 0.1\ \mu A\ cm^{-2}$) (Fig. 2a and b). A Critical threshold of free Chloride ions concentration was reached in the reinforcement steel beginning the atmospheric corrosion. Time to corrosion with induced cracking (t_{cc}) was considered when a high level ($r_{corr} = 0.5–1.0\ \mu A\ cm^{-2}$) and Very high level ($r_{corr} < 1.0\ \mu A\ cm^{-2}$) was reached (Fig. 2a and b). Therefore, Time-to-corrosion initiation (t_i) was one year for RCS of water/cement ratios 0.5 and 0.6 and concrete cover thickness of 20 mm. One and two years for concrete cover thickness of 20 and 40 mm respectively. Time to corrosion with induced cracking (t_{cc}) of three and two years were reached for RCS of water/cement ratios 0.5 and 0.6 in both concrete covering thicknesses. Service life obtained from sum between the two times does not exceed the five years for the reinforced concrete structures built at short distance from the sea exposed to Corrosivity categories of the atmosphere Very high (C5) and Extreme (CX) in

Table 3 Time-to-corrosion-initiation and time to corrosion with induced cracking

		Concrete covering thickness 20 mm			Concrete covering thickness 40 mm		
Specimens	w/c	t_i (years)	t_{cc} (years)	S_l (years)	t_i (years)	t_{cc} (years)	S_l (years)
RCS	0.4	2	12	14	18	~36	~54
	0.5	1	3	4	2	3	5
	0.6	1	2	3	1	2	3

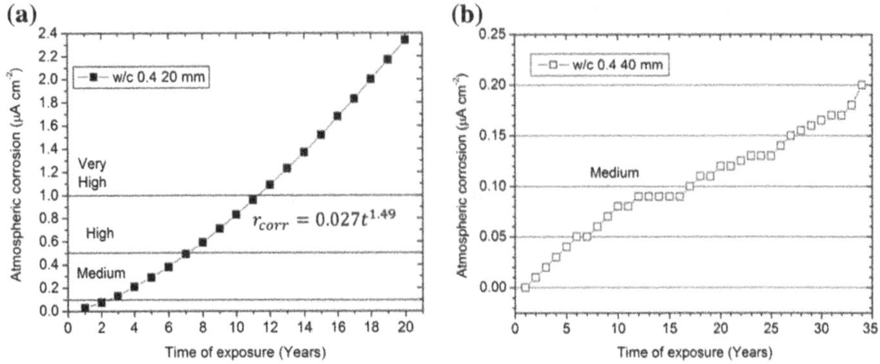

Fig. 3 Prediction of electrochemical corrosion velocity values in RCS of water/cement ratio 0.4

coastal city of Havana (Table 3). Water/cement ratios 0.5 and 0.6, as well as concrete cover thicknesses of 20 and 40 mm, correspond to design conditions most used in the structures build in Cuba.

Time-to-corrosion-initiation (t_i) was two years for RCS of water/cement ratios 0.4 and concrete cover thickness of 20 mm (Table 3). Prediction of electrochemical corrosion velocity values allowed obtaining both times for RCS of water/cement ratio 0.4 in concrete covering thicknesses of 20 and 40 mm (Fig. 3a and b). Values higher than three years until twenty years obtained from regression fitted (1) are shown for RCS of water/cement ratio 0.4 for a concrete covering thickness of 20 mm. Regression fitted obtained was:

$$r_{corr} = 0.027t^{1.49} \quad R^2 = 91\% \tag{1}$$

Time to corrosion with induced cracking (t_{cc}) was reached at twelve years (Fig. 3a). Service life for reinforced concrete structures built with these conditions could be around fourteen years in coastal city of Havana.

Although, electrochemical corrosion velocity as an indicator of the atmospheric corrosion of reinforcement steel presented a slight increase in the time of exposure for RCS of water/cement ratio 0.4 and concrete covering thickness of 20 mm (Fig. 2a). Regression fitted (1) shows the opposite. Constant value b was higher than

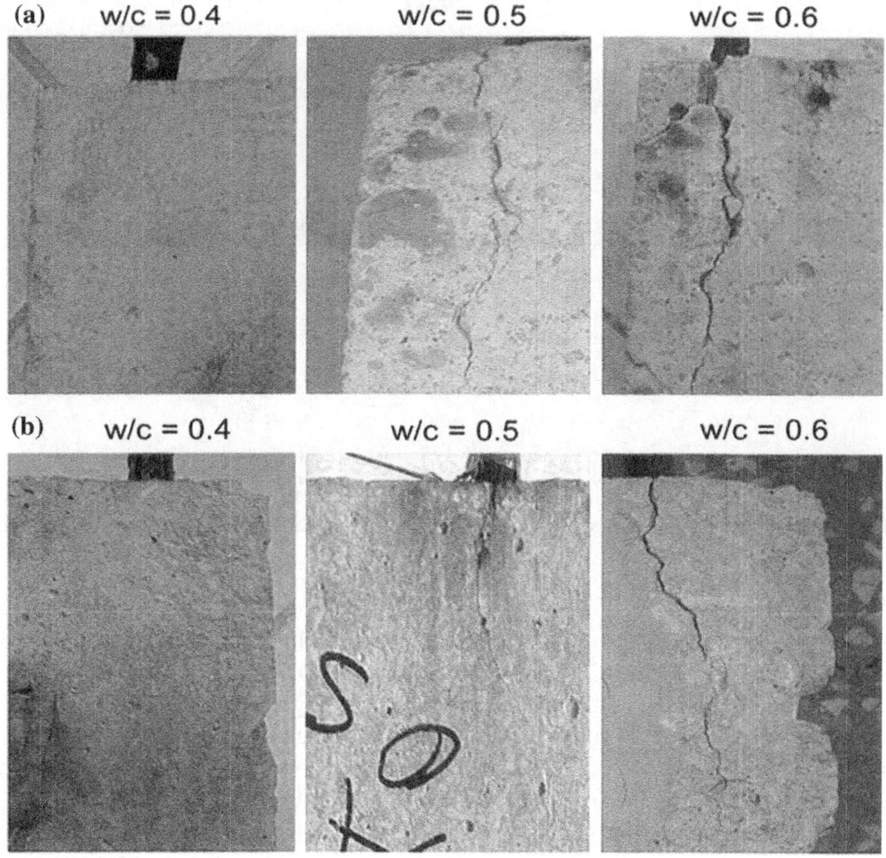

Fig. 4 Visual observation of RCS

1. Therefore, a significant increase is shown also for RCS of water/cement ratio 0.4 and concrete covering thickness of 20 mm. Service life around fourteen years for reinforced concrete structures built with these conditions should not be considered high. Service life term higher than fifty years should be obtained. However, Time-to-corrosion-initiation (t_i) for RCS of water/cement ratio 0.4 for a concrete covering thickness of 40 mm was reached at eighteen years. Time-to-corrosion initiation was nine times higher compared with the obtained for a concrete covering thickness of 20 mm. Influence of concrete covering thickness continue being demonstrated. Electrochemical corrosion velocity values were obtained from statistical method of the Cumulative Sum (Fig. 3b). Considering Time to corrosion with induced cracking (t_{cc}) would be only three times higher than Time-to-corrosion initiation for concrete covering thickness of 40 mm, service life terms higher than fifty years could be obtained (Table 3). Time to corrosion with induced cracking (t_{cc}) could be higher due to concrete covering thickness influence.

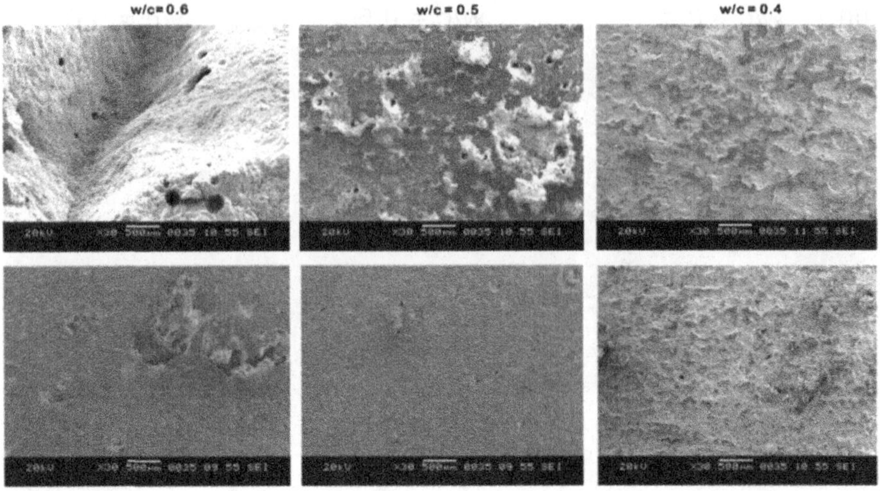

Fig. 5 Scanning electron microscopy tests in small samples of hardened cement paste for the three water/cement ratios

Induced cracking due to atmospheric corrosion of reinforcement steel is observed at third year of study in RCS of water/cement ratios 0.5 and 0.6 for both concrete covering thicknesses (Fig. 4a and b). Induced cracking is more noted for concrete covering thickness of 20 mm. However, induced cracking appeared at two years in RCS of water/cement ratio 0.6 for both concrete covering thicknesses. Induced cracking provoked by the oxides expansion in perpendicular form to RCS surface, was more observed for the RCS of water/cement ratios 0.5 and 0.6 for both concrete covering thicknesses. Induced cracking was more observed too for concrete covering thickness of 20 mm. Influence of water/cement ratio increase, as well as the concrete covering thickness in the atmospheric corrosion of reinforcement steel is confirmed in the outdoor exposure site. Corrosivity categories of the atmosphere in the outdoor exposure site at short distance from the sea were Very high (C5) and Extreme (CX) during first year of study. A major penetration of the aggressive agents, mainly the chloride ions is provoked by induced cracking. Concrete with water/cement ratios 0.5 and 0.6 and concrete covering thicknesses of 20 and 40 mm do not assure a suitable primary protection as physic barrier. Induced cracking appeared confirm that service life for the reinforced concrete exposed to Corrosivity categories of the atmospheres Very high (C5) and Extreme (CX) do not exceed the five years.

Scanning electron microscopy tests in small samples of hardened cement paste for the three water/cement ratios allowed to continue the confirming of the influence of Percentage of effective capillary porosity. Capillary pores formation were less appreciable in small samples of hardened cement paste in concrete with water/cement ratio 0.4 (Fig. 5).

3 Conclusions

1. Corrosivity categories of the atmosphere in the outdoor exposure site at short distance from the sea were Very high (C5) and Extreme (CX). Atmosphere was classified as coastal-industrial. Many reinforced concrete structures in the coastal city as Havana, Cuba are exposed to those conditions.
2. Service life of reinforced concrete structures in a coastal city as Havana, Cuba was demonstrated from behavior of atmospheric corrosion of reinforcement steel in reinforced concrete specimens. A methodology to obtain Time-to-corrosion-initiation and Time to corrosion with induced cracking was demonstrated.
3. Compressive strength and ultrasonic pulse velocity are necessary test but not enough in the concrete quality characterization. Percentage of effective capillary porosity should be determined as a deeper factor for concrete quality characterization before to carry out the building works in coastal cities at short distance from the sea.
4. Designs conditions for the concrete elaboration with a quality high and more durable, before to carry out the building works in a coastal city as Havana City, as well as in all shoreline north and south of Cuba with the purpose to increase the service life in reinforced concrete structures were established.
5. Results obtained were confirmed from visual observation in reinforced concrete specimens exposed to Corrosivity categories of the atmospheres obtained, as well as from Scanning electron microscopy tests in small samples of hardened cement paste for the three water/cement ratios.

References

1. Castañeda, A., Corvo, F., Howland, J.J., Marrero, R.: Penetration of marine aerosol in a tropical coastal city: Havana. Atmósfera **31**(1), 87–104 (2018)
2. ISO 9223: Corrosion of metals and alloys. Corrosivity of atmosphere. Classification, determination and estimation (2012)
3. Troconis de Rincon, O.: DURAR network members, manual for inspecting, evaluating and diagnosing corrosion in reinforced concrete structures. CYTED. ISBN 980–296-541-3 Maracaibo, Venezuela. 1997/1998/2001 (1st Edition, 2nd Edition and 3rd Edition in Spanish) (1st Edition in English) (2000)
4. Cuban Standard 724: Test on concrete. Resistance of hardened concrete (2009)

Preliminary Results on Corrosion Rate in Carbonated LC^3 Concrete

Elizabeth Cabrera, Adrián Alujas, Bernhard Elsener and Jose Fernando Martirena-Hernandez

Abstract The new blended cement, LC^3, can reduce by up to a 30% of the CO_2 emissions per ton of cement, including calcined clay and limestone as main components. However, the low level of clinker content concern about carbonation-induced corrosion of the reinforcing steel. The aim of this project is to measure the corrosion rate and understand the corrosion mechanism of reinforcing steel in concrete made with LC^3 cement. The samples were cast in molds specifically designs in order to minimize the carbonation time and the moisture equilibration time. There is a finer porosity for LC^3 cement and the porosity decrease when the carbonation takes place due to the transformation of the portlandite into calcite. However, when the clinker substitution is higher, the total porosity increase, but there is a coarsening porosity for blended systems after carbonation because the C-S-H are converted into calcite with a smaller volume. In all LC^3 samples, the corrosion rate is higher compared with OPC cement, but there are differences regarding the induced carbonation and natural carbonation especially for LC^3 samples with a more tortuous pore system than OPC assuring the high resistance to the carbonation.

Keywords Carbonation-induced corrosion · LC^3 cement

E. Cabrera (✉)
Faculty of Chemistry, Universidad Central de Las Villas, 54830 Santa Clara, Cuba
e-mail: ecapolinaire@uclv.edu.cu

A. Alujas
Centro de Estudios de Química Aplicada, Universidad Central de Las Villas, 54830 Santa Clara, Cuba

B. Elsener
ETH Zürich, Institute for Building Materials (IfB), 8093 Zurich, Switzerland

J. F. Martirena-Hernandez
Centro de Investigación y Desarrollo de Estructuras y Materiales, Universidad Central de Las Villas, 54830 Santa Clara, Cuba

© RILEM 2020

J. F. Martirena-Hernandez et al. (eds.), *Proceedings of the International Conference of Sustainable Production and Use of Cement and Concrete*, RILEM Bookseries 22,
https://doi.org/10.1007/978-3-030-22034-1_33

1 Introduction

Preliminary evaluation of LC^3 cement, proves a high performance when used in concrete. However, the low level of clinker content concern about carbonation-induced corrosion of the reinforcing steel. Carbonation is the neutralization of the concrete due to the reaction with CO_2 from the atmosphere due to the inherent chemical properties of the blended systems, in particulary the reduced pH buffering capacity due to decreased calcium hydroxide content. First, a lower amount of calcium hydroxide is formed in the hydration reaction [1] as consequence of a reduced amount of CaO with respect to Ordinary Portland Cement (OPC 65% CaO), variably down to 40% [2], depending on the type and amount of substituents; second, the Calcium hydroxide is additionally consumed in presence of SiO_2- and Al_2O_3, rich components in the pozzolanic reactions [3]. For these two reasons, the carbonation penetrates much faster into the concrete matrix in case of blended cements [4–6], and it could reduce the service life of a structure and could limit the practical use of the LC3.

For Portland cement the corrosion rate increases with increasing higher relative humidity of the environment [7] and is because with a higher water demand, the CO_2 can dissolve easily and can form carbonic acid, and reacting with the locally available calcium hydroxide, the calcium carbonate precipitates and allow the pH to fall to a level where steel will corrode (Corrosion of the steel in concrete). In addition, there is a correlation between corrosion rate and resistivity if carbonated concrete made with different cements [8]. At low relative humidity, between 75 and 80% there is no or is negligible the corrosion rate found [9]. However, at 95% of relative humidity was found a maximum of corrosion rate [10] and it was interpreted in terms of a decreasing concrete resistivity.

The initiation and propagation stage of corrosion in concrete (Fig. 1) made with blended cements have to be carefully evaluated in order to reach the required service life. Carbonation induced corrosion becomes much more important, and factors that influence the rate of corrosion of steel in carbonated concrete are by far less investigated than chloride induced corrosion. The faster carbonation process can, in addition, strongly influence also chloride-induced corrosion, by reducing the pore solution pH and, consequently, the critical chloride content. The corrosion rate cannot be predicted due to a lack of laboratory data, field experience, and mechanistic understanding.

2 Materials and Methods

In this work, different mix designs were used, in order to evaluate the influence of the clinker content and water/cement ratio influences. The experimental design is shown in Table 1.

Samples were specifically designed in order to minimize carbonation time and moisture equilibration time [11]. The samples after casting and de-moulding were

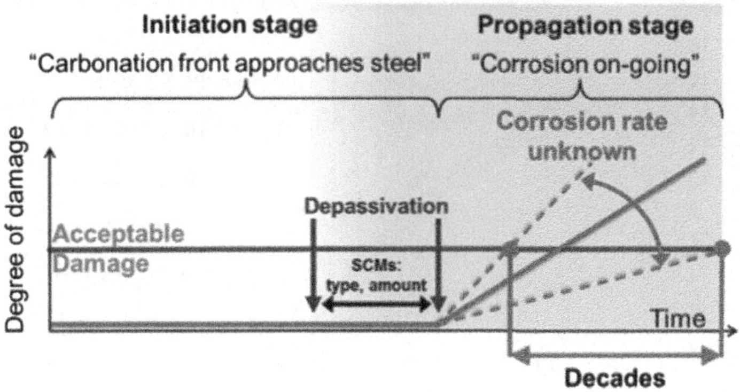

Fig. 1 Schematic representation of the service life of a reinforced concrete structure ("Tuutti diagram") showing the importance of the propagation-stage of corrosion in concrete

Table 1 Experimental design for OPC and LC^3 cements

Mix design	w/c = 0.5	w/c = 0.6
Portland Cement	x	x
LC^3-50 2:1	x	x
LC^3-65 2:1	x	
LC^3-65 1:1	x	x

carbonated at 65% of relative humidity and 4% of CO_2 concentration. Carbonation was checked by phenolphthalein test. Amount of released heat during hydration reaction was determined by isothermal calorimetric study in cement pastes. The porosity and the pore size distribution of mortar samples were obtained by using Mercury Intrusion Porosimetry (MIP) before and after carbonation. SEM analysis were carried out in the carbonated and non-carbonated samples to study the porosity changes. Finally, Corrosion rate was measured by Linear Polarization Resistance (LPR) technique.

3 Results and Discussion

The time for full carbonation (duration until carbonation reached the embedded steel, cover depth 3 mm) was about three times faster for the substituted cement (LC^3) compared to OPC. This can be explained by the lower CaO content in the LC^3 cement, leading to a lower buffering capacity.

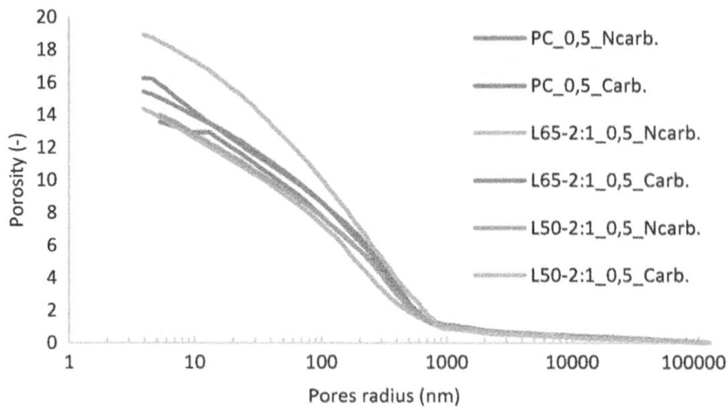

Fig. 2 MIP results for OPC and LC^3 mortars

3.1 Porosity Characterization by MIP and SEM Analysis

OPC samples present finer porosity than LC^3 samples in non-carbonated conditions. However, when carbonation take place, there is a decreasing porosity in case of the OPC mortars due to the conversion of the $Ca(OH)_2$ in $CaCO_3$ with bigger volume. On the other hand, it is observed a decreasing of the total porosity when the clinker substitution is higher but it is evident the finer porosity for blended cement after carbonation because the C-S-H gel into calcite. These possible results can be related to the interfacial transition zone that will be objective of further work of the author (Figs. 2 and 3).

3.2 Corrosion Rate

When the calcined clay/limestone ratio is different, the corrosion rate is very similar. However, when the water/cement ratio change, the corrosion rate increase even if the clinker content is different too, at 95% relative humidity (Fig. 4).

4 Conclusions

- The pozzolanic reaction consumes $Ca(OH)_2$ and thus decreases the pH. This might be beneficial with respect to ASR, but increases the risk of corrosion.
- In case of LC^3 cement, the calcined clay content, as SCMs have positive effects on porosity (denser microstructure) and thus slow down penetration of carbonation.

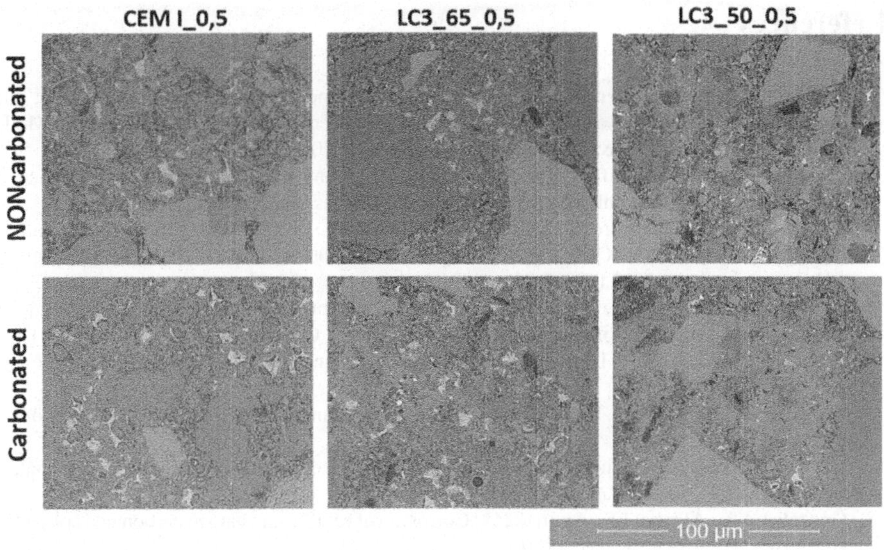

Fig. 3 Coarsening porosity for blend systems after carbonation

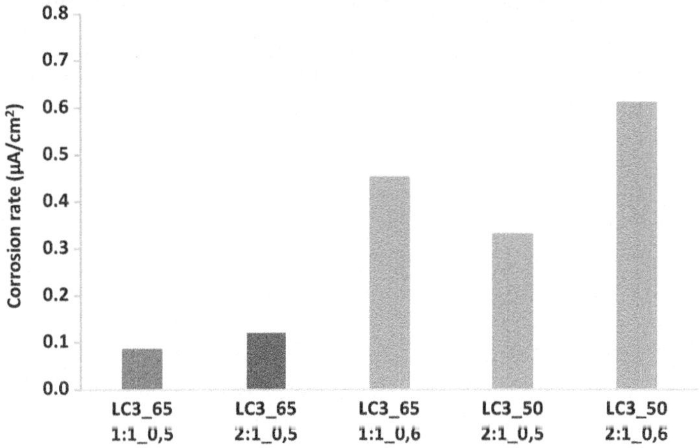

Fig. 4 Corrosion rate for LC^3 mortars at 95% relative humidity

- The difference in corrosion rate at different calcined clay/limestone ratio is negligible, however increase with a higher water/cement ratio in a very high relative humidity.

Acknowledgements The authors would like to acknowledge Dr. Prof. Ueli Angst, Dr. Prof. Bernhard Elsener and the Corrosion Team from ETH Zurich, Switzerland, for the helpfully support.

References

1. Lothenbach, B., Scrivener, K., Hooton, R.D.: Supplementary cementitious materials. In: Cement and Concrete Research, Conferences Special: Cement Hydration Kinetics and Modeling, Quebec City, 2009 & CONMOD10, Lausanne, 2010 **41**(12), 1244–1256 (2011)
2. Papadakis, Vagelis G.: Effect of supplementary cementing materials on concrete resistance against carbonation and chloride ingress. Cem. Concr. Res. **30**(2), 291–299 (2000)
3. Kolani, B., Buffo-Lacarrière, L., Sellier, A., Escadeillas, G., Boutillon, L., Linger, L.: Hydration of slag-blended cements. Cem. Concr. Compos. **34**(9), 1009–1018 (2012). https://doi.org/10.1016/j.cemconcomp.2012.05.007.7
4. Leemann, A., Nygaard, P., Kaufmann, J., Loser, R.: Relation between carbonation resistance, mix design and exposure of mortar and concrete. Cement Concr. Compos. **62**, 33–43 (2015)
5. Lye, C.Q, Ghatoara, G.S., Dhir, R.K.: Carbonation resistance of fly ash concrete. Mag. Concr. Res. **67**(21) (2015)
6. Lye, C.Q, Ghatorara, G.S., Dhir, R.K.: Carbonation resistance of GGBS concrete. Mag. Concr. Res. **67** (2015)
7. Glass, G.K., Page, C.L., Short, N.R.: Factors affecting the corrosion rate of steel in carbonated mortars. Corros. Sci. **32**(12), 1283–1294 (1991)
8. Gonzalez, J.A., Algaba, J.S., Andrade, C.: Corrosion of reinforcing bars in carbonated concrete. Br. Corros. J. **15**(3), 135–139 (1980)
9. Hunkeler, F., Lammar, L.: Requirements for the carbonation resistance of concrete mixes. VSS research report 649 (2012)
10. Parrot, L.J.: Damage caused by carbonation of reinforced concrete. Mater. Struct. **23**, 230–234 (1990)
11. Stefanoni, M., Angst, U., Elsener, B.: Innovative sample design for corrosion rate measurements in carbonated blended mortars. In: 11th International Concrete Sustainability Conference. Washington DC, USA, 15–18 May 2016

Use of Electrical Test Method on Determination Aging Factor of Concrete Incorporating Supplementary Cementitious Materials

Hossein Mosavi, Raid Alrashidi, Mohammed Almarshoud, Mohammed H. Alyami, Kyle A. Riding, Christopher C. Ferraro, Michael D. A. Thomas and Harvey DeFord

Abstract One of the biggest benefits of using supplementary cementitious materials (SCMs) in concrete is the large reduction in the concrete chloride ion diffusion coefficient with age. The formation factor calculated from electrical resistivity measurements is proportional to the concrete effective chloride diffusion coefficient. Traditionally, the aging factor used in service life modeling has been determined from examination of the apparent diffusion coefficient calculated from concrete bulk diffusion experiments, lumping chloride binding and with diffusion. This study proposes to use the formation factor determined from electrical resistivity tests to quantify the concrete aging parameter for service life modeling. To determine the effect of cement chemical composition, four types of cement were tested including type I/II low alkali cement, type IL cement, type V cement and type I/II high alkali cement. Silica fume, slag cement, Class F fly ash, and metakaolin were used as supplementary cementitious materials (SCMs) in binary and ternary blends with different replacement ratios to evaluate the effect of SCMs on the concrete electrical properties with age. The electrical properties were determined at 28, 56, 91, 182 and 365 days. A discussion about the advantages and disadvantages of this method over traditional methods to quantify aging are presented.

Keywords Chloride diffusion · Chloride migration · Aging factor · Service life modeling

---

H. Mosavi · R. Alrashidi · M. Almarshoud · M. H. Alyami · K. A. Riding (✉) · C. C. Ferraro
University of Florida, Gainesville, FL, USA
e-mail: kyle.riding@essie.ufl.edu

H. Mosavi
e-mail: smosavi@ufl.edu

M. D. A. Thomas
University of New Brunswick, Fredericton, NB, Canada

H. DeFord
Florida Department of Transportation, Gainesville, FL, USA

© RILEM 2020

J. F. Martirena-Hernandez et al. (eds.), *Proceedings of the International Conference of Sustainable Production and Use of Cement and Concrete*, RILEM Bookseries 22,
https://doi.org/10.1007/978-3-030-22034-1_34

1 Introduction

Concrete transport properties are some of the most critical parameters that determine the concrete durability. These properties improve as concrete matures. The change in transport properties can be modeled for use in service life models. For example, the concrete diffusion coefficient change with time has been modeled using Eqs. (1) and (2) [1]:

$$D_t = D_{28}\left(\frac{t_{28}}{t}\right)^m + D_{ult}\left(1 - \left(\frac{t_{28}}{t}\right)^m\right) \tag{1}$$

$$D_{ult} = D_{28}\left(\frac{28}{365 \times 100}\right)^m \tag{2}$$

where D_t us the diffusion coefficient (m^2/s) at time t (days), D_{28} is the concrete diffusion coefficient at 28 days (m^2/s), $t_{28} = 28$ days, t is time (days), and m is the decay constant or aging factor. The aging factor determines the change in the concrete properties with time. Data for this parameter is difficult to obtain because it requires aging the samples for a long time, the concrete properties change with age, concrete transport property measurements are often affected by ongoing chemical reactions, and there are multiple unknowns to solve for with chloride ponding tests [2].

Formation factor is an empirical parameter that was originally developed to relate hydraulic conductivity of rock formations to electrical measurements in the oil industry [3]. As a porous material, formation factor can be applied to concrete to calculate the concrete effective diffusion coefficient from electrical resistivity measurements, as shown in Eq. (3):

$$\frac{\rho}{\rho_0} = F = \frac{D_0}{D} \tag{3}$$

where ρ is the concrete bulk resistivity ($\Omega \cdot m$), ρ_0 is the pore solution resistivity, F is the formation factor, D is the effective diffusion coefficient (m^2/s), D_0 is the Ionic diffusion coefficient in an infinitely dilute solution (m^2/s). For chlorides, the self-diffusion coefficient is 20.3×10^{-10} m^2/s at 25 °C [4]. The diffusion coefficient for concrete can be determined using Eq. (4):

$$D = 20.3 \times 10^{-10}\left(\frac{\rho}{\rho_0}\right) \tag{4}$$

In this paper, the formation factor is examined for estimating the aging factor of the concrete.

2 Methodology

The pore solution resistivity was estimated from the constituent material composition and degree of hydration [5, 6]. Concrete mixtures examined included four types of cement that are commercially available and supplementary cementitious materials such as silica fume, slag, fly ash and metakaolin [7]. Mixture proportions are presented in Table 1.

100 × 200 mm concrete cylinders were made for each test. The concrete specimens were removed from their molds 24 h after mixing and stored in an ASTM C511 [8] moist room until the specified testing ages 28, 56 and 365 days for rapid chloride migration testing. The bulk resistivity samples were cured using a simulated pore solution. A different simulated pore solution was made for each mixture using the NIST pore solution calculator [6, 9]. The solution was also saturated with $Ca(OH)_2$. Six liters of solution was made for each mixture and placed with four 100 × 200 mm cylinders in a container with a sealable lid. The sealed containers were then placed in the moist curing room to ensure that all samples tested had the same temperature history.

The bulk electrical resistivity of the concrete was measured according to AASHTO TP119-15 [10]. The concrete electrical resistivity can be determined using Eq. (5):

$$\rho = R_{cylinder}\left(\frac{A}{L}\right) \tag{5}$$

where $R_{cylinder}$ is the cylinder resistance, A is the cylinder cross-sectional area, and L is the cylinder length. The electrical conductivity was determined using Eq. (6):

$$\sigma = \frac{1}{\rho} \tag{6}$$

where σ is concrete electrical conductivity (S/m). The conductivity of the simulated pore solution was measured using a conductivity meter. The diffusion coefficient was determined using Eq. (4). NT Build 492 was used to measure rapid chloride migration [11]. The concrete diffusion coefficient with time was fit to measured data using Eqs. (1) and (2).

Table 1 Concrete mixture proportions and aging factor

Mix ID	SCM	w/cm	Cement type (%)				SCM type (%)				Aging factor	
			I/II	V	1L	HA	FA	SG	SF	MK	F	RCMT
C-100	No SCM	0.35	100	–	–	–	–	–	–	–	0.36	0.14
C-100h	No SCM	0.44	100	–	–	–	–	–	–	–	0.19	0.11
C-F10	Binary	0.35	90	–	–	–	10	–	–	–	0.11	0.38
C-F20	Binary	0.35	80	–	–	–	20	–	–	–	0.11	0.65
C-F10 h	Binary	0.44	90	–	–	–	10	–	–	–	0.40	0.32
C-F20 h	Binary	0.44	80	–	–	–	20	–	–	–	0.62	0.60
C-G60	Binary	0.35	40	–	–	–	–	60	–	–	0.42	0.19
C-S8	Binary	0.35	92	–	–	–	–	–	8	–	0.33	0.31
C-M10	Binary	0.35	90	–	–	–	–	–	–	10	0.29	0.15
C-F10G30	Ternary	0.35	60	–	–	–	10	30	–	–	0.00	0.27
C-F10G45	Ternary	0.35	45	–	–	–	10	45	–	–	0.12	0.35
C-F10G60	Ternary	0.35	30	–	–	–	10	60	–	–	0.15	0.27
C-G55S8	Ternary	0.35	37	–	–	–	–	55	8	–	0.11	0.71
C-G55M10	Ternary	0.35	35		–	–	–	55	–	10	0.14	0.52
CV-100	No SCM	0.35	–	100	–	–	–	–	–	–	0.21	0.20
CV-100h	No SCM	0.44	–	100	–	–	–	–	–	–	0.15	0.08
CV-F10G60	Ternary	0.35	–	30	–	–	10	60	–	–	0.20	0.14
CV-F20S8	Ternary	0.35	–	72	–	–	20	–	8	–	0.17	0.68
CV-M10	Binary	0.35	–	90		–	–	–	–	10	0.00	0.36
CL-100	No SCM	0.35	–	–	100	–	–	–	–	–	0.36	0.10
CL-100h	No SCM	0.44	–	–	100	–	–	–	–	–	0.19	0.08
CL-F10G60	Ternary	0.35	–	–	30	–	10	60	–	–	0.11	0.33

(continued)

Table 1 (continued)

Mix ID	SCM	w/cm	Cement type (%)				SCM type (%)				Aging factor	
			I/II	V	1L	HA	FA	SG	SF	MK	F	RCMT
CL-F20S8	Ternary	0.35	–	–	72	–	20	–	8	–	0.11	0.39
CL-M10	Binary	0.35	–	–	90	–	–	–	–	10	0.40	0.06
CHA-100	No SCM	0.35	–	–	–	100	–	–	–	–	0.62	0.01
CHA-100h	No SCM	0.44	–	–	–	100	–	–	–	–	0.42	0.11
CHA-F10G60	Ternary	0.35	–	–	–	30	10	60	–	–	0.33	0.39
CHA-F20S8	Ternary	0.35	–	–	–	72	20	–	8	–	0.29	0.40
CHA-M10	Binary	0.35	–	–	–	90	–	–	–	10	0.00	0.02

3 Results

The diffusion coefficient was calculated for both methods at 28 days and compared, as shown in Fig. 1a. While there is a significant correlation between the 28-day diffusion coefficients calculated using both methods, the diffusion coefficient derived from rapid chloride migration were greater compared to the diffusion coefficients from *F*. This could be because of the effect of chloride binding or common ion effect.

Figure 1b shows the diffusion coefficients for concrete incorporating 10% fly ash and Slag with replacement percentages of 0, 30, 45 and 60. Incorporating at least 30% slag significantly reduced the diffusion coefficient by refining the pore structure of the concrete, with much of the benefits occurring before 28 days. Figure 1c and d show the diffusion coefficient determined from formation factor for a water-to-cementitious material ratio (w/cm) of 0.35 and 0.44, respectively for the cement types I/II, 1L, V, and high alkali cement only mixtures. Unexpectedly it was seen that for concrete with high diffusion coefficients that after a certain point the diffusion coefficient calculated from the bulk resistivity measurements started to increase. The low value for the aging coefficient is due to the increase in the diffusion at later ages, likely an artifact of the test method. Work is ongoing to understand why the calculated diffusion coefficient would increase, but it is thought to be an artifact of alkali migration from the simulated pore solution into the concrete or a significant increase in the concrete degree of saturation for the samples with higher diffusion coefficients, possibly beginning to fill in air voids. Figure 1e shows the diffusion coefficient from the formation factor for concrete with the type I/II cement and different SCM replacements. The reason that the aging factor reduction was not very high for some of the ternary mixtures is due to the very low initial coefficient at 28

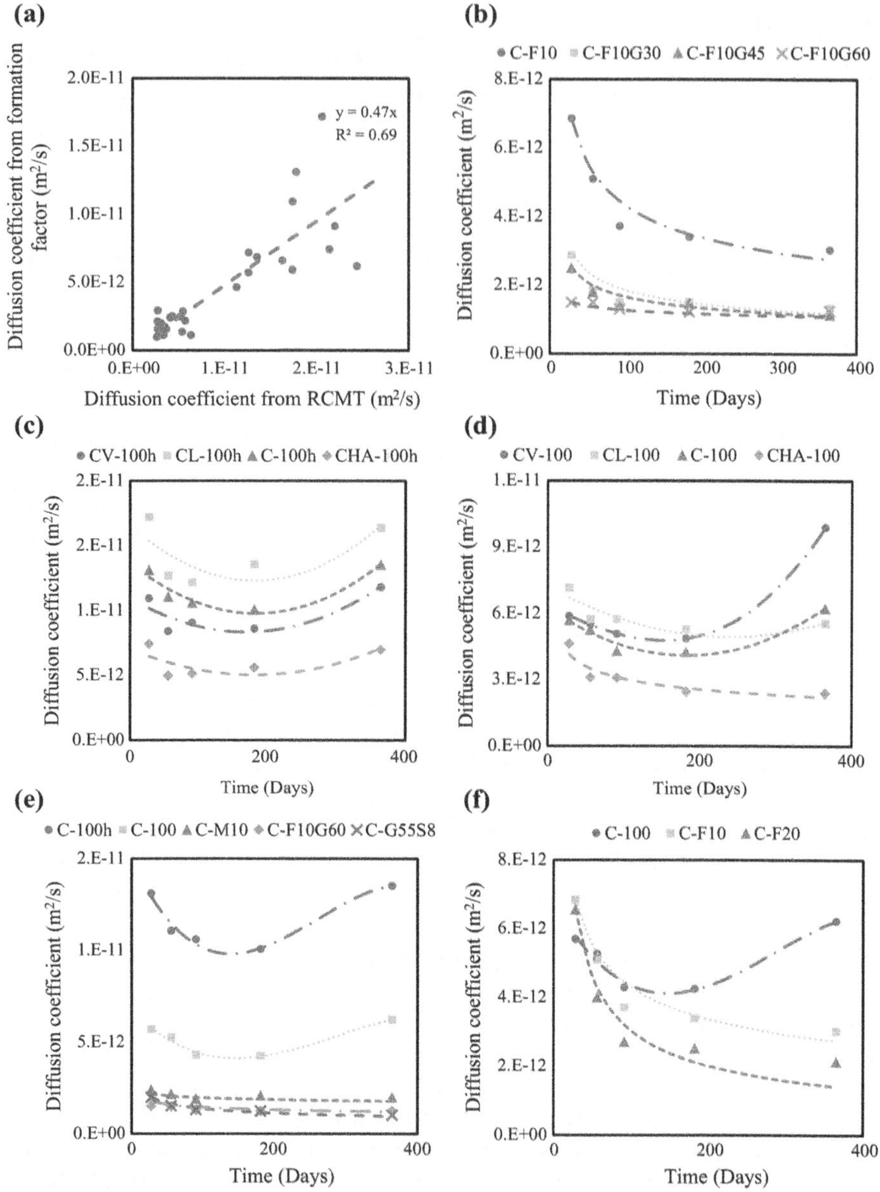

Fig. 1 Diffusion coefficients: (1): formation factor vs. RCMT, (2): slag replacement (3), (4): four types of cement high and low water-cement ratio respectively. (5): effect of different re SCM placement (6): effect of fly ash

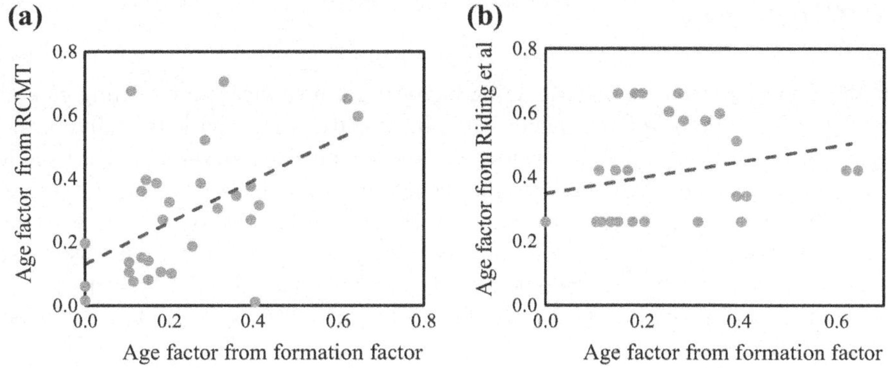

Fig. 2 Aging factor derived from **a** Rapid chloride migration vs aging factor from the formation factor, and **b** calculated from Eq. (7)

days. The ternary mixtures started from a low diffusion coefficient and the reduction in the diffusion coefficient was not significant compared to the initial value. Figure 1f shows the diffusion coefficient for concrete incorporating different replacement ratios of fly ash. Fly ash did significantly improve the diffusion coefficient as the concrete matured. The diffusion coefficient decreased up to 60% for concrete incorporating fly ash compared to control specimens at 365 days.

Table 1 shows the aging factors determined from the diffusion coefficient calculated from the formation factor and rapid chloride migration. Figure 2a shows the aging factors determined from rapid chloride migration vs. the aging factor derived from the formation factor. No relation was found, most likely because of chloride binding, different pore solution conductivities could influence the charge passed, and because the formation factor samples were not vacuum saturated, giving a lower degree of saturation and consequently lower diffusion coefficients.

The calculated aging factor determined from the experiments were compared to the aging factor from literature [1]. Equation (7) was used to determine the aging factor:

$$m = 0.26 + 0.4\left(\frac{FA}{50} + \frac{SG}{70}\right) \tag{7}$$

where *FA* is the percentage of fly ash and *SG* is the percentage of slag among the cementitious materials. Silica fume and metakaolin are assumed to not affect m [1]. Figure 2b shows the aging factor from the formation factor versus the aging factor from literature [1]. The results of this study did not compare well to the equation.

4 Conclusions

In this paper, it is shown that the aging factor used in service life modeling can be measured from the formation factor using the electrical resistivity. In addition, the aging factor was determined from the rapid chloride migration test for comparison. It was observed that a slightly higher aging factor was found with the formation factor, possibly because it did not include the effects of binding. Use of formation factor can be used to more easily quantify the concrete aging factor than currently used test methods, however caution should be used for mixtures without SCMs because of the effects the curing method can have on the results, giving very misleading diffusion coefficients.

References

1. Riding, K.A., Thomas, M.D., Folliard, K.J.: Apparent diffusivity model for concrete containing supplementary cementitious materials. ACI Mater. J. **110**, 705–714 (2013)
2. Mangat, P.S., Molloy, B.T.: Prediction of long term chloride concentration in concrete. Mater. Struct. **27**, 338 (1994)
3. Archie, G.E.: The electrical resistivity log as an aid in determining some reservoir characteristics. Trans. AIME **146**, 54–62 (1942). https://doi.org/10.2118/942054-G
4. Spragg, R., Qiao, C., Barrett, T., Weiss, J.: Assessing a concrete's resistance to chloride ion ingress using the formation factor. In: Corrosion of Steel in Concrete Structures, Elsevier, pp. 211–238 (2016)
5. Moradllo, M.K., Qiao, C., Isgor, B., Reese, S., Weiss, W.J.: Relating formation factor of concrete to water absorption. ACI Mater. J. **115** (2018). https://doi.org/10.14359/51706844
6. Bentz, D.P.: A virtual rapid chloride permeability test. Cem. Concr. Compos. **29**, 723–731 (2007). https://doi.org/10.1016/j.cemconcomp.2007.06.006
7. Riding, K.A., Ferraro, C.C., Almarshoud, M., Mosavi, H., Alrashidi, R., Alyami, M.H.: Durability Evaluation of Ternary Mix Designs for Extremely Aggressive Exposures (2018)
8. ASTM, C511-13: Standard Specification for Mixing Rooms, Moist Cabinets, Moist Rooms, and Water Storage Tanks Used in the Testing of Hydraulic Cements and Concretes (2013)
9. Spragg, R., Villani, C., Snyder, K., Bentz, D., Bullard, J., Weiss, J.: Factors that influence electrical resistivity measurements in cementitious systems. Transp. Res. Rec. J. Transp. Res. Board **2342**, 90–98 (2013). https://doi.org/10.3141/2342-11
10. AASHTO, TP 119-15: Standard Method of Test for Electrical Resistivity of a Concrete Cylinder Tested in a Uniaxial Resistance Test. American Association of State Highway and Transportation Officials, Washington, DC (2015)
11. N.T. Build, 443: Concrete, hardened: accelerated chloride penetration. Nordtest Method (1995)

Evaluation of Carbonation in Specimens Made with LC3 Low Carbon Cement

Yosvany Díaz Cárdenas, Ernesto Díaz Caballero and Jose Fernando Martirena-Hernandez

Abstract The durability is a parameter of great importance to be taken into account in reinforced concrete, this is affected by environmental agents such as CO_2, so the present work evaluates the carbonation behavior in hydraulic concrete specimens made with cement Low carbon content (LC3) with respect to Portland cement, located at three exposure sites (Punta Matamoros and University SEDE) with different environmental conditions, by performing the phenolphthalein test according to NC-355: 2004. To understand this phenomenon, other parameters are taken into account such as: the type and content of cement, the pore structure of the material and the curing time. In this way, it can be concluded that concrete made with LC3 and P-35 cement with a characteristic strength of 25 MPa, the former have a better behavior in tidal areas, reaching maximum values of 2.9 mm; This is because those produced with Portland cement have a greater amount of soluble compounds that leach with sea water and affect the pore structure of the same, removing compounds that provide alkalinity to the material. On the other hand, when evaluating the carbonation in places where the relative humidity varies between 50 and 70%, the structural elements produced with the ternary cement present a more unfavorable behavior, due to the percentage of substitution of clinker with the consequent reduction of the alkalinity, Reaching maximum values of 11 mm.

Keywords Carbonation · Cement · Pore structure · Durability · Concrete

Y. D. Cárdenas (✉) · J. F. Martirena-Hernandez
Center for Research and Development of Structures and Materials (CIDEM), UCLV, Santa Clara, Cuba
e-mail: yosvanyd@uclv.edu.cu

E. D. Caballero
Construction Research and Development Center (CIDC), La Habana, Cuba

© RILEM 2020
J. F. Martirena-Hernandez et al. (eds.), *Proceedings of the International Conference of Sustainable Production and Use of Cement and Concrete*, RILEM Bookseries 22,
https://doi.org/10.1007/978-3-030-22034-1_35

1 Introduction

Cement is one of the essential products for the manufacture of concrete. The contribution of concrete and therefore of this binder to construction makes its production a reflection of a country's economic activity. CO_2 emissions related to concrete correspond mainly to emissions associated with the production of cement, its main constituent. CO_2 emissions occur through the decarbonation of raw materials used in the cement manufacturing process and the use of fossil fuels as direct and indirect energy sources [1, 2].

The Portland cement clinker, which is the main component of this binder, is formed after calcining limestone and clay at a temperature between 1350 and 1450 °C [3] and some of the materials used as one of its substitutes and which react with calcium hydroxide are commonly called Cementitious Supplementary Materials (CSM) [4].

The amount of clinker that can be replaced by MCS depends on the type of material used. The CSM traditionally used are wastes from industrial processes, such as fly ash, microsilica and blast furnace slag, as well as natural pozzolans such as clays, zeolitized tuff and volcanic ash. The availability of industrial wastes is limited by the industrial development of each country and its future, conditioned by the efficiency of the processes that generate them, whose tendency points to the reduction of these wastes [5].

Clays, on the other hand, are minerals that are widely scattered throughout the earth's crust and can be activated by thermal processes that require much less energy than that demanded by the clinker. However, most of the studies reported in the literature refer to the calcination of pure kaolinite clays, commercially called metakaolin. The use of metakaolin as MCS greatly increases clinker replacement levels, but this type of mineral resource is highly demanded by the paper and ceramics industry, which is reflected in its high price [6].

With calcined clay it is possible to replace a large part of the clinker, almost doubling the amount of cement produced from the same amount of limestone, giving rise to the so-called Low Carbon Cements (LC3). The most common proportion of this binder are 50% Portland cement, 30% calcined clay and 20% limestone [7]. According to studies carried out by [8] this new cement makes it possible to reduce CO_2 emissions by 20–30% compared to traditional cement; a great reduction if we consider that cement is responsible for 5–8% of the emissions produced by human activity. It is a low-carbon, low-cost cement that achieves properties similar to or even superior to Portland cement.

One aspect to take into account for concrete and especially produced with low carbon cement LC3 is durability, where the Cuban standard (NC 250:2005) "Durability requirements for the design and construction of buildings and civil works of structural concrete", defines it as the capacity of the concrete to support during the useful life for which it has been designed, the physical and chemical conditions to which it will be exposed. The durability of concretes depends on different factors including: the water/cement (a/c) ratio, the curing process to which the material is subjected, the type of cement and aggregates used; as well as: the actions of chloride ions, porosity and carbonation [9], the latter as the focus of our research.

In most cases, the reaction of CO_2 with the alkaline components of the aqueous phase of the concrete creates a "carbonate front" which, when it reaches the steel, causes it to become depassive, causing widespread corrosion of the reinforcement. The rupture and destruction of the protective film of oxides implies the loss of the passivity and the increase in the speed of corrosion (NC250-2005). Carbonation is the second cause of corrosion in reinforced concrete structures, and in tropical non-marine environments it may be the main mechanism. Researches such as [10] confirm that the a/c ratio is one of the factors that directly influence the carbonation process. In this sense, it is known that the increase in the a/c ratio increases the depth of carbonation.

2 Discussion and Development

On the basis of the foregoing, this research proposes to evaluate the carbonation depth of concrete specimens made with LC3 and P-35, taking samples taken from two different areas of environmental aggressiveness: Punta Matamoros (Very high) and University SEDE (High) manufactured in Villa Clara. In both zones the results of exposure of concretes were evaluated and the influence of the porosity of the concrete before a phenomenon such as carbonation was determined.

2.1 Characteristics of the Assessed Concrete Samples

The dosages chosen for the concretes of this investigation respond to a design resistance of 25 MPa, with a ratio of coarse aggregate/fine aggregate of 57:43 and the inclusion of additives (MAPEI) to achieve a settlement in plant and on site of 12 ± 3 cm by the Cone of Abrams classified according to NC 120:2014 as fluid consistency. The following Tables 1 and 2 show the dosage and proportions adopted.

The amount of water and additive vary at the time of manufacture of each of the mixtures, taking into account that the mixtures respond to different types of concrete, using different cements. In order to achieve the established consistency, a minimum effective a/c ratio of 0.4 and an additive percentage of 0.5% is set as the starting point, as recommended by the manufacturers of this product.

2.2 Experimental Procedure

The experimental work was developed on a laboratory scale in which hydraulic concrete cores were evaluated, manufactured with low carbon cement (LC3) and

Table 1 Mix design used in elements of Punta Matamoros (very high aggressiveness)

Ident.	Dosif.	Resist. (MPa)	Cem. (kg/m^3)	a/c	Sand	Coarse	Super plasticizer (L/m^3)	Date
M32	D10-P35	25	300(P35)	0,40	Palenque	Palenque	2.8	10/2/2014
M25	D4-LCC	25	360 (LC^3)	0,47	Palenque	Palenque	4.04	6/2/2014
M381	DA-LCC	25	360 (LC^3)	0,47	Palenque	Palenque	4.04	2/11/2013

Table 2 Mix design used in elements of University SEDE (high aggressiveness)

Ident.	Resist. (MPa)	Cem. (kg/m^3)	a/c	Sand	Coarse	Date
M2	25	464 (LC^3)	0.45	Purio	Purio	6/5/15
M6	25	411 (LC^3)	0.45	Purio	Purio	27/5/15
M8	25	470 LC^3)	0.45	Purio	Purio	29/5/15
M11	25	421 (LC^3)	0.45	Purio	Purio	1/6/15
P	25	445 (P35)	0.45	Purio	Purio	13/6/15

P-35 as a reference, these were extracted from two zones with different levels of aggressiveness (Very high and High). The dependent variables that were evaluated are: carbonation depth, porosity and resistivity established in NC 355:2004 and NC 345:2011. The following sequence was established as an experimental plan:

1. Witness extraction of the structural elements exposed in the Site of Punta Matamoros and University SEDE subjected to environments of very high and high aggressiveness respectively.
2. Preparation of the sample for the tests of carbonation and water absorption by capillarity, air permeability and resistivity in the Faculty of Constructions of Villa Clara (FC).
3. Application of established methods to determine carbonation depth, porosity and resistivity.

3 Discussion of Results

In order to understand the phenomenon of carbonation, these results are correlated with the tests for air permeability (Torren Method), resistivity (Recipod Proceq) and water absorption by capillarity (NC 45:2011) of these samples in 2017.

3.1 Analysis of Carbonation in Concrete Specimens from Punta Matamoros

If we look at Fig. 1 we can see that samples M25 and M381 (LC3 cement) have lower carbonation values than M 32 (P-35 cement). This result tends to be atypical considering that the alkalinity index of concretes produced with low carbon cement is lower.

In order to understand this behaviour, it is necessary to take into account the main factors that may influence this phenomenon, which are described below.

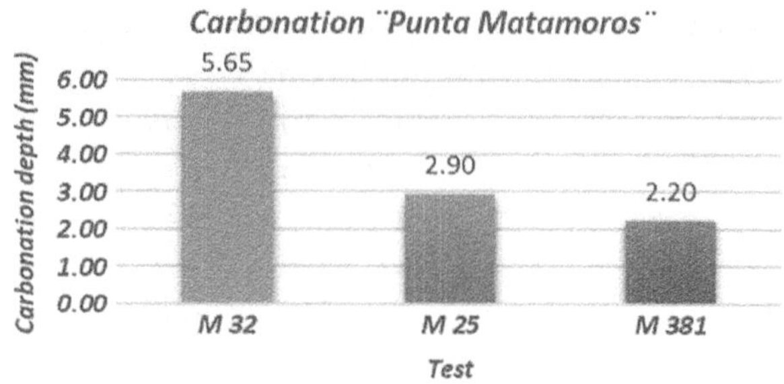

Fig. 1 Carbonation profile of Punta Matamoros

3.2 Influence of Type and Content of Cement on Carbonation

With the objective of achieving a characteristic resistance of 25 MPa in concretes made with a low carbon LC3 cement, a cement content of 360 kg/m^3 was used, with a smaller quantity being used in concretes made with Portland P-35 cement (300 kg/m^3). With the use of a greater quantity of LC3 cement content in the mixture with respect to the content of Portland cement, it is not possible to increase its basicity content, but it does considerably improve the impermeability of the material. This difference in cement content does not represent an express economic disadvantage [11].

Authors such as [11, 12] agree that concretes made with ternary cements are more susceptible to carbonation due to the reduction of the alkaline content. However, in the author's opinion, for the unexpected behaviour of the carbonation depth in this area, the alkaline content does not represent a significant parameter in comparison to the tidal exchange environmental variable.

3.3 Influence of the Environmental Parameter of the Exposure Site

An environmental characteristic to which these specimens are exposed is that they are found in a tropical marine zone, where the constant effect of the tides causes the relative humidity to be very high achieving the saturation of the pores. These conditions make it difficult to transport CO_2 molecules inside the element, which represents a lower aggressiveness to the phenomenon of carbonation in concrete [13, 14], compared to other types of environments.

The atypical behaviour of carbonation may be due to the fact that, in this area, the concretes spend most of the day under water and the high $CaOH_2$ content of elements made of Portland cement has a high solubility, so it may tend to leach, a factor that induces negative indices of porosity during the drying period (low tide), and it is at this moment of the day that CO_2 diffusion occurs inside the structure. However, concretes produced with LC3 low carbon cement are advantageous as the additions of

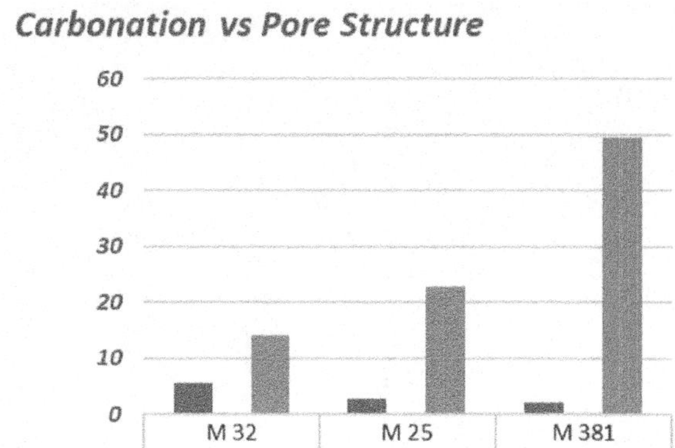

	M 32	M 25	M381
Carbonatación(mm)	5.65	2.9	2.2
Permeabilidad(mm^2x10^{-16})	0.021	0.015	0.0042
Resistividad (kΩ)	14.1	22.9	49.4

Fig. 2 Carbonation versus pore structure, Punta Matamoros

this cement are combined with the reaction products of Clinker to produce insoluble compounds [11].

3.4 Influence of Pore Structure

As has been well explained in the literature, the porous structure of the concrete specimen plays a very important role in the diffusion of CO_2 molecules into the interior of the material, in Fig. 2 the behaviour of carbonation is plotted in comparison with resistivity and permeability.

As can be seen in the graph above, the air permeability index is higher in M32 than in ternary cements, the corrosion rate in M25 and M381 samples is low while that of Portland cement is high, implying that the pore structure of the material is more tortuous in the first case than in the second, if we observe the resistivity values in Portland cement is lower than those of LC3 cements, implying that the former have greater connectivity between pores than the latter.

In this way, it is concluded that concretes made with low-carbon cement have a better behaviour in terms of the microstructure of the material than that of Portland cement, due not only to the advantages provided by the pozzolanic material to the ternary cement but also to the disadvantage caused by the leaching of M32 already explained above.

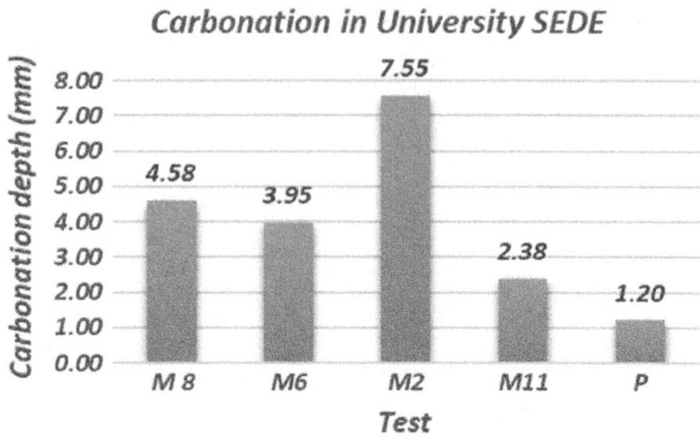

Fig. 3 Carbonation depth in concretes exposed at the University SEDE

3.5 *Analysis of Carbonation in Concrete Specimens from the University SEDE*

Figure 3 shows the carbonation depth of the concrete blocks exposed at the University Campus with high aggressiveness. In this we can note that, unlike the site, Punta Matamoros is the representative sample P of a Portland cement P35 which has a better behavior to the phenomenon of carbonation than the other samples of LC3 (M2, M6, M8, M11).

3.6 *Influence of Type and Content of Cement on Carbonation*

If we examine Table 3, note that in order to reach a characteristic resistance of 25 MPa, the cement content in sample P was higher than in concretes produced with LC3, also guaranteeing the required resistance in the precise time. Although the cement content is greater, the difference is not so considerable [11] when using the same content of LC3 and Portland Cement P-35 in concretes similar resistant properties are achieved, and, although the remaining amount of calcium hydroxide is sufficient to maintain the conditions of high alkalinity inside the concrete, the reserve of alkaline is reduced being more unfavorable for the first case that for the second one as for the depth of carbonation.

It can be observed that, although the percentage of clinker substitution is equal to that of M6, the proportion of pozzolanic material is much higher, so that the reactions of this material with the low percentage of hydroxide further reduce the alkaline content of the concrete, being more susceptible to the phenomenon analysed.

Table 3 Proportions of raw materials used in University SEDE elements

Sample	Characteristics	Clinker (%)	CCL (%)	Limestone	Limestone extra	Gypsum
M2	50% clínquer, 2:1	50	29	14	0	7
M6	50% clínquer, 1:1	50	22	11	11	7
M8	65% clínquer, 2:1	65	19	10	0	6
M11	56% clínquer, 1:1	65	10	10	10	6
P		89	0	0	5	6

3.7 Influence of the Environmental Parameter of the Exposure Site

As we saw in the previous section, the environmental factor plays a decisive role in the behavior of the phenomenon we are dealing with, in this case the specimens analyzed are not affected by the effect of the tides, nor in the continuous wetting and drying that this implies, so that the dissolution of Portlandite is not a parameter that affects in this case.

In question, the relative humidity percentage is between 60 and 70% values suitable for the reactions necessary for the specimens to carbonate [10, 15]. It can be concluded, at the author's discretion, that the ideal conditions exist regarding the humidity of the samples and the susceptibility of specimens made with low-carbon cement, and the behavior shown in Fig. 3 is normal.

3.8 Influence of Pore Structure

As shown in Fig. 4, the porosity and permeability to air in the P sample is greater than that of concretes made with low-carbon cement, suggesting that the material has a higher percentage of effective vacuum volume than in LC3 samples. However, factors such as the connectivity between these pores and the tortuosity in the material can be inferred with resistivity [16, 17].

Note that the lowest value of electrical resistivity is found in P-35 samples where the probability of corrosion is higher than for LC3 samples, implying the high connectivity that exists between the pores of this material.

If we analyze the phenomenon of resistivity among the specimens elaborated with LC3 we will notice that the M2 sample presents a low corrosion probability, so the pore structure in this sample is very good, so we can infer that the high value of carbonate depth in this specimen is due to the low alkalinity index.

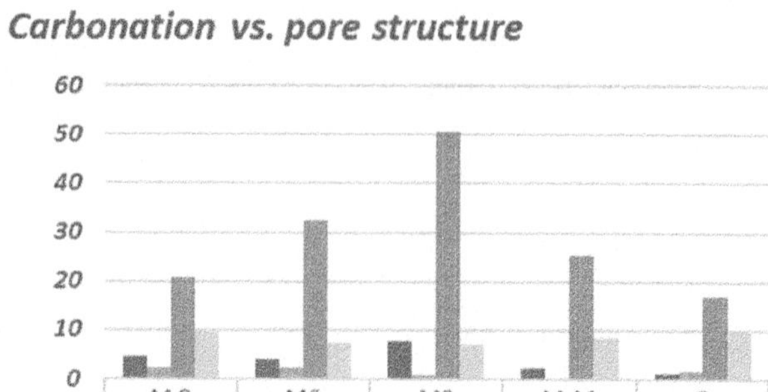

	M 8	M6	M2	M 11	P
Carbonatación(mm)	4.6	4	7.6	2.4	1.2
Permeabilidad(mm²x10⁻¹⁶)	2.4	2.4	0.96	0.38	1.6
Resistividad (kΩ)	20.7	32.5	50.7	25.2	17.1
Porosidad(%)	9.6	7.3	7.1	8.4	10

Fig. 4 Comparison between carbonation and pore structure, University SEDE

4 Conclusions

1. Concretes made with LC3 cement have a good behaviour against carbonation in tidal areas compared to those made with P-35 cement, this is due to the fact that the latter has a greater quantity of soluble compounds that leach with seawater and affects the pore structure of the same, removing compounds that provide high alkalinity to the material.
2. On the other hand, the shallower carbonation depths in places where relative humidity varies between 50 and 70% are found in concretes made with Portland cement, the alkaline content being the parameter that determines this trend.
3. The elements made with LC3 located in areas of medium RH are much more susceptible to the effect of carbonation even when higher impermeability and porosity indices are guaranteed, mainly due to the low alkalinity index in the pores of the elements, a primordial factor in the effects of carbonation.
4. The advance of carbonation in concretes manufactured with LC3 in the first years of exposure is abrupt and as time goes by the growth rate decreases, inferring that it reaches constant values of carbonate depth.

References

1. Gartner, E.M., MacPhee, D.E.: A physico-chemical basis for novel cementitious binders. Cem. Concr. Res. **41**(7), 736–749 (2011)
2. Habert, G., Choupay, N., Montel, J.M., Guillaume, D., Escadeillas, G.: Effects of the secondary minerals of the natural pozzolans on their pozzolanic activity. Cem. Concr. Res. **38**(7), 963–975 (2008)

3. Damidot, D., Lothenbach, B., Herfort, D., Glasser, F.P.: Thermodynamics and cement science. Cem. Concr. Res. **41**(7), 679–695 (2011)
4. Justice, J.M., et al. (2005) Comparison of two Metakaolins and a silica fume used as supplementary cementitious materials. In: Seventh International Symposium on Utilization of High-Strength/High Performance Concrete, to be held in Washington, DC, pp. 1–88 (2005)
5. Tironi, A., Castellano, C.C., Bonavetti, V.L., Trezza, M.A., Scian, A.N., Irassar, E.F.: Kaolinitic calcined clays—Portland cement system: hydration and properties. Constr. Build. Mater. **64**, 215–221 (2014)
6. Sabir, B., Wild, S., Bai, J.: Metakaolin and calcined clays as pozzolans for concrete: a review. Cem. Concr. Compos. **23**(6), 441–454 (2001)
7. Martirena, F., Scrivener, K.: Low Carbon Cement LC3 in Cuba: Ways to Achieve a Sustainable Growth of Cement Production Identification of Suitable Clay Deposits, vol. 2, pp. 3–6 (2018)
8. Martirena, P.F.: LC3. The cuban experience (2017)
9. Tang, S.W., Yao, Y., Andrade, C., Li, Z.J.: Recent durability studies on concrete structure. Cem. Concr. Res. **78**, 143–154 (2015)
10. Baojian, Z., Chisun, P., Caijun, S.: Cement & concrete composites CO_2 curing for improving the properties of concrete blocks containing recycled aggregates. Cem. Concr. Compos. **42**, 1–8 (2013)
11. Andrés, L.M.V.: Cemento de bajo carbono a partir del sistema cementicio ternario clinquer—arcilla calcinada—caliza. Editorial Universitaria (2015)
12. Gettu, P.R. et al.: Life cycle assessment based on CO_2 emissions and energy calculations incorporation of service life in the sustainability estimation. (2017)
13. Young, J.F., Berger, R.L., Breese, J.: Accelerated Curing of Compacted Calcium Silicate Mortars on Exposure to CO, pp. 394–397 (1974)
14. Xu, J., Fan, C., Teng, H.H.: Calcite dissolution kinetics in view of Gibbs free energy, dislocation density, and pCO2. Chem. Geol. **322–323**, 11–18 (2012)
15. Borges, P.H.R., Milestone, N.B., Costa, J.O., Lynsdale, C.J., Panzera, T.H., Christophoro, A.L.: Carbonation durability of blended cement pastes used for waste encapsulation. Mater. Struct. Constr. **45**(5), 663–678 (2012)
16. Farnam, Y., Washington, T., Weiss, J.: The influence of calcium chloride salt solution on the transport properties of cementitious materials. Adv. Civ. Eng. **2015**(1), 1–15 (2015)
17. Helene, P., Medeiros, M.H. F., Real, L.V., Quarcioni, V.A.: Concreto con protección de superfície y expuesto à soluci6n de cloruros: Espesor de recubrimiento equivalente. Rev. ALCONPAT, vol. 5, pp. 219–234 (2015)

Deterioration of Structures Affected by Concrete Leaching

Alberto Hernández Oroza and Dainelys Guerra Bouza

Abstract The present work reflects the results obtained on the study of the concrete leaching in beams and slabs in real structures. Leaching is a phenomenon that promotes the migration of ions from the solution of the pores to the outer side of the structural elements, lowering the pH value of the pore solution and accelerating the corrosion rate of the bars. The objective of this study is to demonstrate how this phenomenon has a remarkable impact on the durability of structures. The investigations were carried out in existing buildings after 5 years of exposure. To evaluate the deterioration rate of the buildings, techniques of polarization resistance, electrical resistivity of concrete and corrosion potentials were applied. Calculation of the remaining section of the reinforced was perform from the direct measurement of the section loss on the steel bars. The research concludes by demonstrating the negative effect of leaching on the durability of the structures.

Keywords Concrete leaching · Durability · Corrosion

1 Introduction

Reinforced concrete structures are conceived to be durable. However, as time progresses, they begin to show signs of degradation in the early stages, mainly caused by exposure to aggressive environments and substances [1]. The sources of deterio ration that have the greatest impact on the durability of the structures are exposure to wetting and drying cycles, strength and direction of horizontal wind flow, thermal variations, chemical composition of soil and water, and the matrix of atmospheric pollutants present in the air [2].

A. H. Oroza (✉) · D. G. Bouza
Department of Diagnostic and Surveying, Restaura, Office of the Historian of Havana, Havana, Cuba
e-mail: ahernandez@proyectos.ohc.cu

D. G. Bouza
e-mail: dguerra@proyectos.ohc.cu

© RILEM 2020

J. F. Martirena-Hernandez et al. (eds.), *Proceedings of the International Conference of Sustainable Production and Use of Cement and Concrete*, RILEM Bookseries 22,
https://doi.org/10.1007/978-3-030-22034-1_36

During a prolonged exposure to environments of high relative humidity, or cycles of systematic wetting and drying, the water, with less ionic charge, begins to percolate through the capillary pores of the concrete. As it passes through the mass of the material, it leaches all the soluble salts of calcium (Ca^{2+}), sodium (Na^{+}) and potassium (K^{+}), until reaching the exposed face of the structural element. Once this happens, the salts react with the carbon dioxide (CO_2) in the air and are transformed to carbonates of calcium, sodium and potassium ($CaCO_3$), ($NaCO_3$) and (KCO_3). This process of chemical deterioration, known as leaching, gradually leads to the loss of the primary protection of the cement, decreases the pH of the medium, increases porosity and permeability, and facilitate the entry of atmospheric pollutants into the interior of the concrete mass [3, 4]. Once the reinforcing steel is unprotected from the alkaline medium, it is fully exposed to initiate the phenomenon of corrosion. The objective of this work is to evaluate the effect of leaching on the corrosion of reinforcing steel in two real structures exposed during 5 years to different environmental conditions. For the evaluation, the techniques of polarization resistance, half-cell corrosion potential and electrical resistivity of concrete were used. The section loss of the reinforcing steel was evaluated by comparison with the nominal section of the rebar. The study showed how the deterioration of the concrete by leaching causes high corrosion rates and a drastic loss of the durability of the structures, which evidences the need to consider in the project, the effect of the leaching according to the class of exposure to extend the service life of reinforced concrete structures.

2 Methodology

2.1 *First Case Study*

The first case study is a building built in the late 1950s. It consists of a rigidly framed system and reinforced concrete slabs. The light of the beams is 6 m in length. The tests were carried out on 8 beams of the intermediate level, isolated from the external environment where an uninterrupted air injection system with temperatures between 22 and 25 °C and relative humidity (RH) between 60 and 80% was installed.

2.2 *Second Case Study*

The second building studied is a metallic structure from the mid-nineteenth century, which currently serves as a cultural fair hall. The property is located in an area of urban/industrial classification [5]. For the restoration project of the roof, it was build slabs of reinforced concrete with 12 cm thickness, strength of 25 MPa and as a reinforced was used 9.5 mm diameter bars spacing are at 25 cm in both directions.

The tests was carried out on a section affected by leaching and corrosion of the bars, in a determined space of 5 × 2 m.

2.3 *Measurement of Polarization Resistance*

To evaluate the corrosion rate (Icorr), an electrochemical measurement was used by means of the GECOR 8 corrosion meter. For the classification of the level of risk the ASTM Standard was applied [6], allowing to determine the probability of corrosion according to the value of the potential (Ecorr).

2.4 *Electrical Resistive of Concrete (ρ)*

For reinforced concrete slabs, it was not possible to use the polarization resistance technique using GECOR 8, due to the impossibility of confining the current. To estimate the corrosion of the reinforcing steels, the technique of electrical resistivity was used, which works based on the Wenner method or the four points. The calculation of the Icorr was made using the equation proposed by RILEM [7].

2.5 *Corrosion Potential*

The method of evaluation of corrosion through the half-cell potential is a technique that allows to measure the possible corrosive activity from the Ecorr value obtained. When Ecorr < 350 mv versus CSE, indicates that there is a probability greater than 90% that there is corrosion of reinforcing bars. Conversely, if the Ecorr > −250 mv versus CSE, the probability is less than 10%. The equipment used for the measurements was Proceq Canin+. This test method was applied to the slabs, the measured spaces were 2 × 1 m and the fixed grid was 50 × 50 cm.

3 Results and Discussion

3.1 *First Case Study*

In beams exposed to the humidity-controlled environment, after 5 years of exposure, the appearance of efflorescence spots began on the lateral faces, mostly on the upper half. Coinciding with the location of the steel bars, longitudinal cracks formed that extended into almost all the light of the beams. Figure 2 shows the results obtained

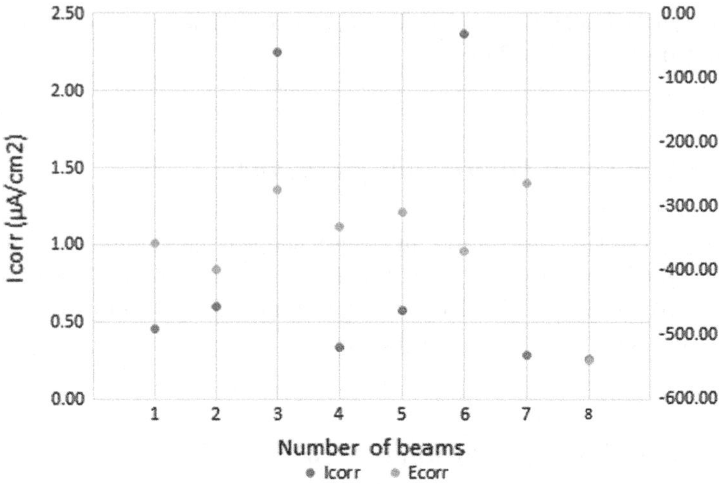

Fig. 1 Results of Icorr versus Ecorr obtained in beams

Fig. 2 General data obtained from Icorr versus Ecorr

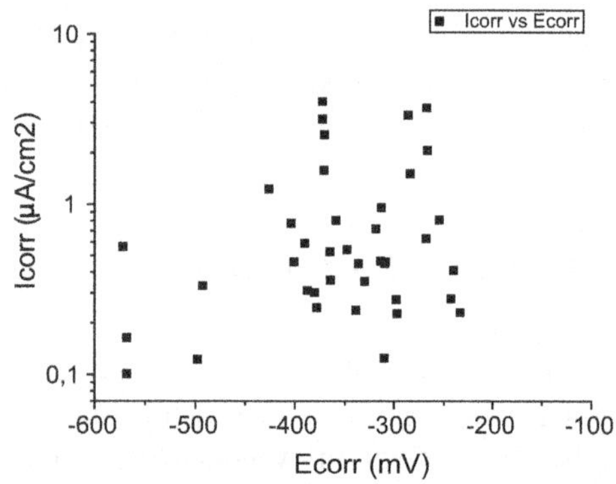

Fig. 3 Deterioration state of the slab

for Icorr and Ecorr. Analyzing the Ecorr obtained (Fig. 3), it is observed that the largest set of values is found with Ecorr between −200 and −400 mV, while the Icorr oscillate between 0.5 and 2.0 μA/cm^2 indicating an active corrosion process.

Table 1 Results of section loss values

Beam	Bar no.	Measured diameter (mm)	Remaining area (mm^2)	Nominal area (mm^2)	Loss section (%)
1	1	23.50	433.74	510	15.0
	2	23.60	437.44	510	14.2
2	1	23.90	448.63	510	12.0
	2	24.10	456.17	510	10.6
3	1	24.15	458.06	510	10.2
	2	22.86	410.43	510	19.5
4	1	20.90	343.07	510	32.7
5	1	20.80	339.80	510	33.4
6	1	21.10	349.67	510	31.4
7	1	23.80	444.88	510	12.8
8	1	20.50	330.06	510	35.3

The migration towards the outer faces of the cement alkalis produces a decrease in the pH of the material that covers the bars, and the necessary primary protection that protects the bars from corrosion is lost. The high RH of the medium causes an increase in the moisture content of the beams, which leads to an increase in the electrochemical activity and the Icorr of the bars. The data obtained show that there is a correspondence between the predicted corrosion based on the measurement of potentials and the kinetics of the process, from the obtained Icorr. The sections measured (Table 1) directly on the reinforcements reflect the existence of a mass loss corresponding to the dissolution and transformation of the steel to the corresponding hydroxides of the corrosion product. In all the reinforced steels measured, the section loss was greater than 10%.

3.2 *Second Case Study*

After 5 years of exposure to the elements, the underside of the slab shows stains of efflorescence caused by leaching with rainwater and exposure to wetting and drying cycles. Due to this element is exposed to an urban environment, it is also under the influence of carbonation, which by acting simultaneously with leaching leads to an increase in the loss of alkalinity and the increase of the porosity of the material. Compared to the first case study, the slabs were more affected with large areas with loss of concrete coating, corroded and fractured steels, concrete unbundling and formation of $CaCO_3$ stalactites (Figs. 3, 4 and 5).

The potential map (Fig. 6) reflects a slab with high electrochemical activity, caused by the water content in the concrete mass of the slab. The highest results were obtained in the area of active corrosion (Ecorr < -350 mV). The resistivities recorded are in

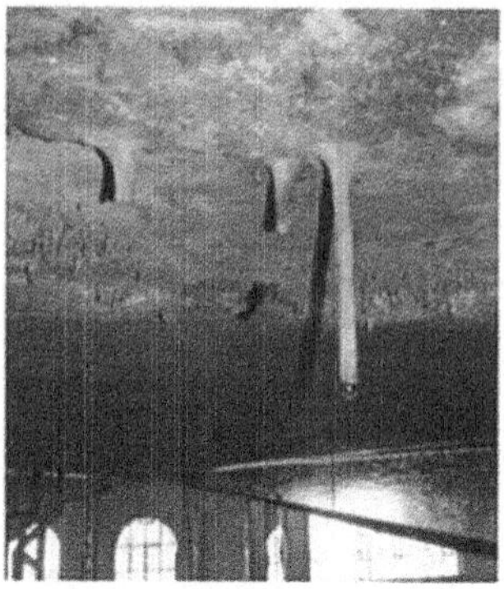

Fig. 4 Formation of $CaCO_3$ stalactites

Fig. 5 Corrosion and fracture of reinforced bars

Table 2 Calculated values of Icorr and Vcorr from measured ρ

ρ (KΩ/cm)	Icorr (μA/cm²)	Vcorr (mm/year)
12.4	2.42	0.028
10.4	2.88	0.033
8.7	3.45	0.040
32.2	0.93	0.011
29.8	1.01	0.012
29.7	1.01	0.012
19.5	1.54	0.018
29.0	1.03	0.012
28.1	1.07	0.012
19.6	1.53	0.018

correspondence with the values of Ecorr, reflecting a favorable environment for the corrosion of the steel bars (Fig. 7). Table 2 shows the results of the calculation made for Icorr and in Fig. 8 the diagram obtained from the relation Icorr versus ρ. The sections measured in the bars demonstrate an active corrosion process with losses greater than 20% for most of the cases (Table 3).

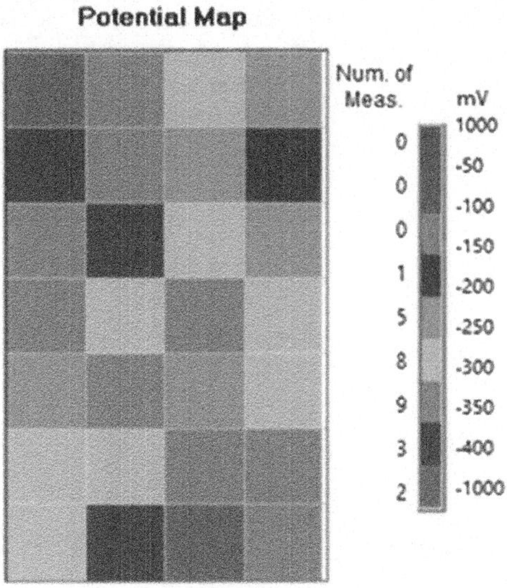

Fig. 6 Potential mapping of an area affected by leaching

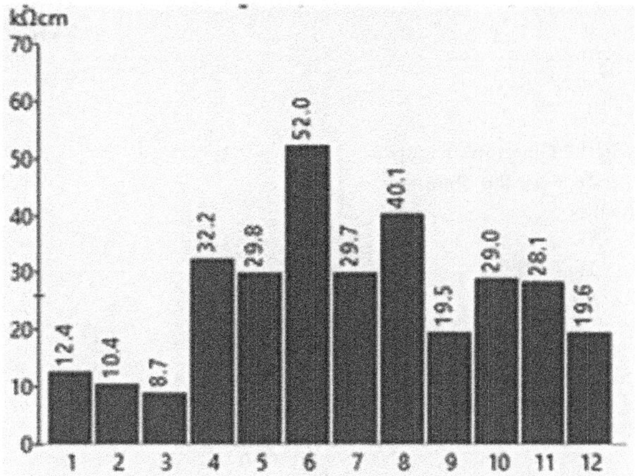

Fig. 7 Concrete resistivity of an area affected by leaching

4 Conclusions

In the structures studied, the deterioration resulting from the 5 years of exposure was reflected in high corrosion rates, loss of concrete cover and efflorescence stains caused by concrete leaching. The effect of humidity on concrete is a factor that must always be taken into account from the design stage to guarantee the service life of the structure. The combined effect of carbonation and leaching significantly increases the rate of deterioration of cement-based materials and the corrosion of the bars, due to a drastic decrease in the alkalinity and an increase in the porosity of the material.

Table 3 Section loss of reinforcing bars of the slab

Axis	Bar no.	Nominal diameter (mm)	Nominal area (mm^2)	Measured diameter (mm)	Resildual area (mm^2)	Section loss (%)
10-11:E-F	1	9.5	71.0	7.21	40.8	42.5
	2			8.11	51.6	27.3
	3			8.43	55.8	21.4
	4			8.72	59.7	15.9
	5			5.96	27.9	60.7
12-13:C-D	1			7.83	48.1	32.2
	2			7.17	40.4	43.2
	3			7.64	45.8	35.5
	4			7.77	47.4	33.2
	5			7.18	40.5	43.0
22-23:C-D	1			8.50	56.7	20.1
	2			9.00	63.6	10.4
	3			8.70	59.4	16.3
	4			8.50	56.7	20.1
	5			9.00	63.6	10.4

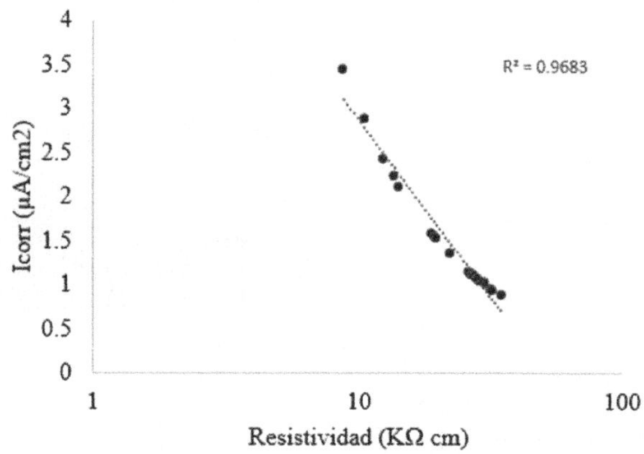

Fig. 8 Diagram of Icorr versus ρ for the obtained results

References

1. CYTED: Manual de rehabilitación de estructuras de hormigón, reparación, refuerzo y protección. In: Red temática XV.B (2003)
2. Brimblecombe, P., Grossi, C.M.: Potential damage to modern building materials from 21st century air pollution. Sci. World J. **10**, 116–125 (2010)
3. Duong, V., Sahamitmongkol, R., Tangtermsirikul, S.: Effect of leaching on carbonation resistance and steel corrosion of cement-based materials. Constr. Build. Mater. **40**, 1066–1075 (2013)

4. Hartwich, P., Vollpracht, A.: Influence of leachate composition on the leaching behaviour of concrete. Cem. Concr. Res. **100**, 423–434 (2017)
5. ISO-9223: Corrosion of metals and alloys. Corrosivity of atmospheres. Classification, determination and estimation. International Standard Organization (ISO), pp. 1–22 (2012)
6. ASTM C876: Standard test method for corrosion potentials of uncoated reinforcing steel in concrete. In ASTM C876-09 (2009)
7. Andrade, C., Alonso, C.: RILEM TC 154-EMC: electrochemical techniques for measuring metallic corrosion. Test methods for on-site corrosion rate measurement of steel reinforcement in concrete by means of the polarization resistance method. In: Materials and Structures, pp. 623–643 (2004)

Atmospheric Corrosion Study of Carbon Steel in Havana Waterfront Zone

Cecilia Valdés, Abel Castañeda, Francisco Corvo, Rigoberto Marrero and Ricardo Montero

Abstract Havana City is located along Cuban North-Western shoreline. Many old building structures are located in the waterfront zone to the West of Havana Bay. This zone is defined as "Traditional". An atmospheric corrosion study of carbon steel CT-3 in the Havana Traditional waterfront zone was carried out. The main aim is to classify the atmospheric corrosivity respecting carbon steel, the main metallic material used in the construction industry. Monthly and cumulative behavior of steel corrosion rate was analyzed, as well as the factors influencing on the kinetics. The influence of climatic factors on chloride and sulfur compounds deposition was also analyzed. Four exposure sites were placed at the same distance from the sea. The results obtained will allow establishing the suitable selection of the primary and secondary protection systems to the structures that are intended to be restored and build in the study area.

Keywords Atmospheric corrosion aggressiveness · Corrosion rate · Factors

1 Introduction

When performing atmospheric corrosion studies, the classification of corrosivity of the atmosphere is only a technical characteristic from the engineering point of view. It provides the basic information toward selecting building materials and protection systems suitable for structures. On the other hand, these results can be extrapolated to the execution of maintenance work, especially related to repair structures located in coastal tropical climate, classified as highly corrosive atmosphere.

C. Valdés (✉) · A. Castañeda · R. Montero
Materials Protection Group, National Center for Scientific Research, Playa, Havana, Cuba
e-mail: cecilia.valdes@cnic.edu.cu

F. Corvo
Corrosion Research Center, Autonomous University of Campeche, Campeche, Mexico

R. Marrero
Technological University of Havana "José Antonio Echeverría", Marianao, Havana, Cuba

© RILEM 2020

J. F. Martirena-Hernandez et al. (eds.), *Proceedings of the International Conference of Sustainable Production and Use of Cement and Concrete*, RILEM Bookseries 22,
https://doi.org/10.1007/978-3-030-22034-1_37

Fig. 1 Havana waterfront: **a** calm sea; **b** waves during the entry of a cold front

Numerous atmospheric corrosion studies have been carried out in Cuba since the 1980s to the present [1, 2]. However, the Havana Waterfront area has not been subject to this type of research. It is worth nothing that these investigations are based on the determination of the corrosive aggressiveness of a given area by selecting several exposure sites at different distances from the coastline. In the present investigation, all sites were placed at the same distance from the sea, because it is a long and narrow area.

The "Havana waterfront zone" (see Fig. 1) is constantly affected by heavy weather (storms, hurricanes, cold fronts, etc.) and its integrity depends on the surrounding buildings, as well as the welfare of its inhabitants. Therefore, the aim of this paper is to study the atmospheric corrosion aggressiveness for the main metallic materials most used in the construction industry in the Havana waterfront "Traditional Zone". The application of the results will guarantee the useful life of the structures will be extended to 50 years.

2 Experimental Part

2.1 Test Sites

Four outdoor exposure sites in the Traditional waterfront were located at a distance of 20 m from the sea (m) (S1, S2, S3 and S4). A long road at the border of the sea called "Havana Malecon" is the first limit between buildings and the sea. The section built from the Paseo del Prado to Belascoaín Street is known as "Traditional waterfront" (see Fig. 2a). The sites were placed from Castillo de la Punta to the vicinity of Antonio Maceo Park (see Fig. 2b).

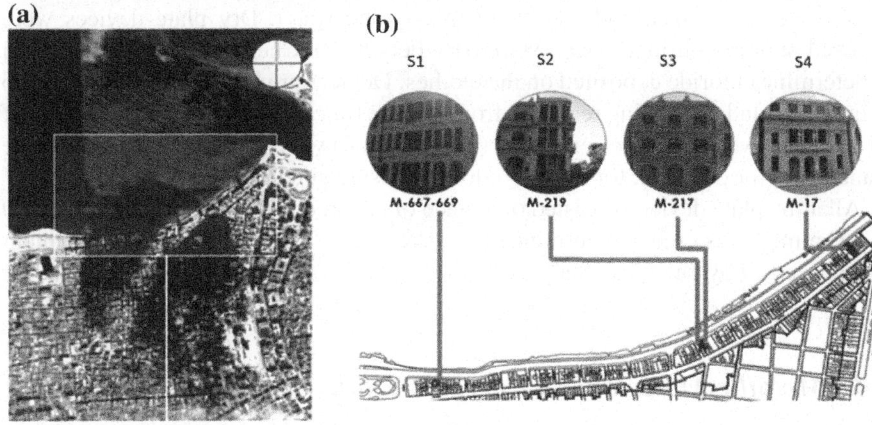

Fig. 2 Havana waterfront: **a** traditional zone; **b** exposure sites

Devices for determination of atmospheric pollutants: chloride deposition (Cl^- *DR*) and sulfur compound deposition (SO_x^- *DR*); [3] together with carbon steel (LCS) specimens (150 × 100 × 0.3 mm) to determine the corrosion rate, were placed in wooden racks on each outdoor exposure site [4]. Dry plate for Cl^- *DR* determination and cellulose filter plates for determination of SO_x^- *DR*. Wooden racks supporting the pollutant collectors were placed under cover to avoid the effect of washing due to rain.

2.2 Meteorological Parameters

Relative humidity (RH), temperature (T) and wind speed (W) data were provided by the Meteorology Institute (INSMET). RH and T data collected during 1 year of study were used to calculate time of wetness (TOW or τ (h y^{-1})) according to ISO standard [4]. TOW was estimated according to ISO 9223 as the time during which RH is over 80% and Temperature is above 0 °C (temperature below zero is never reached in Cuban insular tropical climate).

TOW amount indicates the number of hours per year during which metallic surface should be covered by the water film. However, it is only approximated because under relatively high temperature, evaporation may increase and real TOW decreases.

2.3 Outdoor Pollutants

Cl^- *DR* (mg m^{-2} d) was determined using dry plate device, it consisted of a piece of absorbent cloth sizing 320 × 220 mm. Two pieces of dry plate device (cloth)

were perfectly cleaned and washed with distilled water. Dry plate devices were exposed monthly during 1 year (Mar/2016–Feb/2017). Chemical analysis was used to determine chloride deposited on the clothes. Devices are kept in plastic bags up to of exposure and after being removed from the site for chemical analysis. Two data of Cl^- *DR* for every month of exposure (in the period) were obtained. One for device 1 and other for device 2, for every outdoor exposure site (total = 24).

Alkaline plate device consisted of a piece of absorbent cellulose paper sizing 150 × 100 mm. It was used to determine SO_x^- *DR* (mg m^{-2} d) [3]. Data of this pollutant was processed by chemical analysis.

2.4 Metallic Material and Corrosion Rate

Metallic material exposed (LCS): Carbon steel (C: 0.13–0.2%, Mn: 0.39%, Al: 0.02%, S < 0.05%, and $P < 0.07\%$).

Chemical cleaning previous exposure of samples and procedure for removing corrosion products after exposure to determine corrosion rate (mass loss per area over time) were carried out according to established standard procedures [5, 6]. Specimen's thickness was considered equal and constant. Three samples were exposed for each site and exposure period. Exposure period started on March/2016 up to February/2017.

Determination of the atmospheric corrosion rate (*r*corr g m^{-2}) was carried out and monthly behavior was analyzed, as well as its cumulative behavior over time period of 3, 6, 9 and 12 months. Furthermore, annual data was used to determine corrosive aggressiveness according to the classification intervals required in the ISO standard.

3 Results and Discussion

3.1 Meteorological Parameters Behavior

The behavior of RH–T complex and WS (monthly average) was characterized during the year of study (see Fig. 3). Monthly relative humidity (RH) was always over 70% and the annual average was 75%. Monthly temperature was over 25 °C, with an annual average of 26.6 °C. Higher monthly average RH and temperature was observed in the semester May–October/2016. It corresponds to the wet period when heavy rain occurred. Under these conditions, Cl^- *DR* should occur mainly as saline solution and not as crystals [4].

Monthly data and the annual average show possible the beginning and development of any deterioration process in the structures. Annual time of wetness (TOW) calculated was 3450 h y^{-1}. According to ISO 9223 standard, it corresponds to τ_4

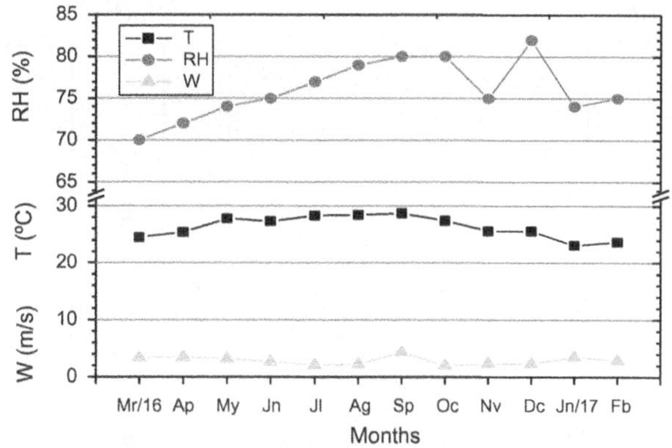

Fig. 3 Monthly meteorological parameters behavior

($2500 < \tau \leq 5500$) category, as it occurs in the coastal insular tropical climate of Cuba [7–9].

In accordance with to the calculated TOW, the metallic surface should be covered by a water film about 40% of total annual time (8760 h y^{-1}). Atmospheric corrosion generally occurs in presence of thin water films on the metallic surface. The water layer may contain chloride ions and dissolved oxygen, factors strongly influencing in atmospheric corrosion rate.

With respect to the wind flow velocity, monthly average data was obtained from the components from the east-northeast to the north directions, i.e. the strength and direction of wind flow from the sea. The wind flow rate remained stable during the period of study, although the highest value corresponded to September month. Annual average obtained was 3.82 m s^{-2}.

Previous studies about atmospheric corrosion showed that, under an annual average higher than 3 m s^{-2} it's enough to ensure not only the origin and marine aerosol formation on the northern Havana coast without shielding conditions, but also sufficient to ensure chloride ions deposition and high values of RH [10].

3.2 *Pollutants Deposition Behavior (Cl^- DR and SO_x^- DR)*

Notable difference is observed between monthly deposition of chloride ion salts (see Fig. 4a) and sulfur compounds (Fig. 4b) during the year of study. The difference is due to the fact that chloride ion salts source is the sea and they are transported in the marine aerosol. While the sulfur compounds come from both the marine aerosol and emissions from mobile sources that circulate in the area.

The highest monthly average was obtained during the winter season (dry season), mainly in March where deposition reached an average more than 900 mg m^{-2} d. Drastic decrease in December in the sites 1, 2 and 3 was due to an excess of collectors surface washing, product of the heavy rainfall recorded in the month.

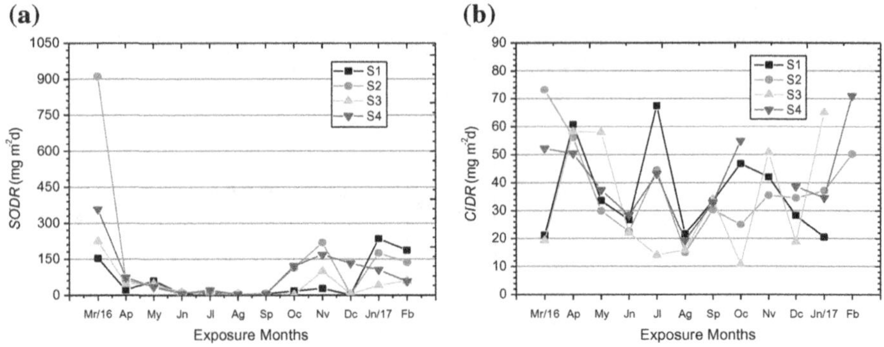

Fig. 4 Monthly pollutants deposition behavior: **a** chloride ion salts; **b** sulfur compounds

Table 1 Average annual deposition for both aggressive agents in each exposure site

Outdoor exposure sites	Cl^- *DR* ($mg\ m^{-2}\ d$)	$SO_x{}^-$ *DR* ($mg\ m^{-2}\ d$)	Classification of the atmosphere
1	42.62	38.58	Coastal-industrial
2	173.27	40.17	Coastal-industrial
3	56.53	31.19	Coastal-industrial
4	82.77	38.47	Coastal-industrial

Average annual deposition for both aggressive agents in each exposure site, allowed to classify the atmosphere type in the area under study (Table 1). In general, the chloride ion salts deposition is higher than sulfur compounds deposition. Therefore, the atmosphere can be classified as coastal-industrial, one of the most aggressive atmospheres for metals and reinforced concrete structures.

3.3 Monthly and Cumulative Corrosion Rate Behavior

Monthly atmospheric corrosion behavior. The behavior of corrosion rate versus time (Months) during the year allows to determine the more and less aggressive periods of the period. Monthly corrosion rate behavior is similar to the chloride ion salts deposition (see Fig. 5a). Therefore, it is expected that deposition of these salts is one of the main factors that influence on the corrosion rate or what is the same, in the atmospheric corrosion of carbon steel in the area of study. This type of monthly behavior is what is known in the specialty as atmospheric corrosion curve.

During the period, the highest corrosion rate was determined in the winter season (dry season). In March/2016 and January/2016 corrosion rate data over 200 g m^{-2} were reported. Kinetics variability of corrosion process (monthly) confirms the difference that exists between the summer and winter stage in terms to the occurrence

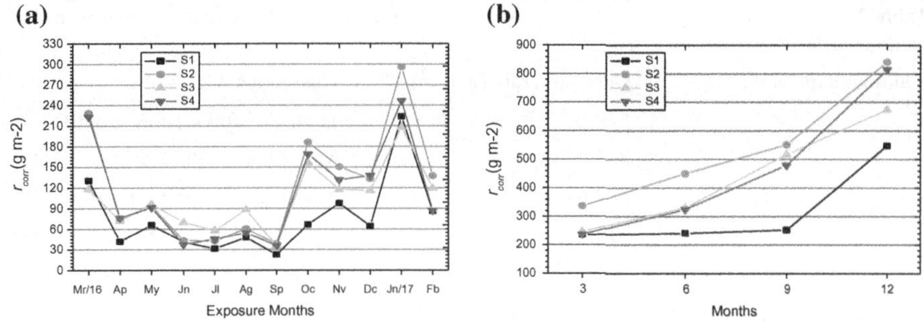

Fig. 5 Monthly and cumulative atmospheric corrosion behavior for each exposure site

of the phenomenon. This is typical for coastal tropical climates with high corrosive aggressiveness as is the case of the northern coast of any region in Cuba [1, 2]. Therefore, this behavior indicates that repair work would be more effective during the summer stage, i.e. when there is a lower level of atmospheric corrosivity.

Cumulative corrosion rate behavior. Increased cumulative corrosion rate over time in the four exposure sites (see Fig. 5b) is presented, which confirms the high corrosive aggressiveness of the zone. This type of analysis of the phenomenon as a function of cumulative time is not required by the specialty regulations, but its interpretation gives very interesting results, since the atmospheric corrosion process evolution over time is known.

After three months of exposure, the average corrosive aggressiveness category is reached (C3 $200 < Vc \leq 400$ g m^{-2}) for the four exposure sites. In site 1, the aggressiveness category increased from medium (C3) to high (C4 $400 < Vc \leq 650$ g m^{-2}) [4]. However, for the other exposure sites the corrosive aggressiveness category of the atmosphere increased from medium (C3) to very high (C5 $650 < Vc \leq 1500$ g m^{-2}).

3.4 *Corrosive Aggressiveness Category Determination of the Atmosphere*

Corrosivity categories of the atmosphere of all outdoor exposure sites were classified based on annual average corrosion rate (g m^{-2}). Classification ranges are established in ISO 9223:2012 (Table 2). Corrosivity category in outdoor exposure sites S2 to S4 was classified as very high (C5). Only in outdoor exposure site S1, corrosivity was classified as high.

The corrosive aggressiveness categories of the atmosphere determined (C4 and C5), not only has a scope in the metallic material deterioration tested (carbon steel). The specialty regulations establish that: the corrosive aggressiveness categories of the atmosphere, are extensive for all types of material that is used for constructive purposes (reinforced concrete, etc.) [4].

Table 2 Corrosivity category of the atmosphere classification for LCS (carbon steel) on each outdoor exposure site

Outdoor exposure sites	Corrosion rate (g m^{-2})	Categories (C)
1	548.0	High C4 (400 < *r*corr < 650)
2	841.0	Very high C5 (650 < *r*corr < 1500)
3	672.67	Very high C5 (650 < *r*corr < 1500)
4	814.78	Very high C5 (650 < *r*corr < 1500)

4 Conclusions

Monthly meteorological parameters behavior and atmospheric pollutants deposition, allowed to demonstrate the existence of a high atmospheric corrosivity in addition to the existence of a coastal-urban atmosphere, which is cataloged as one of the most aggressive to structures in coastal tropical climates.

Cl^- *DR* and SO_x^- *DR* occurs in saline solution and not as solid crystals form. It facilitates to increase corrosion rate.

The monthly and cumulative behavior over time, confirmed the predominance of an atmospheric corrosive aggressiveness category of very high (C5) in the zone under study. The result was demonstrated through visual observation of the carbon steel specimens placed up to one year of study in each exposure sites.

References

1. Corvo, F., Pérez, T., Dzib, L.R., Martín, Y., Castañeda, A., González, E., Pérez, J.: Outdoor-Indoor corrosion of metals in tropical coastal atmospheres. Corros. Sci. **50**, 220–230 (2008)
2. Corvo, F., Pérez, T., Martin, Y., Reyes, J., Dzib, L.R., González, A.J., Castañeda, A.: Corrosion Research Frontiers. Atmospheric corrosion in tropical climate. On the concept of Wetness and its interaction with contaminants deposition. In: Electroanalytical Chemistry: New Research. Nova Science Publishers, New York(2008)
3. ISO 9225:2012 standard. Corrosion of metals and alloys—Corrosivity of atmospheres—Measurement of environmental parameters affecting corrosivity of atmospheres
4. ISO 9223:2012 standard. Corrosion of metals and alloys—Corrosivity of atmosphere—Classification, determination and estimation
5. ISO 9226:2012 standard. Corrosion of metals and alloys. Corrosivity of atmospheres. Determination of corrosion rate of standard specimens for the evaluation of corrosivity
6. ISO 8407:2009. Corrosion of metals and alloys. Corrosivity of atmospheres. Removal of corrosion products from corrosion test specimens
7. Corvo, F., Pérez, T., Martin, Y., Reyes, J., Dzib, L.R., González-Sánchez, J., Castañeda, A.: Time of wetness in tropical climate: considerations on the estimation of TOW according to ISO 9223 standard. Corros. Sci. **1**(50), 206–219 (2008)
8. Corvo, F., Perez, T., Martin, Y., Reyes, J., Dzib, L., Gonzalez, J., Castañeda, A.: Electroanalytical Chemistry: New Research. Nova Science Publisher, New York, (2008)
9. Corvo, F., Torrens, A., Martín, Y., González, E., Pérez, J., Valdés, C., Castañeda, A., Portilla, C.: Indoor atmospheric corrosion of steel. Its particularities in Cuban tropical climate. Revista Metalurgia **44**(2), 129–137 (2008)

10. Valdés, A.C.: Estudio de la corrosión atmosférica del acero de refuerzo embebido en el hormigón armado. Revista CENIC Ciencias Químicas **45**, 77–80 (2014)

Studies Carried Out on Concretes Produced with LC3 According to Cuban Standard NC 120: 2014

Eilys Valdés Alemán, Yosvany Díaz and Jose Fernando Martirena-Hernandez

Abstract In order to implement the low carbon cement (LC3) as part of the Cuban regulations, a study of durability in concretes manufactured with this cement was developed, in different conditions of aggressiveness according to the performance parameters established in the NC 120: 2014 "Hydraulic Hydraulic-Specifications", the taking of witnesses of the H1, H2, and H4 concretes elaborated in 2016 was carried out. Carbonation depth tests, air permeability, chloride ion migration, were carried out. electrical resistivity, effective to determine the changes experienced by the series 1.5 years versus the series 6 months and the concretes made with cement under carbon LC3 compared to those made with Ordinary Portland Cement (CPO). Subsequent to the analysis of the results, it can be verified that the concretes made with LC3 showed to be in a range of 11–16 times superior in their behavior to the electrical resistivity that those made with Portland cement and 15–22 times more resistant to the chloride ion penetration that those made with CPO, also show a more favorable performance with the passage of time.

Keywords Durability · Concrete · Low carbon cement (LC3) · Cuban norm · Exhibition sites

1 Introduction

Concrete is man most used material after water. However, it did not prove to last as much as expected due to durability problems [1]. Most concrete is durable, but some aggressive conditions pose problems, particularly with corrosion of reinforcing steel due to chloride and carbonation leading to steel corrosion [2]. Recent studies have

E. V. Alemán (✉)
Faculty of Construction, Marta Abreu de las Villas Central University, UCLV,
Santa Clara, Cuba
e-mail: evaleman@uclv.cu

Y. Díaz · J. F. Martirena-Hernandez
Center for Research and Development of Structures and Materials (CIDEM), UCLV,
Santa Clara, Cuba

© RILEM 2020

J. F. Martirena-Hernandez et al. (eds.), *Proceedings of the International Conference of Sustainable Production and Use of Cement and Concrete*, RILEM Bookseries 22,
https://doi.org/10.1007/978-3-030-22034-1_38

Table 1 Properties of the cements used in this protocol

Cement	% >90 μm	Strength 2 d Mpa	Strength 7 d Mpa	Strength 28 d MPa
LC3-50 2:1	94.5	17.50	31.00	48.12
P35		18.20	40.00	58.90

proven that concrete under aggressive conditions, such as marine environments, can have problems with chloride and carbonation.

For the implementation of a novel cement a standard for cement is required but in order to be able to use the cement, a durability standard should prove that concrete made with this cement stands aggressiveness conditions considered for the country/region [3, 4].

In the present investigation, a durability study is carried out in concrete made with a new kind of low carbon cement made with a ternary combination of Portland cement, calcined clay and limestone (known as LC3). Concrete blocks made with LC3 and Portland cement (as a reference) have been produced and laid at different exposure sites with different conditions of aggressiveness according to the performance parameters established in the NC 120: 2014 "Hydraulic Hydraulic-Specifications" [5]; and their durability has been assessed at different ages (6 and 18 months).

The testing protocol to be followed included: air permeability tests, electrical resistivity, resistance to chloride ion penetration by ASTM 1202, carbonation and water absorption.

2 Materials and Experimental Protocol

Concretes subject of study were produced with a ternary blend calcined clay-limestone-Portland cement, name LC3. A reference series was made with Cuban PC P35. Table 1 presents the properties of both cements.

Cylindrical specimens with size 35 × 50 cm were made with these cements. They were placed at different locations resembling the exposure conditions demanded by the Cuban standard NC 120:2014. Every 6 months cores were taken for studies at CIDEM, with the experimental protocol described in this paper. Concrete mix design were made considering prescriptive parameters of the Cuban standard. Results are presented at Table 2. Table 3 presents compressive strength results.

Table 2 Mix design used for PC and LC3 concrete (Medina Sanchez, Ernesto)

Mix	Exposure class NC 120:2014	Strength (Mpa)	W/C	Cem (kg)	SP (kg) PC	SP (kg) LC3	Sand (kg)	Middle (kg)	Coarse (kg)	Water (kg)
H1	Very high	35.0	0.40	430	3.87	8.6	634	352	775	172
H2	High	30.0	0.45	405	3.65	8.1	651	362	796	182
H4	Low	20.0	0.55	345	3.1	6.90	690	384	844	190

Table 3 Compressive strength in concrete made with LC3 and PC

Series	Cement (kg)	SP (%)	A/C	Slump (cm)	Compressive strength (MPa)					
				P35/LC3	3 d LC3	3 d P35	7 d LC3	7 d P35	28 d LC3	28 d P35
H-35	430	0.65	0.4	20	14.00	23.33	20.60	30.00	36.34	34.00
H-30	405	0.65	0.45	20	10.33	17.33	16.30	23.00	27.00	28.00
H-25	370	0.65	0.5	20	7.66	14.33	15.00	18.60	22.33	23.00
H-20	340	0.65	0.55	20	7	12.00	12.00	14.70	18.57	19.66

2.1 Exposure Sites

Punta Matamoros (Cayo Santa María)

This site is located at the central region of Cuba, on the northern coast of Villa Clara [6]. It has two concrete platforms each of approximately 40 m^2 for exposition. This is the most aggressive environment in Cuba (Fig. 1).

Sede Universitaria (Cayo Santa María)

This site is situated at some facilities of the university in the area. It has a 20 m^2 platform and is situated at approximately 1500 m from the sea side. This is considered high aggressiveness (Fig. 2).

Faculty of Constructions (UCLV)

It is located at CIDEM's facilities at the university, at more than 20 km from the seashore. This área is considered low aggressiveness. (Series H4) (Fig. 3).

Fig. 1 Exposure site Punta Matamoros, Cayo Santa María

Fig. 2 Exposure site Sede Universitaria Cayo Santa María

Fig. 3 Exposure site Faculty of Constructions

2.2 Testing Protocols

- Measurement of surface resistivity of fully saturated concrete cores. A four-point probe was used—Wenner probe—four equally spaced electrodes are placed on the concrete surface. An AC current is injected through the outer two electrodes, and the voltage drop is measured between the two inner electrodes [7].
- Measurement of air permeability of concrete cores using the Torrent method: it measures the coefficient of air-permeability kT with the PermeaTORR instrument [8]. The coefficient of air-permeability kT (10–16 m^2) is calculated as function of the increase in pressure recorded in the measuring chamber.
- Carbonation depth, measured by spraying the exposed surfaced with phenolphthalein. Cores of the elements were taken. The cores were split in two parts so have the carbonated part exposed [6, 9].
- Rapid Chloride permeability test (RCPT) according to the procedure described at ASTM 1202, which determines the total charge passing [10].

3 Discussion of Results

3.1 Air Permeability Tests

In Fig. 4 it can be seen that the specimens of the series 1.5 years show better behavior than those of the series 6 months, unlike the concretes of the series H2 CP and H1 LC3 the latter with irrelevant differences.

The above comparisons allow us to determine that after its exposure to the medium, the passage of time has a positive influence on the permeability of the concrete when

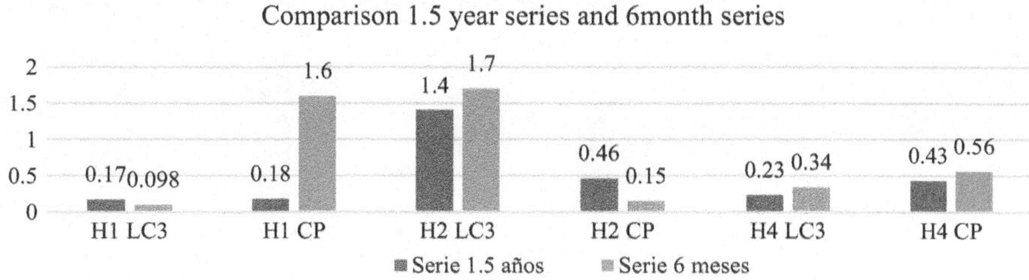

Fig. 4 Comparison of air permeability results (series 1.5 years and series 6 months)

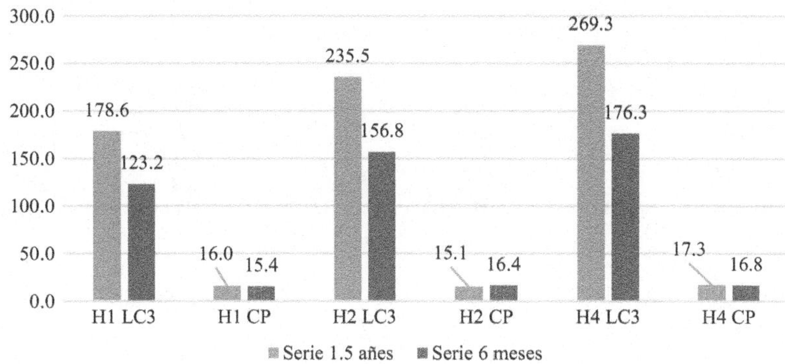

Fig. 5 Comparison of resistivity results (series 1.5 years and 6 months)

the hydration reactions are completed, since the best results were obtained with the series 1.5 years, on the other hand, the concretes made with LC3 showed to have better permeability behavior than those made with Portland cement.

3.2 *Resistivity Tests*

Figure 5 shows, concretes with LC3 increase their electrical resistivity considerably over time. In the case of concretes made with CPO, no noticeable increase in electrical resistivity is observed.

The above comparisons allow to affirm that the passage of time has a positive influence on the electrical resistivity of the concretes made with LC3, however, in those manufactured with CPO this property does not show variations with the passage of time (series 6 months to series 1.5 years), in addition, the concretes made with LC3 showed to be in a range of 11–16 times better in terms of their behavior to resistivity than those made with Portland cement.

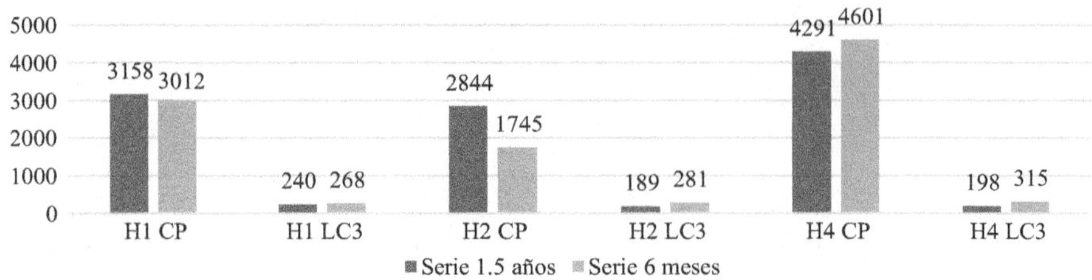

Fig. 6 Comparison of chloride ion penetration resistance results by ASTM 1202 (series 1.5 and series 6 months)

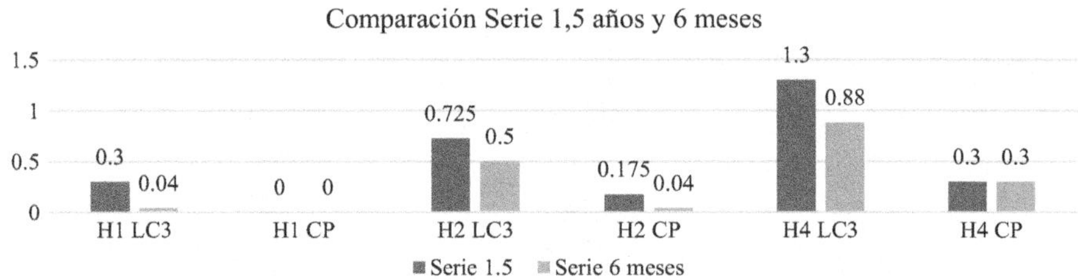

Fig. 7 Comparison of carbonation depth results (series 1.5 and series 6 months)

3.3 Chloride Ion Penetration Resistance Tests According to ASTM 1202

Figure 6 shows that the 1.5-year series decreases the chloride ion permeability of concretes made with LC3 compared to the 6-month series. In concretes made with Portland cement, only improvements of 310 °C in the chloride ion permeability for the H4 samples are observed.

With the passage of time, the chloride ion permeability of the concretes that used LC3 for its elaboration decreases since the series 1.5 years presents the lowest values for the evaluation of this parameter. The concretes produced with LC3 are more resistant to chloride ion penetration than those made with CPO in a range of 15–22 times.

3.4 Analysis of Carbonation Results

Figure 7 reveals that in the series 1.5 years there is a greater depth of carbonation than in the series 6 months, both for specimens made with LC3 cement and those made with Portland cement, this is justified by the longer time of exposure to the medium.

It is appreciated that LC3 specimens have a greater depth of carbonation than those of CPO, according to previous research, it is because the carbonation in concretes produced with LC3 advances quickly in the first months and then stabilizes until presenting a practically constant over time, while in concretes made with CPO the advance of carbonation starts at 0 and increases in a measured way throughout the life of the concrete.

4 Conclusions

- Concrete made with LC3 can be used in any of the aggressiveness conditions prescribed in the Cuban standard NC 120:2014, since all results are favorable.
- Concrete made with LC3 show a better performance in terms of chloride ingress and permeability, but they have a slight increase in carbonation rate.
- At late ages, LC3 concrete continues to improve its properties.

References

1. Idorn, G.M.: Innovation in concrete research—review and perspective. Cem. Concr. Res. **35**, 3–10 (2005)
2. The European cement Association: The role of CEMENT in the 2050 LOW CARBON ECONOMY (2014)
3. Hooton, R.D.: Current developments and future needs in standards for cementitious materials. Cem. Concr. Res. **78**, 165–177 (2015)
4. Hooton, R.D.: Bridging the gap between research and standards. Cem. Concr. Res. **38**, 247–258 (2008)
5. NC-120: Hydraulic concrete -specifications Cuba (2014)
6. Cebey Liquor, A.: Evaluation of carbonation in concretes made with low carbon cement LC3. Universidad Central de las Villas (2016)
7. Angst, U.M., Elsener, B.: On the applicability of the Wenner method for resistivity measurements of concrete. ACI Mater. J. **111**(6), 661–672 (2014)
8. S. 262/1:2913: Bauwesen Betonbau – Ergänzende Festlegungen, pp. 1–52 (2013)
9. Nazco, K.: Evaluation of Cuban multicomponent clays as source of raw material for the production of supplementary cement materials. Universidad Central de las Villas (2014)
10. ASTM-C1202: Standard Test Method for Electrical Indication of Chloride's Ability to Resist Chloride (1994)

Monitoring and Interpreting the Early Properties of Alkali-Activated Materials by Electrical Conductivity Measurement

Luping Tang, Jun Liu and Emma Zhang

Abstract Alkali-activated materials are attracting more and more attention worldwide thanks to their low carbon footprint. Different from the binders based on ordinary Portland cement, the early properties such as flowability, setting, and early strength development, of alkali-activated materials can vary very much and are very sensitive to the compositions of concrete mixture. Conventional methods for testing such early properties are often time-consuming and laborious, whilst the measurement of electrical conductivity is relatively simple and can easily be monitored in an automatic manner. This paper presents a simple monitoring technique based on Winner's resistivity method and some results measured from mortar samples with various compositions of alkali-activated materials as well as ordinary Portland cements. The preliminary results show some reasonable correlations between monitored conductivity curves and measured/observed early properties. Some interpretations and theoretical considerations to these measured conductivity curves in relation to the early properties and possible chemical reactions are discussed.

Keywords Alkali-activated materials · Early properties · Conductivity

1 Introduction

In the past decades, alkali-activated materials are attracting more and more attention worldwide thanks to their low carbon footprint as well as their versatility in applications and adoptability to locally available raw materials [1]. Different from the binders based on ordinary Portland cement, the early properties such as flowability, setting, and early strength development, of alkali-activated materials can vary very much and are very sensitive to the compositions of concrete mixture. Conventional methods for testing such early properties are often time-consuming and laborious,

L. Tang (✉) · E. Zhang
Chalmers University of Technology, 41296 Gothenburg, Sweden
e-mail: tang.luping@chalmers.se

J. Liu
Shenzhen University, Shenzhen, China

© RILEM 2020

J. F. Martirena-Hernandez et al. (eds.), *Proceedings of the International Conference of Sustainable Production and Use of Cement and Concrete*, RILEM Bookseries 22,
https://doi.org/10.1007/978-3-030-22034-1_39

whilst the measurement of electrical conductivity or resistivity is relatively simple and can easily be monitored in an automatic manner. At present, there is at least one standard method available for testing the surface resistivity of hardened concrete [2], but up to now no standard method available for testing the resistivity of fresh concrete, probably due to the contact problem between electrodes and concrete. Li et al. [3] developed an innovative non-contacting device for measuring resistivity of fresh concrete without the contact problem. Their method has been used for determination of concrete setting time [4] and for improving understanding of cement hydration mechanism [5]. Wei et al. [6] used this technique to investigate the effect of superplasticizer on the setting time. However, this non-contacting method need a special device which is not available in ordinary laboratories. On the other hand, the study of fresh alkali-activated materials using the technique of electrical resistivity is limited. This paper presents a simple monitoring technique based on Winner's resistivity method for monitor the resistivity profiles of mortar samples with various compositions of alkali-activated materials as well as ordinary Portland cements.

2 Experimental

2.1 Setup for Resistivity Measurement

The experimental setup for each sample consists of a test cell, a galvanostatic pulse generator, and a potential recorder (datalogger). The test cell is made of PVC plastic with four stainless steel screws equally distributed ($a = 25$ mm) on the bottom (Fig. 1) for resistivity measurement based on Winner's four-electrodes method, which can effectively eliminate the influence of contact resistance when compared with two-electrodes method. Instead of alternative current, a galvanostatic pulse of 0.2 mA at about 0.1 s was used in this study for avoiding the effect of polarization on the potential measurement. The potentials were recorded by a datalogger (Campbell CR10) at an interval of 1–2 min in the first 24 h and 10 min for the hydration period over one day.

2.2 Mixtures and Test Procedure

In this study Swedish cement named "Bascement Slite" (CEM II/A-V 52.5 N according to European standard EN 197-1) and ground granulated blast furnace slag named "Merit 5000" (specific surface area 5000 cm^2/g and density 2920 kg/m^3) were used as cementitious materials. The main chemical compositions of the cementitious materials are listed in Table 1. Natural quartz sand of 0–2 mm was used as aggregate to improve the homogeneity of mixes. The water glass ($Na_2O{\cdot}nSiO_2$) with initial modulus $n = 3.35$ (molar ratio SiO_2/Na_2O) (supplied by Sibelco Nordic) was adjusted

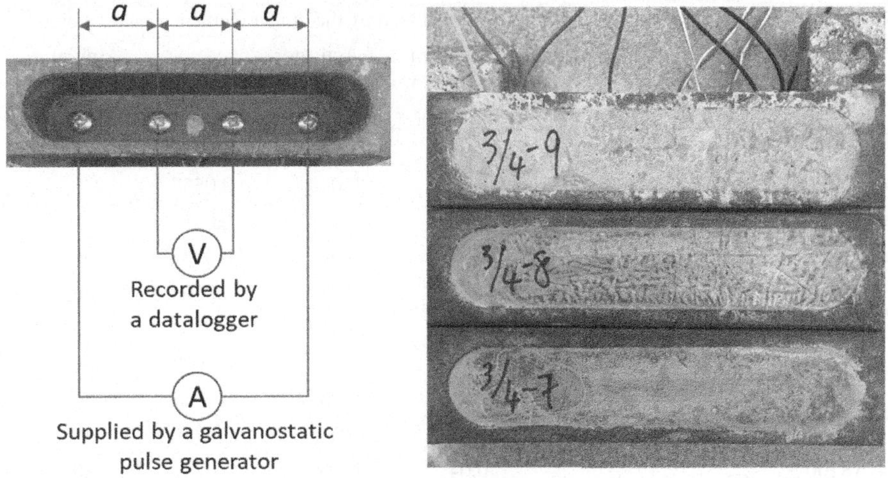

Fig. 1 Setup for resistivity measurement (left) and samples after hardened (right)

Table 1 Main chemical compositions of slag Merit 5000

	CaO	SiO_2	Al_2O_3	Fe_2O_3	MgO	TiO_2	Mn_2O_3	S^{2-}	SO_3	Na_2O-eq
Cement	56.7	22.7	6.1	3.6	2.9				3.4	1.1
Slag	31	34	13		17	2.4	0.8	1.3	0.25	0.9

by addition of NaOH (analytical grade, purchased from Fisher Scientific) to $n = 1.2$ and used as alkali-activator. Gypsum ($CaSO_4 \cdot 2H_2O$, puriss grade, purchased from Fisher Scientific) was used with the purpose for reducing shrinkage in some mixes. The mixes tested in this study are listed in Table 2. In all the mixes the ratio of water to binder (cement + slag) was fixed at 0.5 and the ratio of sand to binder fixed at 1. The mixing procedure was in accordance with EN 196-1 and the initial setting time was measured by a Vicat (Toni Technik) with a calibrated weight of 300 g and a cylindrical needle with a flat tip area of 1 mm^2. The mortar sample for the initial setting time test was 40 mm thick and filled in a plastic cup of diameter about 50 mm. The penetration depth was measured in a certain time interval of 5–10 min until the reading (6 ± 3) mm was reached when compared with the initial reading (zero).

3 Results and Discussions

According to Winner's four-electrodes method the resistivity ρ is calculated by the following equation:

Table 2 Mix proportions of mortar samples normalized by the binder (cement + slag) weight

Mix No.	Binder		Gypsum	Activator		Initial setting time (min)
	Cement	Slag		Na_2O	SiO_2	
BasC	1					150[a]
AAS80-1	0.2	0.8		0.048	0.0557	220
AAS80-2	0.2	0.8	0.02	0.048	0.056	190
AAS80-3	0.2	0.8	0.04	0.048	0.056	90
AAS80-4	0.2	0.8		0.052	0.0604	220
AAS80-5	0.2	0.8	0.02	0.052	0.0604	180
AAS80-6	0.2	0.8	0.04	0.052	0.0604	160
AAS100-1		1		0.065	0.0755	40
AAS100-2		1	0.02	0.065	0.0755	35
AAS100-3		1	0.04	0.065	0.0755	30

[a]Setting time supplied by the manufacturer

$$\rho = \frac{2\pi \Delta U}{I} \tag{1}$$

where a is the distance between the electrodes, U is the recorded potential, and I is the impressed galvanostatic current. To facilitate the discussion, conductivity σ—the inverse of resistivity, was used in this study for expressing the results.

3.1 Conductivity Curves of the Samples with Ordinary Portland Cement

Figure 2 shows the conductivity curves recorded from the samples mixed with ordinary Portland cement. It can be seen from Fig. 2 that, although there is a certain deviation in the absolute values of conductivity, the curves expressed by $\sigma \cdot t$ from three samples are generally comparable, indicating an acceptable repeatability of the measurement method. It seems difficult to correlate the setting time to the changes in conductivity, although the latter dramatically decreased with the time after mixed for 6–12 h, with the first peek at 7–8 h and the second peak at 11–12 h. These peeks indicate structural changes in the matrix due to the accelerated hydration.

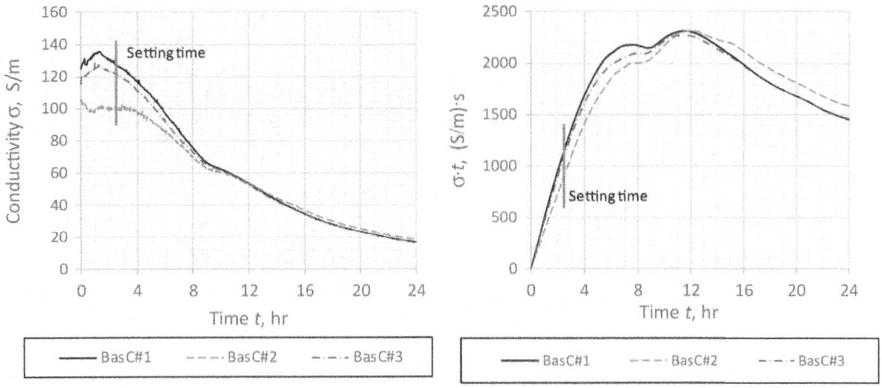

Fig. 2 Conductivity curves from the ordinary Portland cement samples

3.2 Conductivity Curves of the Samples with Alkali-Activated Slag

Figure 3 shows the conductivity curves recorded from the samples mixed with 100% slag. The conductivity curve for AAS100-3 markedly decreased in the first hour and increased again in the followed hour. The addition of gypsum probably changed the ionic concentrations in the mixing water. The measured setting time in this group of samples is relatively short, only 30–40 min. It was observed that, in the first 20–30 min the mixes were very flowable, but suddenly became hardened. The peaks of $\sigma \cdot t$-curves are, however, in the range of 8–9 h, implying the reduction of pore solutions due to the formation of solid network. This means that the early setting may not be due to the formation of hydration products but possibly be caused by some type of gelling. It seems that the addition of gypsum has no clear effect on the hydration process in the mixes with 100% slag, probably due to the less availability of calcium ions for the formation of ettringite or less solubility of gypsum in the high alkaline condition.

Figure 4 shows the conductivity curves recorded from the samples mixed with 80% slag and 20% cement. For samples AAS80-3 and AAS80-6, the peak of conductivity curve coincides with the initial setting time, but not for the other samples. The peeks of $\sigma \cdot t$-curves are, however, always behind the respective initial setting time by 1–2 h except for sample AAS80-6, in which the peak is behind its initial setting time by only 10 min. Without addition of gypsum (samples AAS80-1 and AAS80-4), the peak appeared earlier when higher amount of Na_2O was used (in sample AAS80-4), implying a stronger activation. Addition of gypsum in the mixes significantly changed the hydration process and accelerated the formation of solid network in the matrix, especially when lower amount of Na_2O was used (in samples AAS80-1 to AAS80-3). This effect should be attributed to the formation of ettringite because in these mixes the 20% cement can supply adequate calcium ions. The effects of

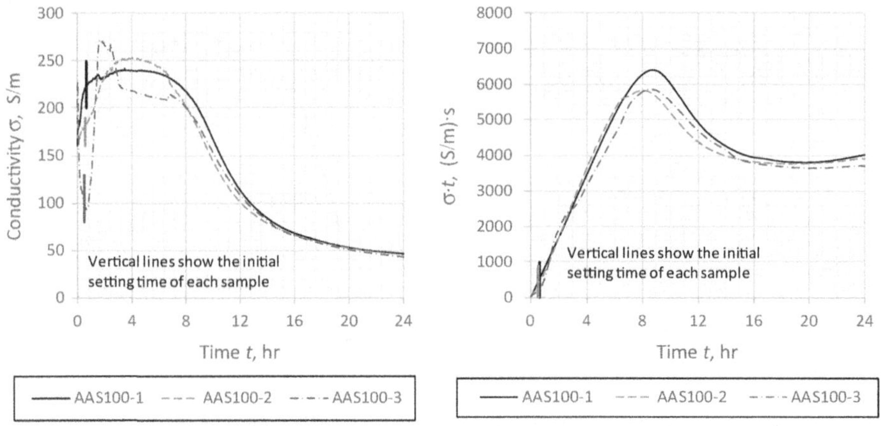

Fig. 3 Conductivity curves from the samples mixed with 100% slag

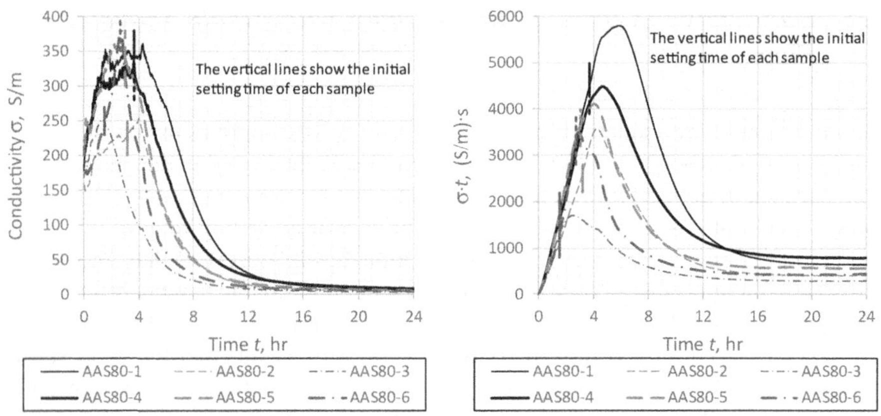

Fig. 4 Conductivity curves from the samples mixed with 80% slag and 20% cement

gypsum in samples AAS80-5 and AAS80-6 are not big as in samples AAS80-2 and AAS80-3, probably due to a lower solubility of gypsum in the former samples with higher amount of Na_2O.

4 Concluding Remarks

From the preliminary results presented above the following concluding remarks can be given:

- The electrical resistivity or conductivity monitoring by means of Winner's four-electrodes method indeed supplies useful information related to the hydration process.
- The test setup proposed in this study is relatively simple and can easily be built up in ordinary laboratories.
- The addition of gypsum in mixes with 100% slag has no clear effect on the hydration process, whilst the addition of gypsum in mixes with 80% slag and 20% cement significantly changed the hydration process and accelerated the formation of solid network in the matrix.

Acknowledgement This paper is part of the projects financed by Vinnova (2016-03367) and EU-H2020 (777823).

References

1. Provis, J.L.: Alkali-activated materials. Cem Concr Res. https://doi.org/10.1016/j.cemconres.2017.02.009 (2017)
2. AASHTO TP 95: Standard test method for surface resistivity of concrete's abil-ity to resist chloride ion penetration, American Association of State Highway and Transportation Officials, USA (2014)
3. Li, Z., Wei, X., Li, W.: Preliminary interpretation of Portland cement hydration process using resistivity measurement. ACI Mater. J. 253–257 (2003)
4. Li, Z., Xiao, L., Wei, X.: Determination of concrete setting time using electrical resistivity measurement. J. Mater. Civ. Eng. **19**(5), 423–427 (2007)
5. Xiao, L., Li, Z.: New understanding of cement hydration mechanism through electrical resistivity measurement and microstructure investigations. J. Mater. Civ. Eng. **21**(8), 368–373 (2009)
6. Wei, X., et al.: Hyperbolic method to analyze the electrical resistivity curve of Portland cement with superplasticizer. J. Wuhan Univ. Tech-Mater. Sci. 245–248 (2008)

Case Study on Concrete Durability Field Exposure Station in Hangzhou Bay Bridge, China

Weiliang Jin, Jianghong Mao, Jinquan Wang, Dawei Zhang and Weijie Fan

Abstract The Hangzhou Bay Bridge is located at the typical marine corrosion environment. Research on the durability of reinforced concrete acts as an important role to evaluate the service life of the bridge. In order to evaluate the durability of Hangzhou Bay Bridge, the field exposure station of Hangzhou Bay Bridge was built during the construction of Hangzhou Bay Bridge in 2005. Hundreds of concrete specimens were placed at the station for at least 10 years, the chloride ions inside the concrete were tested twice per year. This paper summarized the experience of constructing and maintaining for the field exposure station. Meanwhile, the application of information construction and management of the exposure test station to evaluate the durability of the bridge is introduced.

Keywords Concrete structure · Durability · Field exposure station · Hangzhou Bay Bridge

1 Introduction

The test method for reinforced concrete durability can be divided into field exposure test and simulation test [1, 2]. The field exposure test owns the advantages of non-destruction, convenient operation, and reliable test results, thus it is always adopted for important project durability evaluation. Numbers of field exposure stations have been built in China. They are located at typical sea area, such as Jinzhou Port Station, Donghai Bridge Station, Zhanjiang Port Station, and Hangzhou Bay Bridge Station [3–6]. Some of them have accumulated research data of more than 30 years, and many achievements have been adopted by Chinese Construction Standards [7].

W. Jin (✉) · D. Zhang
Institute of Structural Engineering, Zhejiang University, Hangzhou 310058, China
e-mail: jinwl@zju.edu.cn

W. Jin · J. Mao · W. Fan
Ningbo Institute of Technology, Zhejiang University, Ningbo 315100, China

J. Wang
Ningbo Hangzhou Bay Bridge Development Co., Ltd, Ningbo 315317, China

© RILEM 2020

J. F. Martirena-Hernandez et al. (eds.), *Proceedings of the International Conference of Sustainable Production and Use of Cement and Concrete*, RILEM Bookseries 22,
https://doi.org/10.1007/978-3-030-22034-1_40

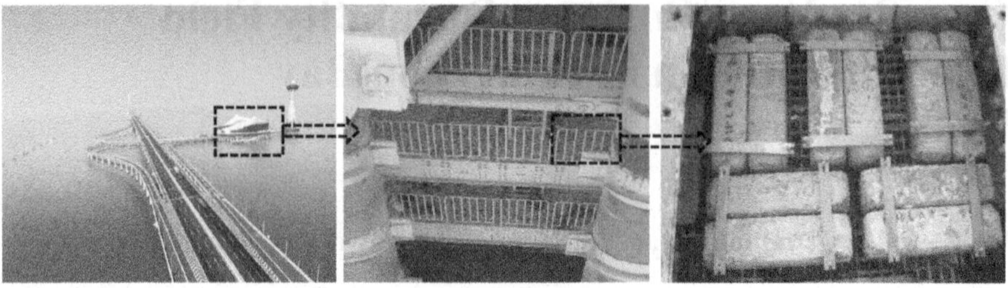

(a) Hangzhou Bay Bridge (b) Field Exposure Station (c) Steel grid

Fig. 1 Field exposure station of Hangzhou Bay Bridge

Hangzhou Bay Bridge Station was built during the construction of Hangzhou Bay Bridge in 2005. It is one of the earliest field exposure stations in China. It is used for evaluating the concrete durability of Hangzhou Bay Bridge. Hangzhou Bay Bridge is the first cross-sea infrastructure project in China with the design service life of 100 years. The environment in Hangzhou Bay is complicated with strong wind, large wave, high tidal range, rapid ocean current and high chloride ion content. In view of this, field exposure stations have been built under the platform located in the sea [5].

The Hangzhou Bay Bridge Station acts as an important concrete durability research base, including durability evaluation and new technology testing platform. Zhejiang University established long-term cooperative relationship with the bridge maintenance company. Several meaningful researches have been carried out in the last ten years, including annual chloride ions data detection and durability evaluation, application of concrete durability monitoring system. In this paper, concrete durability evaluation process is introduced and some results are presented.

2 Construction and Maintenance of the Field Exposure Station

2.1 Construction of Field Exposure Station

The structural design of the field exposure station of Hangzhou Bay Bridge was described in the literature [5]. The station was built at the platform substructure of Hangzhou Bay Bridge. The platform superstructure is a tourist spot of Hangzhou Bay Bridge now. The location and real picture are shown in Fig. 1.

The field exposure station locates at the middle of the bridge. The distance of the station from the off-shore is 18 km. Thus, it has the advantages of good environmental representation and convenient transportation. There are five layers including an atmosphere zone, a splash zone, two tidal zones, and a submarine zone. The test specimens of different elevation zone represent the concrete of different elevation parts of Hangzhou Bay Bridge.

2.2 Maintenance of Field Exposure Station

In order to obtain enough chloride ions data for evaluating, the structural health of field exposure station and the integrity of specimens are quite important. Based on above consideration, following measures are taken for maintenance:

(1) Structural anti-corrosion. The field exposure station is a steel structure. The anti-corrosion coating could be destroyed by specimen's transportation and the hitting of floating objects. Thus, it is necessary to carry out regular maintenance of anti-corrosion coating of structural main members, steel grids and connect joints.
(2) Specimen fixation. The wave forces applied on the specimens are very large. The specimens would be destroyed under long-term wave force attacking. Therefore, specimens should be tightly fixed inside the steel grid with stainless steel strips.
(3) Data detection: The durability database is updated four times per year by detecting the specimens' chloride ions. The sample powder which is used to test the chloride ion are collected using a drill with a diameter of 10 mm. The powder extraction depth for each layer is 5 mm. The migration degree of chloride ion is obtained by testing the content of chloride ion using rapid chloride ion test (RCT).

2.3 Informatization of Data Management

The data management system of the Hangzhou Bay Bridge field exposure station was compiled based on the database. The management system includes three levels, including structural level, grid layer level and test block level, as shown in Fig. 2.

Structural levels are divided into atmospheric zone, splash zone, tidal zone 1, tidal zone 2, and submerged zone. Grid level has 16 grids for each structure layer. Specimen level presents different specimens including mix ratio, specimen dimension, and exposure time.

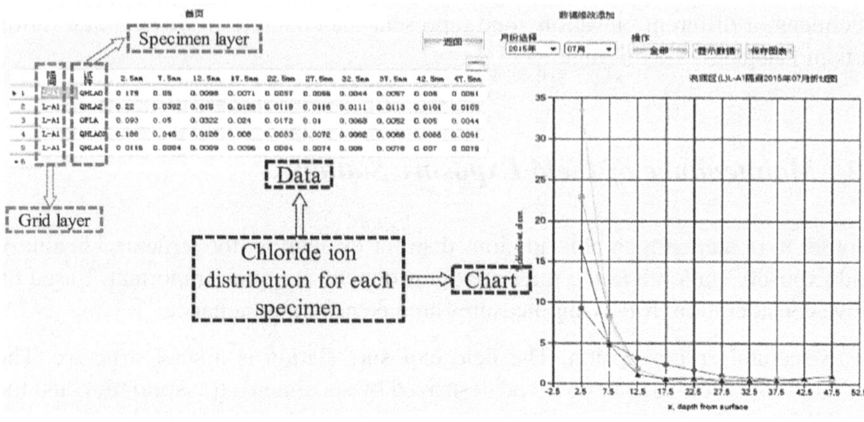

Fig. 2 Software interface for data informatization

3 Utilization of the Field Exposure Station

3.1 *Multi-environmental Time Similarity*

Multi-environmental time similarity (METS) [8–10] method solves the problem by introducing one Reference structures (RS) in the same or similar environment of Objective structures (OS). Generally, the service time of RS should be more than 20 years. Similarity relationship is established on the analysis of inspection data of RS and indoor accelerated tests with specimens of OS and RS. Then deterioration parameters' information of OS in outdoor natural environment can be deduced. Figure 3 shows the methodology of METS schematically.

3.2 *Life Prediction Technology Roadmap*

For Hangzhou Bay Bridge, there exists big difficulties to establish the deterioration parameters' relationship between the outdoor natural environment and the indoor accelerated environment. Moreover, the correlation between the results of artificial climate simulation test and the results of field test is the key to apply the results of artificial climate simulation test to the field practice and to evaluate the durability. In response to the above-mentioned key issues, Zhejiang University proposed METS (the principle is shown in Sect. 3.1) to carry out durability evaluation and life prediction of Hangzhou Bay Bridge.

The method introduces the Zhapu Port, which has similar environmental conditions as the Hangzhou Bay Bridge and a certain number of years of military service as reference objects (environment similarity). Similarity relation of durability parame-

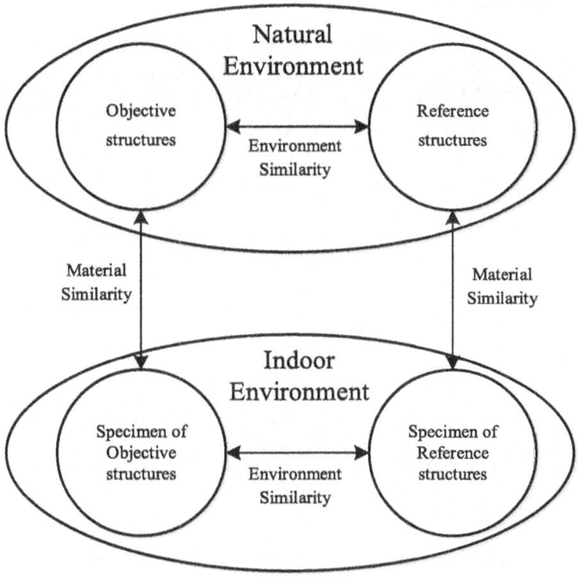

Fig. 3 Schematics of METS method

ters of concrete structures with time changes in two environments (time similarity) can be established through the on-site tested results of the third-party reference objects and the results of corresponding concrete specimens indoor accelerated test. Utilizing the above similarity relation and the results of indoor acceleration test of the same mix ratio concrete specimen of Hangzhou Bay Bridge, the durability evaluation of the concrete structure of Hangzhou Bay Bridge can be realized. In the above evaluation system, concrete specimens set in the field exposure station can provide a large amount of measured data without destruction, so constant detection and evaluation of the durability of the concrete structure of the Hangzhou Bay Bridge can be realized.

3.3 Life Prediction Process

The durability information of different structural components of the Hangzhou Bay Bridge is obtained through the indoor acceleration test and the field test results. Combined with the Bayesian update method and the chloride ions data collected from field exposure station, the key parameters (e.g. surface chloride ion concentration C_s and the diffusion coefficient D) for life prediction keep updating. And then the METS theory is used to evaluate and predict the durability of the different structural parts of the bridge. Based on life prediction method, the concrete durability warning system for Hangzhou Bay Bridge is developed (Fig. 4).

The warning system consists of three modules:

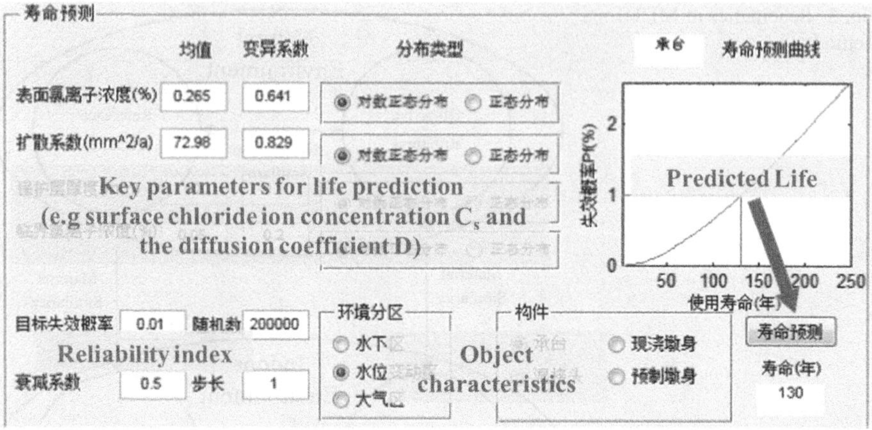

Fig. 4 Hangzhou Bay Bridge concrete durability warning system

(1) Database module. Using the chloride ion concentration data into the bridge durability database, the surface chloride ion concentration C_s and the diffusion coefficient D are fitted by Fick's second law.

(2) Parameter update module. The statistical data of the surface chloride ion concentration C_s and the diffusion coefficient D are obtained by detecting and fitting every year's data. "Update information" combined with "prior information" of previous years by the Bayesian update method can obtain the "posterior information" of the surface chloride ion concentration C_s and the diffusion coefficient D.

(3) Life prediction module. The data of surface chloride ion concentration C_s, diffusion coefficient D, concrete cover thickness C, and critical chloride ion concentration C_{cr} are entered into Hangzhou Bay Bridge concrete durability warning system. Finally, the durability life of the bridge is obtained by Monte-Carlo numerical simulation method.

3.4 Life Prediction Results

The time-varying relationship curve of posterior failure probability of each part of the bridge is shown in Fig. 5.

Figure 5 shows that the time-varying curve of failure probability of each part of the bridge from 2015 to 2018. The information was updated to the end of 2018 and it can be seen from the figure that the service life of each part of the bridge exceeded the design life of 100 years.

In Fig. 5, it can be found that the predicted life of cast-in-place joint is less than that of other parts, while the predicted life of cast-in-place body, precast body and

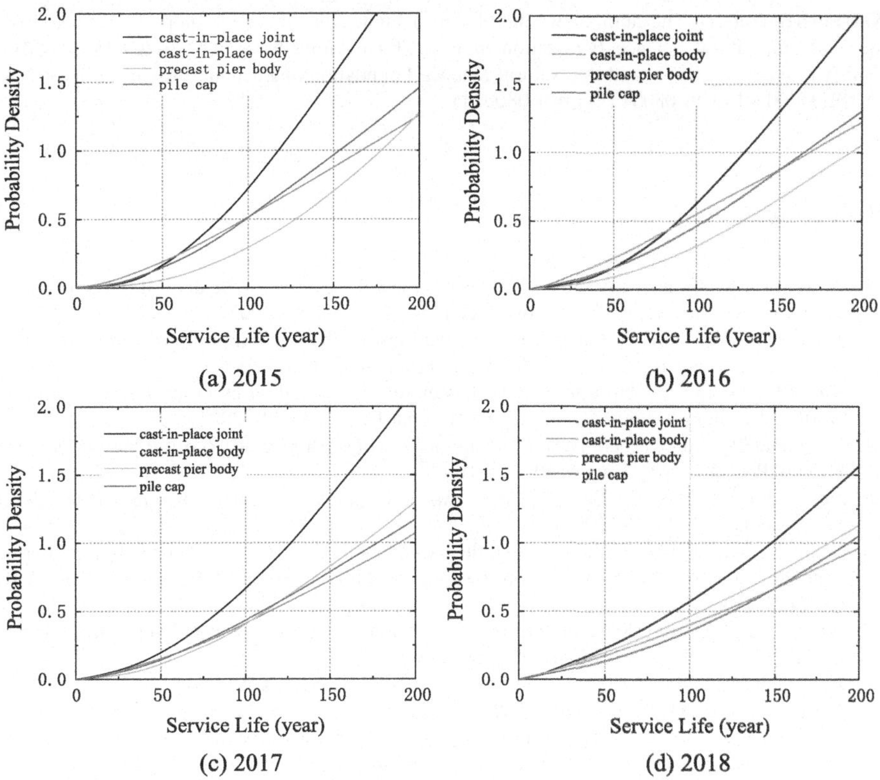

Fig. 5 Concrete durability life of Hangzhou Bay Bridge

pile cap are closer. The predicted results of pile cap and cast-in-place joint are stable in four years, and the range of change is within 15 years.

4 Conclusion

The exposure test station consists of atmospheric area, splash area, tidal range area and underwater area. In order to ensure the continuity, validity and accuracy of the data, this paper puts forward some problems that should be paid attention to in field exposed station management, such as the self-preservation of the structure and the protection of concrete specimens. In order to manage and maintain the field exposed station effectively, the specimen management system is introduced. Based on METS theory and reliability theory, the Concrete Durability Warning System is developed. The numerical simulation shows the service life of the components meeting the requirements of 100 years designed service life.

Acknowledgements The authors would like to acknowledge financial support received from the National Natural Science Foundation of P.R. China (Grant Nos. 51878610, 51820105012, 51638013, and 51778577) and the Natural Science Foundation of Zhejiang Province (Grant Nos. LY18E080003, LQ19E080012, LQ19E080011).

References

1. Li, G., Dai, K., Yuan, Y.: A study on the experimental methods for the durability performance of reinforced concrete. J. Huaihai Inst. Technol. **11**(3), 56–59 (2002)
2. Jin, L., Jin, W., Zhao, X.: Optimization of field test method on durability of coastal concrete structure. J. Southeast Univ. (Nat. Sci. Ed.) **36**(Sup(II)), 61–67 (2006)
3. Kang, B.: Design and building of CHEC's engineering material exposure testing station in Northeast China (Port of Jinzhou). China Harbour Eng. **2**, 35–38 (2004)
4. Li, S., Fan, W., Xu, N., et al.: Design and application of offshore field exposure station(Donghai bridge). Port Waterway Eng. **6**, 30–32 (2011)
5. Jin, L., Jin, W., Chen, T.: Design of field exposure station on coastal concrete structure. Port Waterway Eng. **2**, 14–18 (2008)
6. Wang, S., Tian, J., Fan, Z.: Research on theory and method of service life prediction of marine concrete structures based on exposure test and field investigation. China Harbour Eng. **A01**, 68–74 (2010)
7. Zhao, T., Li, Q.: High Strength and High Performance Concrete and Its Application. China Building Materials Industry Press, Beijing (2004)
8. Jin, L., Jin, W., Wang, H., Xia, J.: Multi-environmental time similarity theory and its application. J. Zhejiang Univ. (Eng. Sc.) **44**(4), 789–797 (2010)
9. Jin, W., Li, Z., Xu, C.: Life prediction method of concrete structures based on relativistic information entropy. J. Zhejiang Univ. (Eng. Sci.) **46**(11), 60–66 (2012)
10. Jin, L.: Multi-environmental Time Similarity Theory and Its Application in Durability of Coastal Concrete Structures. Zhejiang University (2008)

Evaluating Carbonation-Induced Corrosion in High-Volume SCM Mixtures Through the Square Root Model

Federico Aguayo, Anthony Torres, Omkar Thombare and Thano Drimalas

Abstract The dominant cause of premature deterioration and reduced service life of reinforced concrete structures is corrosion. While chloride ingress has typically been the leading cause of corrosion, carbonation-induced corrosion has become an increasingly serious durability concern as higher supplementary cementing material (SCM) replacements have become more commonplace in sustainable concrete mixtures. This study investigated the carbonation resistance for a series of fly ash concrete mixtures placed in an outdoor exposure site to assess the influence of climatic conditions on the propagation of natural carbonation. The carbonation front was determined at 1, 2, and 5 years of exposure for each mixture. In addition, results are compared to similar mixtures placed in an accelerated carbonation chamber at 4% CO_2 concentration. The results highlight the efficacy of the square root model to predict the carbonation front and provide an indication of the initiation phase of reinforcement corrosion induced by natural CO_2. The carbonation coefficient using the square root model at various ages as well as in accelerated conditions are compared for various mixtures.

Keywords Natural carbonation · Accelerated carbonation · Square root model · Fly ash concrete

1 Introduction

The corrosion of reinforcing steel in concrete is considered one of the most serious forms of durability mechanisms plaguing concrete structures throughout the world. According to the National Academy of Corrosion Engineers (NACE), they estimated the global cost of corrosion at a staggering \$2.5 trillion in 2013 [1]. These costs include corrosion incurred in the infrastructure and transportation sectors which are

F. Aguayo (✉) · A. Torres · O. Thombare
Texas State University, San Marcos, TX 78666, USA
e-mail: fred.aguayo@txstate.edu

T. Drimalas
University of Texas at Austin, Austin, TX 78759, USA

© RILEM 2020

J. F. Martirena-Hernandez et al. (eds.), *Proceedings of the International Conference of Sustainable Production and Use of Cement and Concrete*, RILEM Bookseries 22,
https://doi.org/10.1007/978-3-030-22034-1_41

the domain of reinforce concrete structures, and is the main reason for degradation and reduced service life of concrete structures [2]. In reinforced concrete structures, the surface of the steel rebar is normally protected from corrosion by a thin passive layer of iron oxide formed by the high alkalinity of the concrete's pore structure. This phenomenon will protect the steel from corrosion indefinitely unless the passive layer is destroyed either by the presence of an excessive amount of chloride at the location of the steel, and/or by the reduction of the pH in the vicinity of the steel due to the diffusion of gaseous carbon dioxide (CO_2) from the environment reacting with the cement hydrates, a process known as carbonation.

When used appropriately SCMs can increase the resistance to chemical threats such as alkali-silica reaction, sulfate attack, and chloride penetration. However, the carbonation resistance of such cementitious systems is found to be reduced, especially as the volume of cement clinker replacement is increased [3–5]. It is one of the few cases where high dosages of SCMs do not improve the durability of reinforced concrete and in some cases, exacerbate the issue. Mixtures containing high amounts of SCMs tend to have low calcium hydroxide contents lowering the alkalinity of the concrete, and the lack of calcium hydroxide increases the rate of carbonation of other calcium-bearing hydrates in the concrete. This decrease in carbonation resistance could be attributed to the reduction in CaO buffer capacity of the cementitious system with respect to increase in the partial substitution of clinker fraction. In addition, increase in atmospheric CO_2 levels due to increase in man-made industrial and vehicular emissions in urban areas increase the potential threat to reinforced concrete structures by accelerating carbonation. Thus, the estimation of potential carbonation resistance based on prevailing exposure conditions, as well as type and composition of materials used in the structure is necessary to construct a building with the intended life expectancy. This paper presents a unique study on the carbonation resistance of concrete for a series of plain and blended fly ash mixtures placed in both accelerated and natural carbonation exposure. Carbonation coefficients, determined using the square root model in sheltered and unsheltered conditions after 5 years of exposure, are compared to those specimens evaluated under accelerated conditions to determinate the efficacy of the model to predict the carbonation front and provide an indication of the initiation phase of reinforcement corrosion induced by natural CO_2.

2 Experimental Program

2.1 Materials and Sample Preparation

The cement used was an ordinary portland cement (OPC) conforming to an ASTM C150 Type I/II specification. A total of 15 concrete mixtures were produced with the concretes divided into three series of mixtures designated LC, HC, and CONT. The first two series consisted of mixtures with either a low-calcium (LC) or high-calcium

Table 1 Details of concrete mixture proportions and Compressive Strengths

Mix series	# of mix	W/CM	Total binder (OPC + Fly Ash) (kg/m^3)	SCM (%)	f′c @ 28 days (MPa)
CEM-CONT	3	0.45–0.50	310, 340, 400	–	34–43
CEM-LC	6	0.45–0.50	310, 340, 370, 400	15, 30, 50	18–46
CEM-HC	6	0.45–0.50	310, 340, 370, 400	15, 30, 50	18–41

(HC) fly ash used as cement clinker replacement ranging from 15 to 50% by mass of total cementitious content. The last series included the control mixtures with no fly ash. Each series of mixtures included two water to binder (W/binder) values (0.50 and 0.45), and four cement contents of 310, 340, 370 and 400 kg/m^3. The mixture proportions along with compressive strength data are given in Table 1.

2.2 *Testing Procedure*

Natural and accelerated tests were performed under different storage conditions and cast at two different times for each testing condition. Each mixture included a set of 100 × 200 mm concrete cylinders for determining compressive strength at 28 days, and prismatic specimens of 100 × 100 × 400 mm to determine the carbonation depth.

A first set of samples was produced to determine the carbonation resistance under natural exposure. The natural test consisted of a pair of concrete prisms in which one specimen was outside and unprotected from direct rainfall (unsheltered), and another one protected from rainfall (sheltered) through the use of Stevenson Screens. All specimens were stored and cured under wet burlap and plastic for 24 h at 23 ± 3 °C. After 24 h of curing, specimens were then labeled and placed outdoors either in sheltered or unsheltered environments until the age of testing. The carbonation depth was measured after 1, 2, and 5 years of storage.

A second set of prisms was produced to investigate the carbonation resistance through accelerated testing. The concrete production was identical to the first set of samples however, a 1 and 7 day curing treatment was also explored to evaluate the influence of curing age on the rate of carbonation. The prisms were removed from the moist curing chamber at an age of 1 and 7 days and were allowed to air dry in a controlled environment (23 °C, 50% RH) for a minimum of 3 h. Thereafter, prisms were vertically placed into the carbonation chamber with CO_2 level of 4.0 ± 0.5% by volume and a temperature of 23 ± 2 °C. The relative humidity (RH) of the chamber was controlled and monitored at 57 ± 5% through the use of an industrial dehumidifier attached to the chamber. The ages of measuring carbonation depth were 28, 56, 63, 70, and 105 days after placing them in the carbonation chamber.

Table 2 Lowest and highest monthly yearly averages for Texas, USA

	Texas, USA
Temperature, °C	30–12
RH, %	71–46
CO_2, ppm	303–460
Precipitation, mm	873.33

For measuring the carbonation depth, a slice (approximately 80–100 mm thick) was broken off each prism. The depth of carbonation was measured perpendicular to the face by spraying the freshly split face with a 0.5% phenolphthalein ethanol solution with a 0.5 g phenolphthalein and 90 ml 95 V/V% ethanol diluted in water to 100 ml [6]. In the noncarbonated part of the specimen, where the concrete was still highly alkaline, a purple-red color was obtained. In the carbonated part of the specimen where the alkalinity of concrete is reduced, no coloration occurred. The depth of the uncolored zone was measured to the nearest 0.5 mm (0.02 in) made at five points along each side of the square cross-section prism slice for a total of 20 values and average presented. The carbonation coefficient K was determined by calculating the regression of the carbonation depth as a function of the square root of time (Eq. 1):

$$K = \frac{d_k}{\sqrt{t}} \tag{1}$$

where K is the carbonation coefficient in $mm/\sqrt{yr}$, d_k the carbonation depth in mm and t the time in years.

3 Results and Discussion

3.1 Environmental Conditions

Table 2 illustrates the yearly average temperature and relative humidity at the natural exposure site in Texas, USA. These data were taken from a local weather station near the research laboratory [7]. While the data in Table 2 provides a summary of the yearly environmental conditions at the site, the daily minima and maxima for relative humidity and temperature can than the values presented below. It should be noted however, the average relative humidity observed at the site is optimal for progression of carbonation as considered by many researchers [8]

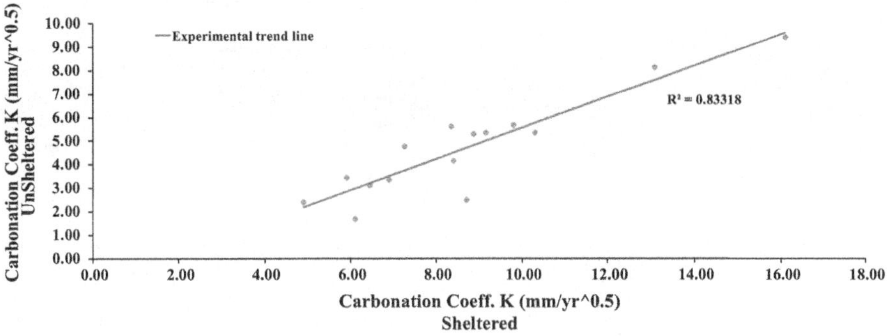

Fig. 1 Accelerated carbonation chamber

3.2 Comparison Between Sheltered and Unsheltered Conditions

The comparison between carbonation rate with respect to both sheltered and unsheltered conditions is presented in Fig. 1. It can be seen that the carbonation rate is significantly lower in the unsheltered condition than in the sheltered condition as the saturated concrete surface prevents CO_2 from penetrating into the concrete. In most cases, the carbonation coefficient is approximately 1.2–1.6 times higher in the sheltered condition compared to the unsheltered specimens; however, there were a few extreme cases where the carbonation rate was more than three times in the sheltered condition; many of these larger differences were found in blended fly ash mixtures. Nonetheless, no systematic change compared to concrete OPC with regard to binder type or W/binder is evident going from the sheltered to unsheltered exposure.

3.3 Comparison Between Accelerated and Natural Carbonation

Figures 2 and 3 presents the accelerated carbonation coefficients between sheltered and unsheltered condition, respectively. There is a strong correlation of the carbonation coefficient between the accelerated and sheltered samples ($R^2 = 0.91$). However, the relationship decreased with the extended 7-day curing age before exposure for samples evaluated under accelerated condition ($R^2 = 0.80$). The observed results however, are expected as the samples placed outdoors were also only cured for 1-day prior to exposure. In general, the trend is fairly consistent regardless of the different compositions and W/binder ratios evaluated in the study. In addition, it can be seen that replacing the portland cement with fly ash at various replacement levels reduces the resistance to natural carbonation; where the higher replacement level, the higher the carbonation depth is observed. In fact, the carbonation rate for higher volume fly

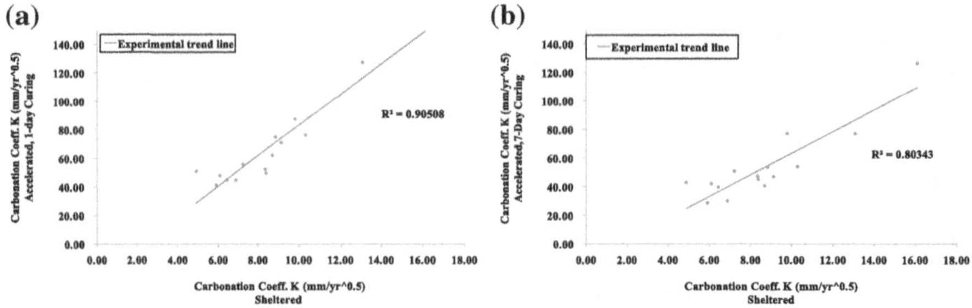

Fig. 2 Comparison of accelerated condition for specimens wet cured for 1-day cure (**a**) and 7-day (**b**) as a function of the natural carbonation in sheltered condition

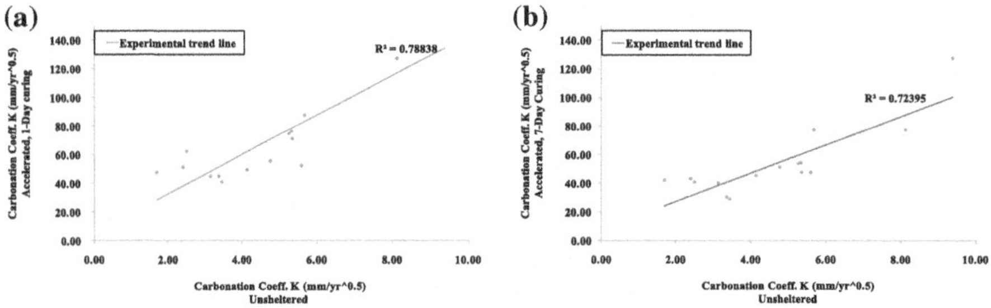

Fig. 3 Comparison of accelerated condition for specimens wet cured for 1-day cure (**a**) and 7-day (**b**) as a function of the natural carbonation in unsheltered condition

replacement were in many cases severely exacerbated to the point where carbonation depth measurements were no longer possible when exposed to the accelerated carbonation chamber.

On the other hand, a decreasing correlation was observed of the carbonation coefficient between the accelerated and unsheltered samples ($R^2 = 0.79$ and 0.73). This may be attributed to the differences between constant RH and dynamic conditions as present in natural exposure, especially for specimens that may be in direct contact with the rain. Early age curing should impact the region close to the concrete surface where hydration will be stopped due to drying. However, the huge impact of early age curing on carbonation as observed here may not remain valid in unsheltered condition. Indeed, rain water penetrating into concrete can result in a recovery of hydration within the dried depth of concrete. In this case, the difference of carbonation depth between early age and advanced age curing is expected to be reduced significantly.

4 Conclusions

From the results presented herein, the following conclusion can be made:

- Higher carbonation coefficients were observed for mixtures stored in sheltered condition (roughly 1.2–1.6 times higher); many of these larger differences were found in blended fly ash mixtures
- The carbonation coefficients for those evaluated in the accelerated carbonation chamber at 4% CO_2 and 57% RH correlated well with similar samples placed outdoors evaluated after 5 years of exposure. The results showed that the carbonation resistance of concrete can be assessed through the use of an accelerated chamber. However, care must be taken when higher volumes of fly ash are incorporated as these mixtures observed full carbonation in short period of time and were no longer measureable. This is likely attributed tot to the poorly cured specimens exacerbating the results.

Acknowledgements The authors wish to acknowledge the financial support from Lafarge Centre de Recherche. In addition, the research team at the University of Texas in helping gather these data and maintain carbonation exposure site.

References

1. NACE: International measure of prevention, application, and economics of corrosion technologies study (IMPACT). NACE International, 15835, Park Ten Place, Houston, TX 77084, p. 216 (2016)
2. Reis, R., Malheiro, R., Camões, A., Ribeiro, M.: Carbonation resistance of high volume fly ash concrete. Paper presented at the Key Engineering Materials (2015)
3. Khunthongkeaw, J., Tangtermsirikul, S., Leelawat, T.: A study on carbonation depth prediction for fly ash concrete. Constr. Build. Mater. **20**, 744–753 (2006)
4. Sisomphon, K., Franke, L.: Carbonation rates of concretes containing high volume of pozzolanic materials. Cem. Concr. Res. **37**(12), 1647–1653 (2007)
5. Wierig, H.: Longtime studies on the carbonation of concrete under normal outdoor exposure. In: Proceedings RILEM Seminar on the Durability of Concrete Structures Under Normal Outdoor Exposure, pp. 239–249 (1984)
6. Chang, C., Chen, J.: The experimental investigation of concrete carbonation depth. Cement Concrete Res. (Elsevier Ltd.) **36**, 1760–1767 (2006)
7. Cedar Lake Ventures, Inc. weatherspark.com. December 1, 2012. https://weatherspark.com/averages/29672/Austin-Texas-United-States. Accessed 26 May 2015
8. Andreas, L., Fabrizio, M.: Carbonation of concrete: the role of CO_2 concentration, relative humidity and CO_2 buffer capacity. Mater. Struct. **50**, 30 (2017). https://doi.org/10.1617/s11527-016-0917-2

Diagnose on the State of Deterioration of the Materials of a Ship of the Company of IT Transports of Combustible Transcupet Caibarien

Santiago V. Sánchez Pérez, Gilberto J. Quevedo Sotolongo and Claudia Rodríguez Rodríguez

Abstract In our country the industrial constructions are very use in different economy sectors, many of them were built in the last century, for that reason is necessary their continuous maintenance and restoration. Some of them present a bad function due of the high grade of deterioration of their structural materials, such is the case of one of the buildings of the fuel transportation company (TRASCUPET) of Caibarién city, in which their main structural elements are in advanced deterioration state, due to: their proximity to the sea, the insufficient protection of their reinforce elements or a high exposition grade of the structural materials to a highly aggressive atmosphere, the absence of maintenance, and for the application of big technological loads due a change in their function. In the work was made: a diagnosis of the physical and mechanical state of the structural elements materials; the geometric model and the structural analysis of the building, the review of the structural capacity of the reinforced concrete elements, and structural steel elements, as well as its unions, taking into a account the concrete deterioration, and the high grade of corrosion of the reinforcement steel and of the structural one. In order to propose solutions to the intervention of the most compromised elements from the point of view of their resistance, and the limitation in terms of work areas and maximum loads to be use in overhead crane.

Keywords Diagnosis · Materials · Corrosion · Deterioration · Structural capacity

S. V. S. Pérez (✉)
Civil Engineering Department, Faculty of Constructions, Central University "Marta Abreu" of the Villages, Santa Clara, Cuba
e-mail: santiago@uclv.edu.cu

G. J. Q. Sotolongo · C. R. Rodríguez
Hydraulic Engineering Department, Faculty of Constructions, Central University "Marta Abreu" of the Villages, Santa Clara, Cuba
e-mail: quevedo@uclv.edu.cu

C. R. Rodríguez
e-mail: claudiamrr@uclv.cu

© RILEM 2020

J. F. Martirena-Hernandez et al. (eds.), *Proceedings of the International Conference of Sustainable Production and Use of Cement and Concrete*, RILEM Bookseries 22,
https://doi.org/10.1007/978-3-030-22034-1_42

Photo 1 Facade of the reparations building, TRANSCUPET, Caibarién (built in 1922)

1 Introduction

The old buildings, due to the normal deterioration and aging of their materials require a constant and continuous maintenance so that they can continue working after a long period of being built (96 years), moreover, when these buildings are in a very aggressive environment due to its proximity to the sea (approximately 100 m), and are subject to much higher and completely more different loads in nature to those that were used in their design. Such it is the case of the reparations building of the Transport of Fuel Company (TRANSCUPET Caibarién), building that were originally conceived and built to store goods, and that for some time have been used to repair the truck cistern used for fuel transportation (Photo 1).

To be able to repair the fuel transportations truck cistern, were carried out a series of transformations to the structure of the edifices to install three bridge cranes of different capacities (2.0, 3.5 and 5.0 tons) (Photo 2).

Equipment that transmit to the structure, big dynamics vertical and horizontal loads which for the structure was not designed.

The above-mentioned, together with the aggressiveness of the environment and the humidity due cover filtrations, and breaks of the plumbing downspouts, caused the deterioration of some of the structural steel elements (corrosion, large deformations and vibrations), as well as, the failure of some of the structural reinforcements concrete elements which presented: Detachment of the reinforcements clear cover, great corrosion of the same ones, excessive cracking, large deformations, and great vibrations during the bridge cranes work, situations that compromise, the resistance and the structural stability, and impaired or prevented the proper functioning of the offices area (frontal and lateral part of the building) see photo below.

Due the critical condition of some of the structural elements of the edifice that constitute the construction, of the TRANSCUPET company of Caibarién requested the International Center of Havana, S.A, CIH, the professional scientist technicians services of the CIDEM specialists of the Constructions Faculty of the Central Univer-

Photo 2 Central building where a bridge crane of 2.5 tons was added

sity "Martha Abreu" of the Villages, to carry out a study that included the structural "Analysis and the assessment of the pathologies of the repair building, with the objective to establishing the measures that should be taken immediately, to avoid the possible failure of other structural elements, and to be able to continue exploiting the edifices, and also, to assess the possibility of increase the bridge cranes capacities in the edifices 3 and 4.

For the realization of this task we aid ourselves of one of the CAE tools, tool that facilitating the creation of the construction geometric model, and also allows to integrate its properties, conditions that is exposed, materials, etc. allowing to calculate how the structure will behave in the reality, in so diverse aspects as: deformations, resistance, thermal characteristic, vibrations, etc.

Nowadays it is normal to use professional software (with the integration of the "Finite Element Analysis"—FEA, together with the "Computer Aided Design"—CAD), with the objective of reduce the project times. This way, important advantages like the elimination of unnecessary tests in prototypes, reductions of time and money, increase in the perception of the response to the fatigue load of the product, and optimization of the fatigue design.

If the geometric model of the construction represents authentically the real structure, can be carried out a many experimental or virtual tests, that facilitate the study and the validation of the causes that cause the pathologies, to study the exploitation situations that are not have previously analyzed, and to settle down the possible variants of solution. All of the above-mentioned facilitates to reduce considerably the structures diagnostic studies because the kindnesses of the modeling are exploited to the maximum always that exists a good correspondence between the real structure and the geometric model of the construction.

2 Development

To the realization of the work the CIDEM specialists, assumed the following tasks:

- General study of the problem and definition of the activities to carry out.
- Geometric rising of the edifices.
- Pathology Survey of the construction and establishment of the initial hypotheses.
- Modeling the geometry of the edifices
- Modeling of the materials that make up the structural elements of the edifices.
- Modeling of the states and the combinations loads (For current conditions and for a possible bridge crane of 10 tons in the edifices #3 and #4).
- Evaluation of the runs results models for the current technological loads, and for a possible bridge crane of 10 tons.

Survey of the edifices pathologies

The edifice #4 presents a better technical conditional, and with the present bridge crane, doesn't exist any important structural problem, but, it's observed the absence of total maintenance that could deteriorate the steel structure and compromise its stability (see Photo 3).

The situation of the edifice #3 from the technical point of view is a little more compromise than the previous one, in its first two distances among columns have a reinforced concrete structure composed of slab, beam and columns. The technical state of the second distance among columns, that has the bathroom above, is enough critic, with one beam practically in failure, and the other with a very high grade of the steel corrosion with damage of the clear cover, and the other concrete beams and concrete slab, equally with presence of corrosion in the reinforcement steel (see Photos 4 and 5).

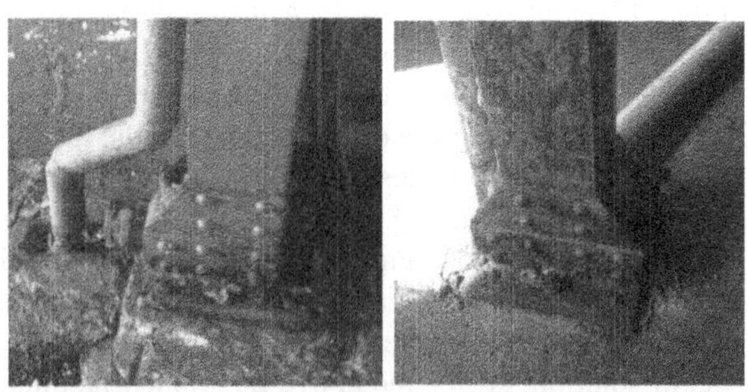

Photo 3 Corrosion in the lower part of the columns and in the base plate

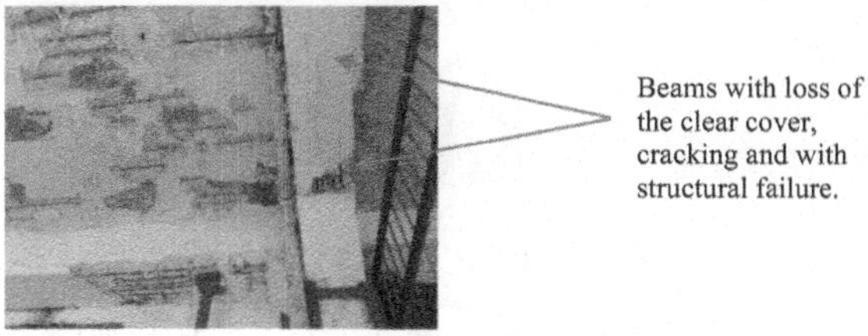

Photo 4 Loss of the clear cover reinforcement in concrete slabs and beams, corrosion of the reinforcement steel, cracking of the concrete and large deformations in the reinforced concrete elements

Photo 5 Reinforced concrete beam with loss of the clear cover reinforcement steel, corrosion of the reinforcement steel, cracking of the concrete and large deformations, element with structural failure

The lower part of the external columns shows a high grade of deterioration, evidencing in the same way the absence of total maintenance that could deteriorate the steel structure and to compromise its stability (see Photo 3).

The external columns of the edifice #2 (Columns of the left), are in the same technical state as the external columns of the edifice #3 (see Photo 3).

The bridge crane of this edifice, on the right side is supported by steel columns that are connected to the main structure of the concrete edifice (columns there are common with the edifice #1, see Photo 6), solution that undoubtedly has been the cause of many of the problems that today presents this concrete structure with some elements in imminent failure (Photo 6).

The edifice #1 have a similar structural solution to that of edifice #2, it is formed by a main structure of concrete with slab, beams and columns, and connected to this structure there is a steel structure without ground connection which supports the bridge crane work. The general technical condition of the main structure of concrete is not good and in some areas (the first distance among columns, see previous Photo) the elements are in critical conditions.

Numeric modeling of the structure

In order to study the structural response of the building structure under the action of the acting loads in terms of efforts, rotation, and displacements, a model was made by means of the employment of professional software STAAD Pro 2008, with the main factors that influence the behavior of the structure.

Photo 6 Reinforced concrete structures of the edifice #1, beams and columns with loss of the clear cover, corrosion of the reinforcement steel, cracking of the concrete, beam with big deformations, and structural failure

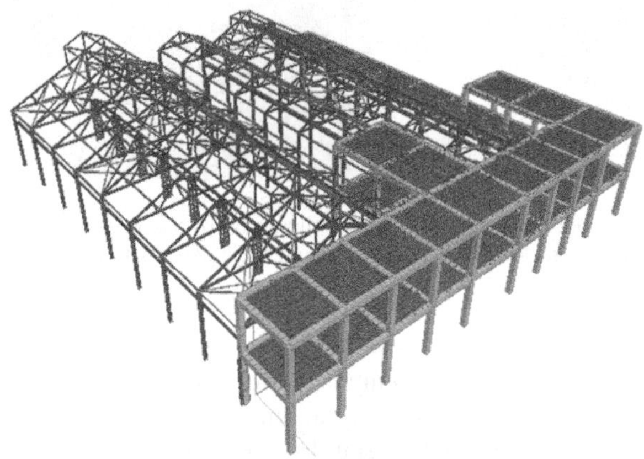

Fig. 1 Geometric model of the construction

Geometric modeling

To make the geometric model of the structure were taking a count the dimensions of the different structural elements that conform each one of the construction parts: The columns, the beams and the slabs of armed concrete of the front part and of the lateral right part of the construction. The three types of columns of steel of the three edifices, the beams of the three bridge cranes, as well as the elements that conform the struts of the covers (Fig. 1).

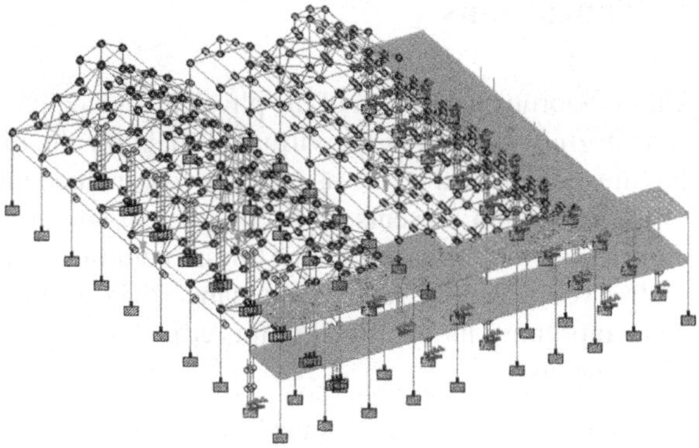

Fig. 2 Live and technological loads (Floor, roof and bridge cranes)

Modeling of the material

When not having the real data of the materials it was considered to use for the reinforced concrete elements a resistance to the compression for the concrete of 17.5 MPa, and for the reinforcement steel a resistance to the fluency of 240 MPa, while for the structural steel elements of a resistance was also used to the fluency of 240 MPa. For both materials it is considered an elastic behavior [1].

Modeling of the loads

For the run of the numeric model three types of loads were considered acting on the structure, which were calculated according to established in the Cuban loads standard for the calculation of structures [2], they were:

- The dead load, coming from the self-weight of the structural and not structural elements (floor solution and roof solution).
- The alive load, in intermediate floors and in the roof due to the use of the construction.
- The technological load, loads due to the use of the bridge cranes (Fig. 2).

In the runs of the virtual model several load combinations [3] were used in order to evaluate the behavior of the structure under the most unfavorable work conditions. Two load situations were considered with the corresponding combinations whit different technological loads positions (**Situation A** and **Situation B**).

A. **Current situation**: considering that bridge cranes of 5.0 tons operate on edifices #4 and #3, and bridge cranes of 3.5 and 2.0 tons operate respectively on edifices #2 and #1:

B. **Future situation**: Analyze the runs of the virtual model considering that bridge cranes of 8.0 tons operate on the edifices #4 and #3, and in the bridge cranes of 3.5 and 2.0 tons operate respectively on edifices #2 and #1.

3 Conclusions

After a rigorous inspection of the physical and technical state of the edifices, as well as analyzing the runs of the virtual models for the different load combinations for the Situations A and B, it was possible to conclude the following:

The edifice #4 presents a better technical conditions, and for the current bridge crane (Situation A) doesn't exist any structural problem. For the external columns is observed a lack of total maintenance that could deteriorate the steel structure, and compromise its stability. For this edifice it propose to carried out immediate the following tasks:

- Pass the top of the bridge crane beam to the second column, to avoid the vibrations in the structure of reinforced concrete of the offices for the bridge crane braked.
- A general maintenance must be given to the metallic structure, with emphasis on the lower parts of the external column, included the plate bases.

For the Situation B: To change the current bridge crane for one of 10 tons, in addition to the previous measures it would be necessary to change the bridge crane beam and the externals columns for larger ones, as well as, the placement of crossbeams, type crosses of San Andrés in the initial and the final distance between columns, and to check the columns foundations.

The edifice #3 the external columns have the same situation that the external columns of the edifice #4, being evidenced a lack of maintenance and so they should be intervened in the same way that it was explained previously. For this edifice should take the following additional measures:

- Not to allow the operation of the bridge crane in the first three distance between columns, to avoid the transmission of vibrations to the concrete structure that exists in that area, structure that have an appreciable deterioration
- To put the end of the bridge crane above the fourth column and to eliminate the bridge crane beam in those first three distance between columns.
- To continue using the locals of the assembly hall and the bathroom that are in the area of this first two distance between columns, should be taken the following measures:
- To bolster the concrete slab (the one that has the bathroom above), in order to discharge the last beam of that area which is in critical state.
- To carry out an anticorrosive treatment to the steels of the armed concrete beams and the slab.
- To bolster the 4 beams that supports the concrete slab.
- For the concrete slab that has the assembly hall above, should be carried out to the same anticorrosive treatment described previously.

The edifice #2, their structural elements in general present a not good physical and technician state, and should be taken the following measures to continue their exploitation:

- To give a general maintenance to the metallic structure, with emphasis in the inferior parts of the columns, included the plate bases.
- To eliminate the work of the bridge crane in the first five distance between columns where the concrete structure is more affected,
- To put braces type San Andrés cross in the first and the last distance between columns in where use the hook crane.
- To bolster the entrance reinforce concrete beam, and in the future to put a new piazza of concrete.

The edifice #1, his main concrete structure also have a not good physical and technician state, and in some areas it is critical, so is necessary to take the following measures to continue his exploitation:

- A general maintenance to the metallic structure.
- To limit the work of the bridge crane at the first five distance between columns, area where the concrete structure is more affected.
- To put braces type San Andrés cross in the first and the last distance between columns in use by the bridge crane.
- To closes the locals of the technical department boss and the juridical in immediately way, closing the access to this areas.
- All the beams should be bolstered in the first five distance between columns and give there the anticorrosive same treatment described previously to the concrete slabs.
- In the first five distance between columns should be retire the metallic structure of the bridge crane, the columns and the bridge crane beam that are connected to the columns of reinforced concrete common to the edifices #2 and #1.
- Due to the deplorable state of many of the structural elements in the edifices #2 and #1, and as a result of the analysis of the different loads states, we can affirm that it is not possible in these last two edifices to change the bridge crane to one of more capacity.
- Due to the high grade of deterioration of several of the structural elements is not allow for the future operation of the edifices the simultaneous use of the bridge cranes in the edifices #3 and #4, and in the edifices #1 and #2.

References

1. Density of natural and artificial materials and of the construction elements like load design, NC XX (2002)
2. Constructions load factors and combinations, NC 450 (2006)
3. Constructions use load, XX NC (2002)
4. McCormac, J.C.: Steel Structural Design LRFD Method, 2nd edn
5. Manual of the Construction in Steel, AISC-LRFD, 2nd edn
6. Segui, W.T.: Steel Design, Load Resistance Factors Design (LRFD), 2nd edn

Part IV
Recycled Concrete Aggregate RCA

Prediction of Mechanical Properties of Concrete Made with Recycled Concrete Aggregates Using Statistical Analysis of Data Available in Literature

Anuruddha Jayasuriya, Tola Chen, Emily Shibata and Matthew P. Adams

Abstract The concrete design and construction industry has been eagerly adopting sustainability initiatives over the last 20 years to help reduce its carbon footprint, improve economic sustainability, and ensure the important natural resources that support the industry remain available for years to come. Despite these advances, recycled concrete aggregates (RCA) have been significantly underutilized even though they are becoming an increasingly available sustainable resource that can be used to reduce the amount of natural aggregates used in concrete. Use of RCA has been limited, in part, because of a lack of standardized guidance on creating mixture designs and how aggregates may impact concrete properties. Presented in this paper are the results of a statistical analysis of data from over 80 peer reviewed articles that can be used to predict performance in mixtures containing RCA. A large database consisting of 850 compressive strength, 682 elastic modulus, 197 flexural strength, and 462 splitting tensile strength results was created from this literature review with data ranging from 1988 to 2018. Non-linear regression analyses determined that the empirical trends related to compressive strength, elastic modulus, flexural strength, and splitting tensile strength had a linear relationship with the RCA replacement levels depending on absorption capacity, RCA replacement level, effective water-to-cement ratio, maximum aggregate size, and strength of the RCA's parent concrete. Overall, the RCA replacement ratio and the water-to-cement ratio had significant impacts on concrete mechanical strengths, while other factors such as mix proportioning parameters and aggregate properties had minor effects on the mechanical properties of RCA systems over the large dataset. Additionally, significant reductions in strength were not observed until replacement levels of RCA increased above 20%.

Keywords Recycled concrete aggregate · Mechanical properties · Database analysis · Statistical analysis

---

A. Jayasuriya · T. Chen · E. Shibata · M. P. Adams (✉)
John A Reif, Jr. Department of Civil and Environmental Engineering, New Jersey Institute of Technology, Newark, NJ, USA
e-mail: matthew.p.adams@njit.edu

© RILEM 2020

J. F. Martirena-Hernandez et al. (eds.), *Proceedings of the International Conference of Sustainable Production and Use of Cement and Concrete*, RILEM Bookseries 22,
https://doi.org/10.1007/978-3-030-22034-1_43

1 Introduction

Recycling of demolished concrete as coarse aggregates in production of new concrete has been considered as an innovative method to improve material sustainability. Currently, concrete systems including recycled concrete aggregates (RCA) are being extensively studied at laboratory levels experimentally while monitoring the material behavior based on aggregate properties and aggregate substitution quantities (i.e., full or partial). Despite the large number of experimental studies that have already been completed using RCA, the practical application has still been limited to non-structural concrete, and on average, the aggregate substitution is recommended not to exceed 0.20–0.30 of natural aggregates to recycled concrete aggregate ratio [1]. However, the amount of aggregate substitution is extremely important for desired concrete strength properties considering the quantity of adhered mortar content attached to the natural aggregate. The adhered mortar phase in RCA is commonly believed to be the main reason for concrete with RCA exhibiting poorer material properties compared to standard concrete systems [2–4]. Adhered mortar content impacts water absorption, porosity, crushing value, Los Angeles abrasion value, and specific gravity. The properties of RCA obtained from different sources vary significantly due to the fluctuations in compositions of the parent concrete. A majority of the studies that are available at present explain the statistical trends on how mechanical properties of RCA systems would perform under given materials and prescribed loads [5–9], instead of studying the material variability based on different RCA sources. Therefore, this paper is motivated by the need to develop a large-scale database to establish a fundamental statistical evaluation considering the variabilities of mechanical properties of RCA systems covering a substantial scope of literature.

2 Experimental Database

The database for concrete systems with RCA was compiled from over 80 peer reviewed research articles and the data was collected considering four types of RCA mechanical properties; compressive strength (850 data), elastic modulus (682 data), flexural strength (197 data), and splitting tensile strength (462 data). In addition to the size effects of RCA specimens, mixture proportion factors and aggregate characteristics were included in the database analysis, considering the potential effects that may induce on the mechanical properties of RCA systems. Among those mix proportion factors, effective water-to-cement ratio, total aggregate-to-cement ratio, RCA replacement level were considered whereas maximum aggregate size and absorption capacity were used as the aggregate characteristics in the study.

3 Statistical Analysis Results

3.1 *Initial Findings of the Literature Study*

Effective water-to-cement ratio followed a right-skewed distribution of the histogram distribution with an average ratio of 0.48 bounded between 0.20 and 0.87. The effective water-to-cement ratio was primarily controlled by the absorption capacity of the aggregates used in the mixture proportions. The average absorption capacity of the RCA particles that had been used in concrete mixtures was 5.27% with a standard deviation of 2.30%. Since there is a lack of available guidelines on the proper use of optimum RCA replacement content, the database contained a scattered distribution of RCA replacements from 0 to 100%. For analysis purposes, the RCA replacement levels were sub-divided into six levels consisting of 0, 1–20, 21–40, 41–60, 61–80, and 100%. Eventually, nonlinear regression statistical tools were adopted to study the mechanical trends observed within each of the RCA replacement level against the effective water-to-cement ratio levels.

3.2 *Compressive Strength*

The compressive strength data at 28 days for RCA systems were inversely proportional to the effective water-to-cement ratio regardless of the shape of the specimens that were used in the experiments. The dataset trends were able to match an exponential decaying function for 0 and 100% respective RCA replacement levels using cube and cylinder specimens. Based on the nonlinear curve fitting, it was found that when the replacement levels were increased, the compressive strength trends got more linear with the increasing effective water-to-cement ratio. The highest compressive strength was recorded at 1–20% RCA replacement levels for both specimen types. Considering the 100% RCA replacement level systems, it showed a potential increase in the compressive strengths than that of 61–80% replacement level. This was due to the high contents of adhered mortar in RCA systems and the resulting compatibility of the material properties contributed to show an improved strength performance [6, 10]. In a perspective of engineering mechanics, higher natural aggregate content in RCA has a tendency to induce localized stresses during material testing and eventually results in lower compressive strengths [10–12].

3.3 *Elastic Modulus*

The scattered data points were fitted with a decaying exponential function against the effective water-to-cement ratio, and it was observed that, as the RCA replacement level was increased from 0 to 100%, the exponent constant was reduced. Therefore, those concrete systems that had higher RCA replacement levels showed higher linear trends of elastic moduli in the fitted curves. This behavior was characterized by the increase of adhered mortar content within the concrete where the RCA system behaved as a homogeneous concrete section during testing. It was indicated an approximately 70% reduction of the exponent constant compared to a conventional concrete system that showed linear trends of the elastic modulus properties in RCA systems. Although the RCA systems displayed linear behaviors with high adhered mortar contents, the elastic modulus was decreased with minor fluctuations in RCA systems with replacement levels higher than 20%. Over the selected 682 elastic moduli data, highest magnitudes were shown when the adhered mortar was used within 1–20% range.

3.4 Flexural Strength

For 0 and 100% RCA replacements, the flexural strength decayed with the increasing effective water-to-cement ratio. It was observed that when the RCA replacement level was increased from 0 to 100%, the exponent of the nonlinear fit decreased exhibiting a linear material behavior of the flexural strength and the dispersion of the dataset was closer to the general trend. Similar reasons can be drawn that with high RCA replacement levels, the material homogeneity increases with the corresponding attached mortar content increases. Further, it was clearly noticed that the peak flexural strength response was within the 1–20% range.

3.5 Splitting Tensile Strength

The splitting tensile strength was observed to be inversely proportional to the effective water-to-cement ratio, and it was found that 0 and 100% replacement levels were in good correlation compared to the other replacement level ranges with a fitted decaying exponential function. Although the tensile strength capacity in RCA systems is supposedly lower than conventional concrete systems, 100% RCA replacement level showed a considerable tensile strength capacity which was approximately 14% less than that of the natural aggregate concrete system. This type of higher capacity was expected due to the decrease of the material heterogeneity with higher adhered mortar contents included in the RCA system. The splitting tensile strength also showed similar trends in exhibiting highest strength capacities when 1–20% replacement levels were used.

4 Conclusion

Based on the statistical analysis results, it was seen that the RCA replacement levels less than 20% showed the highest mechanical performance providing an obvious indication to optimize the RCA content at 20% or less. Higher replacement levels were adopted in the concrete mix yielded an increased linear behavior on the material where the mechanical performance was linearly decayed with increasing effective water-to-cement ratio at high RCA replacement levels. This was due to the homogenization of material phases through increase of adhered mortar with increasing RCA replacement level. The 100% RCA replacement levels showed good strength performance due to the higher contents of adhered mortar included in the aggregate system.

References

1. Gonçalves, P., de Brito, J.: Recycled aggregate concrete (RAC) – comparative analysis of existing specifications. Mag. Concr. Res. **62**, 339–346 (2010)
2. Etxeberria, M., Vázquez, E., Marí, A.: Microstructure analysis of hardened recycled aggregate concrete. Mag. Concr. Res. **58**, 683–690 (2006)
3. Butler, L., West, J.S., Tighe, S.L.: Effect of recycled concrete coarse aggregate from multiple sources on the hardened properties of concrete with equivalent compressive strength. Constr. Build. Mater. **47**, 1292–1301 (2013)
4. Oikonomou, N.D.: Recycled concrete aggregates. Cem. Concr. Compos. **27**, 315–318 (2005). https://doi.org/10.1016/j.cemconcomp.2004.02.020
5. Xiao, J., Li, J., Zhang, C.: Mechanical properties of recycled aggregate concrete under uniaxial loading. Cem. Concr. Res. **35**, 1187–1194 (2005)
6. Jayasuriya, A., Adams, M.P., Bandelt, M.J.: Understanding variability in recycled aggregate concrete mechanical properties through numerical simulation and statistical evaluation. Constr. Build. Mater. **178**, 301–312 (2018)
7. Rahal, K.: Mechanical properties of concrete with recycled coarse aggregate. Build. Environ. **42**, 407–415 (2007)
8. Hoffmann, C., Schubert, S., Leemann, A., Motavalli, M.: Recycled concrete and mixed rubble as aggregates: influence of variations in composition on the concrete properties and their use as structural material. Constr. Build. Mater. **35**, 701–709 (2012)
9. Lovato, P.S., Possan, E., Molin, D.C.C.D., Masuero, Â.B., Ribeiro, J.L.D.: Modeling of mechanical properties and durability of recycled aggregate concretes. Constr. Build. Mater. **26**, 437–447 (2012)
10. Jayasuriya, A., Bandelt, M.J., Adams, M.P.: Simulation of cracking susceptibility in recycled concrete aggregate systems. In: Proccedings of Computational Modeling of Concrete Structurs. EURO-C 2018 (2018)
11. Li, W., Xiao, J., Sun, Z., Shah, S.P.: Failure processes of modeled recycled aggregate concrete under uniaxial compression. Cem. Concr. Compos. **34**, 1149–1158 (2012)
12. Liu, Q., Xiao, J., Sun, Z.: Experimental study on the failure mechanism of recycled concrete. Cem. Concr. Res. **41**, 1050–1057 (2011)

Evaluation of Different Treatment Techniques in Recycled Aggregates for Use in the Production of Concrete

Giselle B. Limonte Morales, Jose Fernando Martirena-Hernandez and Stefan Linsel

Abstract The results of 5 months of research in the topic Concrete produced with 100% recycled aggregates are presented. We worked with a mixed type recycled material from Madrid, Spain and the tests were carried out in the laboratory (OBP) of the University of Applied Sciences of Karlsruhe, Germany. The recycled material was characterized and separated into three fractions for use as aggregates in the production of concretes (i) fine fraction 0–4 mm (ii) intermediate fraction 4–9 mm (iii) coarse fraction of 9–16 mm. To improve the properties of the recycled aggregates, different treatments were used for each of the fractions (i) accelerated carbonation of the fine fraction, (ii) encapsulation of the intermediate fraction with steam curing, (iii) use of MSC, in this case an active mineral addition of calcined clay-limestone base, denominated by its acronym in English LC2. Established as state parameters for accelerated carbonation of the fine fraction Relative Humidity = 75%, CO_2 concentration = 20% and residence time of 24 h, which resulted in an increase of 0.7% in the formation of Calcite with respect to the fine fraction without carbonation, the absorption of water decreased by 0.63% and the resistance in mortars after 28 days of curing exceeded by 0.89% the strength of the fraction without carbonation, 28 Mpa and 31, 43 Mpa respectively. For the steam curing of the encapsulated fraction, it was worked under conditions of temperature lower than 60 °C and a residence time of 30 min, obtaining a decrease of 0.75% in the absorption of water with respect to the material without encapsulating. In the concretes produced with the treated aggregates, the best performance series was that which uses a combination of fine carbonated aggregate, encapsulated intermediate aggregate with steam curing and Portland cement 32.5 Mpa.

Keywords Recycled aggregates · Steam curing · Accelerated carbonation

G. B. Limonte Morales (✉) · J. F. Martirena-Hernandez
CIDEM, Universidad Marta Abreu de las Villas, Santa Clara, Cuba
e-mail: glmorales@uclv.cu

J. F. Martirena-Hernandez
e-mail: f.martirena@enet.cu

S. Linsel
University of Applied Sciences of Karlsruhe, Karlsruhe, Germany

© RILEM 2020

J. F. Martirena-Hernandez et al. (eds.), *Proceedings of the International Conference of Sustainable Production and Use of Cement and Concrete*, RILEM Bookseries 22,
https://doi.org/10.1007/978-3-030-22034-1_44

1 Introduction

An alternative to traditional methods to improve the quality of coarse aggregate is to encapsulate the 4/8 mm RCA fraction with a layer of cement paste. The results of resistance indicate that the new technique of coating the surface of recycled aggregate with pozzolanic dust contributes significantly to better viability and greater resistance to compression and bending. The interfacial transition zone of the concrete with recycled aggregates is benefited with the new technique [1]. To accelerate the curing process of the encapsulated fraction, water vapor is used, it is known that if we generate enough water vapor under ideal conditions of humidity and temperature, we would be creating favorable conditions for the cement used in the encapsulation of the fraction of the recycled aggregate is hydrated, cured and thus guarantees the curing of the encapsulated aggregates. The substitution of Portland cement for a percentage of addition in concretes produced with recycled aggregates allows to compensate the loss of strength and durability of the concretes, the substitution is effective to increase the resistance to the penetration of chloride ions in the body of the concrete [2]. Fly ash used as a mineral addition in concrete mixtures produced with recycled aggregate improves the resistance to the penetration of chloride ions, but also increases the carbonation depth of the concrete. The compressive strength decreases with the increase in the recycled aggregate content, as well as the increase in the fly ash mixture. The compressive strength of recycled aggregate concrete incorporating fly ash with standard water curing increases significantly over the years [3]. Another method to improve the properties of the recycled aggregate is based on the accelerated carbonation of the fine fraction, it was first disclosed in 1990 when Seifritz exposed a method in which a high concentration of CO_2 is used in a closed chamber that would react with the silicates sprayed in order to trap the gas and allow a safe disposal of this [4]. According to the literature it is recommended to use concentration values of up to 10% without notable differences in the microstructure with respect to the natural carbonation process [5] which allows accelerating the reaction on a time scale of several days to a few minutes and hours [6], otherwise the process would take years or decades due to the low concentration of gas in the atmosphere [7].

2 Materials and Methods

Two recycled M1 and M2 materials from a treatment plant in Madrid, Spain, both with a grain size of 020 mm, were characterized. Lugo to be characterized is chosen the best properties to perform accelerated carbonation treatments and encapsulated with steam curing. After a simple characterization of the two materials, it is decided to work with the M1.

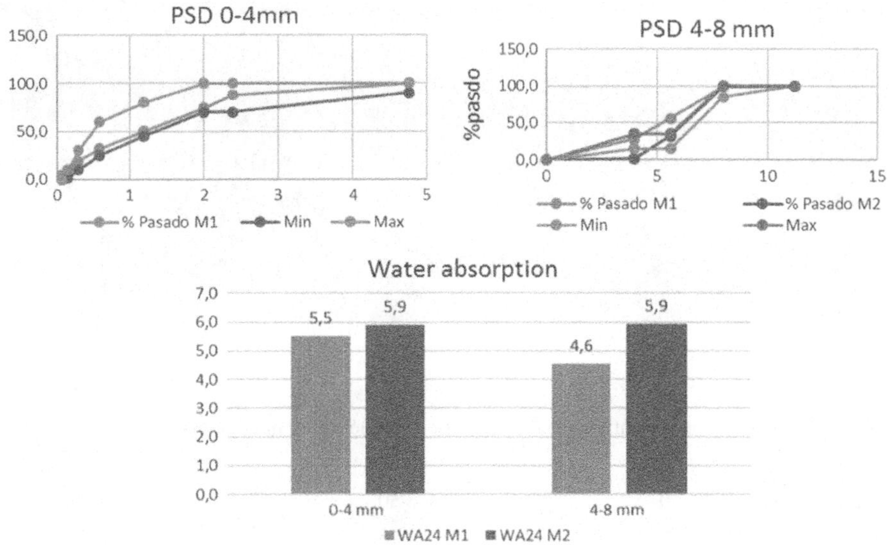

Graph 1 Characterization of recycled aggregates

Table 1 Parameters to be measured

Carbonation process	Temperature	Ambient
	RH	55–75%
	Concentration of CO_2	10–20%
	Carbonation time	Total carbonation

Source Own elaboration

2.1 *Characterization of Recycled Aggregate*

See Graph 1.

2.2 *Quickly Carbonation of Fine Fraction of ARC Recycled Fine Aggregate*

For the carbonation of the fine ARC, different environmental conditions were used, in order to find the parameters of the carbonation process that would allow an improvement of the properties of the recycled aggregate. The environmental variables considered are listed in Table 1.

For the carbonation process, a hermetic glass chamber with inlet and outlet for CO_2 was used, the state parameters were controlled with the help of a weather station. The fine ARC was inside the chamber for 24 h, the state parameters were worked in

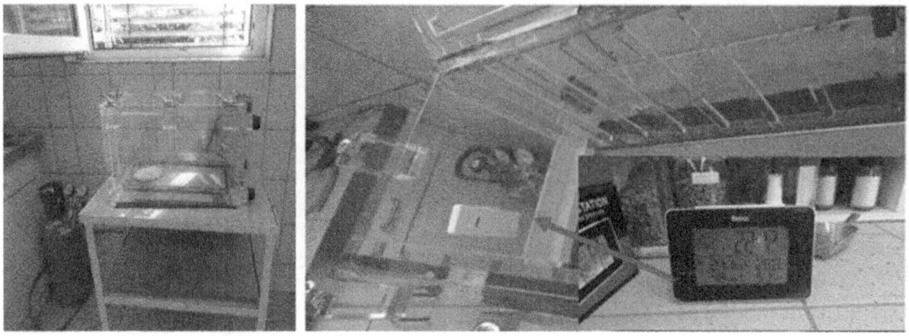

Fig. 1 Carbonation chamber

Table 2 Results of the weight loss trial by indiction for the different combinations

	CO_2	Time	RH	% CH	% calcite
Original sand	0	0	50	7.41	13.64
Series 1 CO_2 10%	**10**	**30**	**55**	**2.47**	**14.77**
Series 1 CO2 15%	15	30	55	2.06	14.55
Series 2	10	60	55	2.88	14.55
Series 3	20	30	65	2.47	13.86
Series 4 CO2 10%	10	60	65	0.82	15.00
Series 4 CO_2 15%	**15**	**60**	**65**	**2.06**	**15.91**
Series 4 CO2 20%	20	60	65	4.53	12.73
Series 5	**10**	**30**	**75**	**2.06**	**17.05**
Drum	**20**	**90**	**75**	**2.06**	**23.16**

different combinations as shown in Table 1. To check the carbonation state reached by the fine RCD, the one tested with phenolphthalein was used and the test was carried out. Loss on ignition (LOI) (Fig. 1).

Taking into account the amount of fine carbonate ARC necessary to perform the different mixtures of mortars and concrete proposed in this research, an accelerated carbonation test was carried out using a drum as a carbonation chamber and maintaining the best combinations achieved in the carbonation chamber, highlighted in bold in Table 2.

2.3 *Encapsulation of the Coarse Fraction of Recycled Aggregate with Steam Cure*

For the encapsulation process, a Portland Cement CEM1-22.5 MPa was used, with an a/c ratio between 0.5 for 1 kg of material to be encapsulated. Once encapsulated, the

Table 3 Parameters to be measured

Parameters			
Cement for encapsulation	40 kg/m^3		60 kg/m^3
Temperature	40–60 °C		
Residence steam	20 min	30 min	35 min

Source Self-made

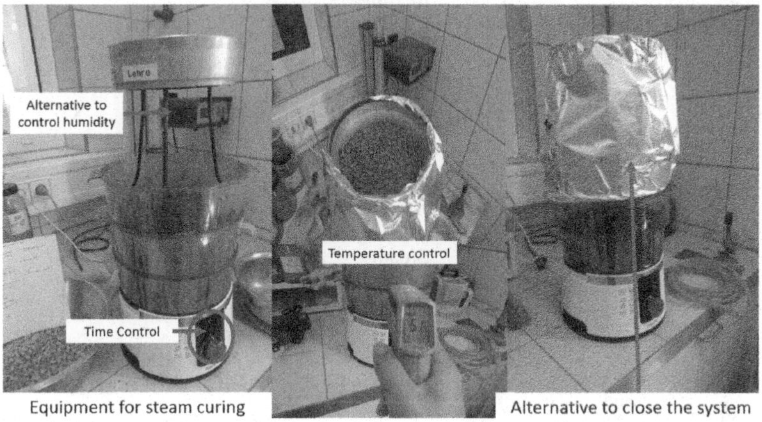

Fig. 2 Steam curing system

recycled aggregate was placed for 35 min in a vaporizer with controlled temperature and relative humidity parameters (Table 3; Fig. 2).

2.4 Production of Concrete Mixtures with 100% Recycled Aggregate

Different mixtures of concrete were prepared by combining the two treatments carried out on the recycled aggregate using Portland Cement (CEM1 35 MPa) and an active mineral addition based on calcined clay, limestone and gypsum LC2. The Touffar program was used for the design of mixtures (Table 4).

Table 4 Concrete mix designs

Samples	Aggregate		Cement	
	Fine	Coarse	CP	LC^3
H1 [8] 100% recycle	Without treating	Without treating	X	
H2 Recycle fine	Carb.	Without treating	X	
H3 Recycle coarse	Without treating	Steam	X	
H4 [1] Recycle + LC^3	Without treating	Without treating		X
H5 Recycle combinado	Carb.	Steam		X
H6 Recycle combinado	Carb.	Steam	X	

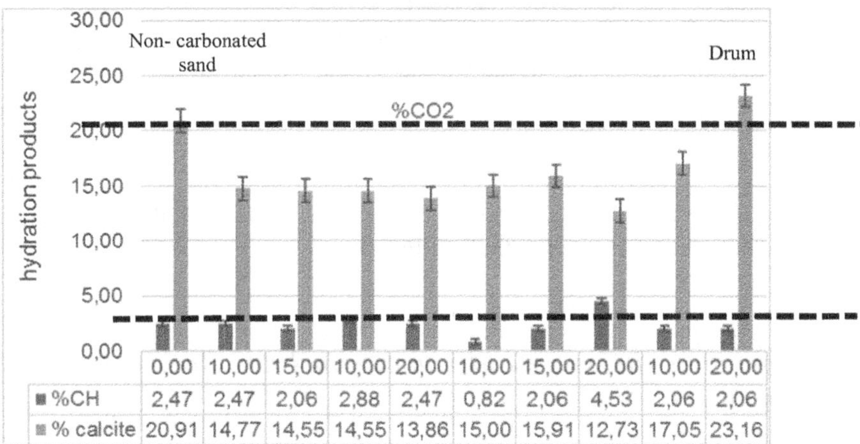

Graph 2 Hydration product formation versus CO_2 concentration

3 Analysis of Results

3.1 Influence of the Degree of Carbonation of the Fine Recycled Aggregate in Mortars

With the analysis of Graph 1 were established as state parameters for the accelerated carbonation of fine recycled aggregate: RH = 75%, CO_2 = 20% for a residence time of 24 h with which 23.16% of Calcite is reached (Graphs 2, 3 and 4).

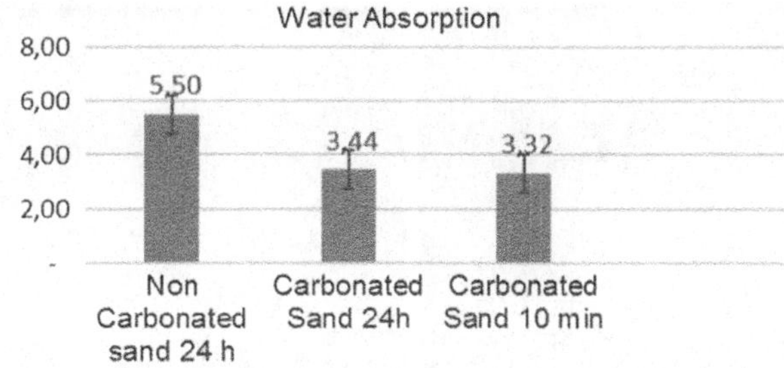

Graph 3 Results of the water absorption test to the carbonated fine recycled aggregate

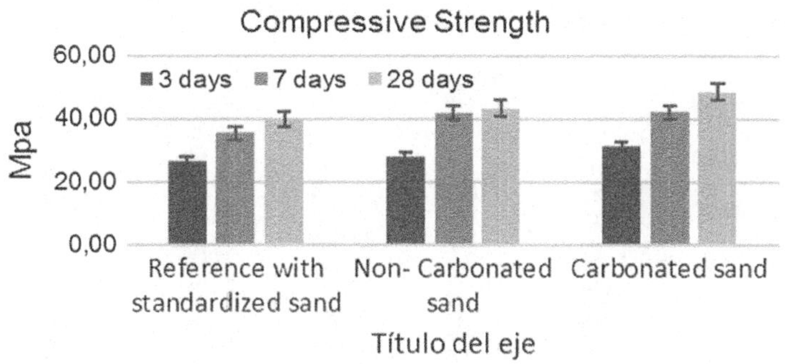

Graph 4 Results of the resistance test in mortars with carbonated fine recycled aggregate

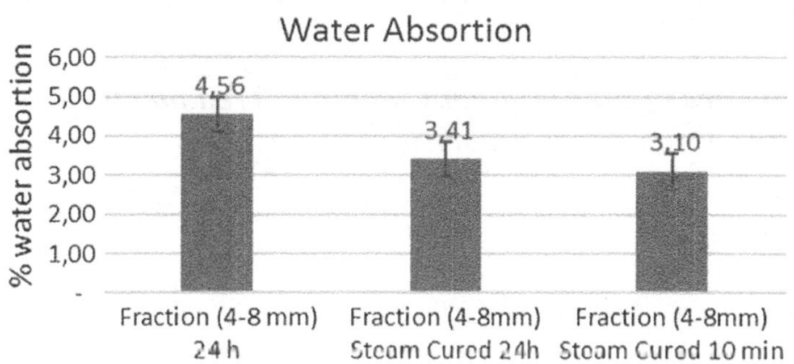

Graph 5 Results of the water absorption test to encapsulated recycled aggregate steam-cured

3.2 Influence of Steam Curing on the Properties of Encapsulated Recycled Aggregate

See Graph 5 and Fig. 3

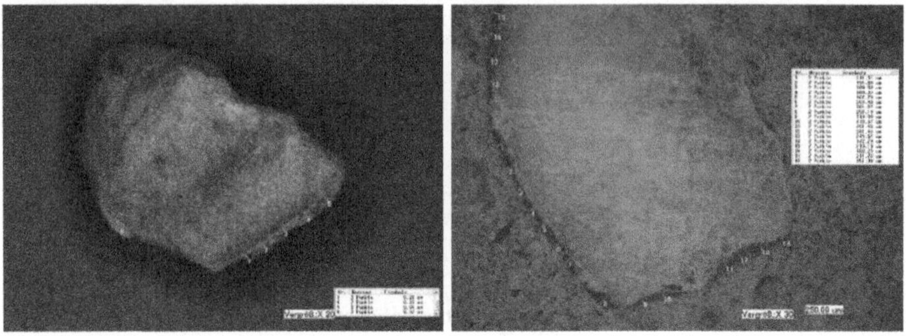

Fig. 3 Encapsulated Arid viewed with an Optical Microscope

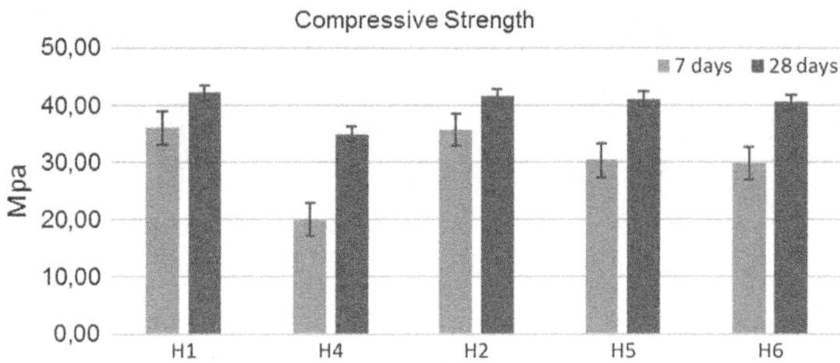

Graph 6 Results of the resistance test to understanding of the concretes produced

3.3 Influence of the Different Combined Treatments on the Mechanical Properties of Concrete

See Graph 6

4 Conclusions

The accelerated carbonation of the fine fraction of recycled aggregate allows to lower the values of water absorption and does not compromise the resistance of the mortars and concretes produced.

The technique of steam curing of the coarse fraction of recycled arid must be technically improved.

References

1. Li, J., Xiao, H., Zhou, Y.: Influence of coating recycled aggregate surface with pozzolanic powder on properties of recycled aggregate concrete .Constr. Build. Mater. **23**, 1287–1291 (2009)
2. Ann, K.Y., Moon, H.Y., Kim, Y.B., Ryou, J.: Durability of recycled aggregate concrete using pozzolanic materials. Waste Manage. **28**, 993–999 (2008)
3. Kou, S.-C., Poon, C.-S.: Long-term mechanical and durability properties of recycled aggregate concrete prepared with the incorporation of fly ash. Cement Concr. Compos. **37**, 12–19 (2013)
4. Cuellar, M., Felipe, A.: Carbonatación acelerada del agregado fino reciclado y su influencia en morteros. Tesis de Master (2017)
5. Morandeau, A., Thiery, M., Dangla, P.: Investigation of the carbonations mechanism of CH and C-S-H in terms of kinetics, microstructure change and moisture porperties. Cem. Concr. Res. **56**, 153–170 (2014). https://doi.org/10.1016/j.cemconres.2013.11.015
6. Johnson, D.C.: Sci Lecture Papers Series Accelerated carbonation of waste calcium silicate materials. Constr. Mater. 1–10
7. Gunning, P.J.: Accelerated carbonation of hazardous wastes (February) (2011)
8. Sánchez, 2011; Chauveinc, 2011, Navarro, 2014; CEDEX Materiales, 2016

Durability of Recycled Aggregate Concrete Under Real Conditions in Tropical Ambience

Sandra Lichtblau, Stefan Linsel and Jose Fernando Martirena-Hernandez

Abstract The substitution of natural resources by the use of recycled aggregates from construction and demolition waste is a huge advantage for the sustainability in the concrete production. The presented paper considers different processing methods of recycling material evaluated with respect to their advantages and disadvantages in sustainability. But once the recycled aggregates are used in concrete, there is a lag of knowledge in the field of its durability behavior. Durability results from a long-time study on 100% recycled aggregate concretes (RAC) suspended the real conditions in a tropical island are presented with respect to their chloride ingress. As reference a natural aggregate concrete (NAC) with the same w/c-ratio was used. The compressive strength of the RAC reached 30 MPa and is only 6% less than that for natural concrete. Samples of each concrete are exposed to three sites of different aggressiveness depending on the distance to the sea, which involves for one site wet-dry cycles by tides or changes in the relative humidity and the chloride content of the ambience. After 5 month exposure the chloride ingress of samples exposed to the sea shore is up to 39% lower in RAC than in NAC. The changes of chloride ingress with time are observed with a second measurement after 17 month exposure. Results in the carbonation show similar trends in the durability behavior with a lower carbonation depth in RAC. However, remaining tests such as electrical resistance and air permeability show better results for NAC.

Keywords Recycled aggregate concrete · Durability · Chloride ingress

1 Introduction

Amounts of construction and demolition wastes, C&DW, are increasing in industrialized countries due to the continue process of upgrading infrastructure and refur-

---

S. Lichtblau (✉) · S. Linsel
University of Applied Sciences, Karlsruhe, Germany
e-mail: smarina@uclv.cu

J. F. Martirena-Hernandez
Central University "Marta Abreu" of Las Villas, Santa Clara, Cuba

© RILEM 2020
J. F. Martirena-Hernandez et al. (eds.), *Proceedings of the International Conference of Sustainable Production and Use of Cement and Concrete*, RILEM Bookseries 22,
https://doi.org/10.1007/978-3-030-22034-1_45

bishing of existing buildings whereas in developing countries sources of C&DW are related to the occurrence of frequent disasters such as hurricanes or earthquakes that destroy part of the existing infrastructure [1].

The scarcity of natural aggregates is prompting for solutions to increase the share of recycled materials in concrete, and also to broaden the use of concrete made with recycled material to mainstream applications [2–4].

A research team assembled between scientists of the Universidad Central de las Villas, Cuba and the University of Applied Sciences in Karlsruhe has developed an alternative to traditional methods to improve the quality of coarse aggregate by an encapsulation of the fraction 5/9 mm of coarse RCA with a layer of cement [5]. Concrete produced with encapsulated aggregate increased roughly 30% of strength and the air permeability is reduced to a half [6].

This paper presents the results of a research project to assess the durability of concrete made entirely with RCA, with the fraction 5–9 encapsulated with cement.

2 Experimental Work

The present study compares mechanical and durability properties of a natural aggregate concrete, NAC, and a recycled aggregate concrete, RAC, where 100% of the aggregates were substituted by crushed concrete waste from a ready mix concrete plant. This plant normally produces concrete having strength around 25–35 MPa, and the wasted material does not comply with standards, so it is supposedly under these strength values. The parent concrete made with calcareous aggregates commonly consists Portland cement of type I, 35 MPa at 28 days, according to EN 196. Table 1 presents the characterization of the RCA and NA.

Table 1 Physical properties of aggregates

	RCA 0–5 mm	RCA 5–9 mm	RCA 9–19 mm	NA 0–5 mm	NA 5–9 mm	NA 9–19 mm
Current density (kg/dm^3)	2.15	2.0	2.17	2.55	2.63	2.54
Water absorption (%)	9.21	10.78	7.48	2.07	3.16	1.72
Bulk density (kg/dm^3)	1.45	1.21	1.11	1.63	1.50	1.53
Fines <0.074 mm (%)	7.26	1.0	0.31	4.27	1.74	2.86

Table 2 Concrete mix designs

	kg/m³						
	Cement	0–5 mm	5–9 mm	9–19 mm	w/c	SP (%)	Slump (cm)
NAC	349	583	438	801	0.45	1.0	10
RAC-E	385	628	425	402	0.45	0.7	16

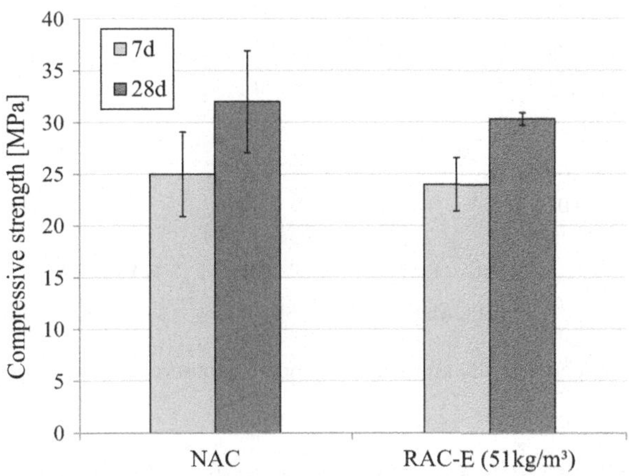

Fig. 1 Compressive strength in MPa at 7 and 28 day

To minimize the negative influence of RCA in concrete properties, the fraction of 5–9 mm of RCA was treated with an improvement method where the material was coated with a thin layer of cement. The cement used was a Portland Pozzolan Cement with 15% Pozzolan PZ-35 (CEM II 32.5 N). The total amount of cement used for coating was 51 kg/m^3 of concrete. Table 2 presents the mix designs for both concretes, NAC and recycled aggregate concrete with the coated fraction, RAC-E. Figure 1 shows their compressive strength values of 7 and 28 days.

The durability study was carried out in Cuba, taking into account three different exposures of aggressiveness according to NC 120:2014 [7] which differ depending on the distance to the sea. That means very high aggressiveness until 500 m to the seashore (Punta Matamoros), high aggressiveness from 500 to 3000 m (SEDE) and low aggressiveness more than 20.000 m away from the sea (UCLV). To be able to study the durability behavior for a few years, two cylinders with the size of 320 mm diameter and a height of 220 mm of each concrete and exposure site were fabricated.

3 Discussions of Results

Durability test series were performed two times since their exposure. After 5 month and 17 month respectively, air permeability, electrical resistivity, carbonation depth

Table 3 Quality classes of cover concrete

Quality of cover concrete	Very bad	Bad	Normal	Good	Very good
kT (10–16 m^2)	>10	1.0–10	0.1–1.0	0.01–0.1	<0.01

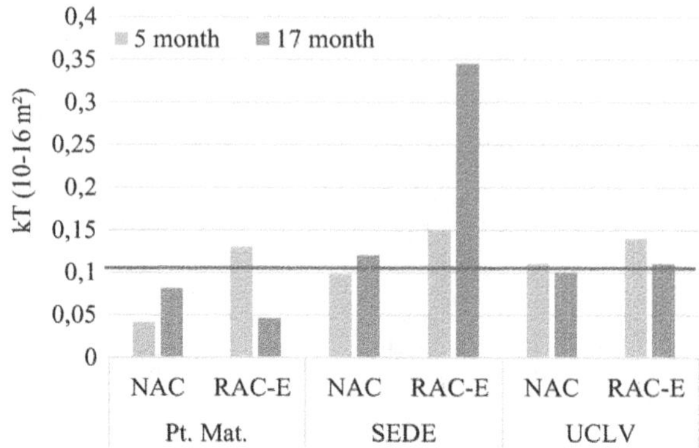

Fig. 2 Results of air-permeability after 5 and 17 month exposure in kT (10^{-16} m^2)

and chloride profile were executed. Therefore cores of 100 mm diameter and 50 mm diameter for chloride profile were taken on the sites.

3.1 Air-Permeability

The air-permeability was measured on site directly on the concrete cylinders just before taking the cores. By this method, the quality of the cover concrete is determined using the coefficient kT and after judged in five quality classes (Table 3) [8].

The results of air-permeability obtained after an exposure time of 5 and 17 month are presented in Fig. 2. The values of air-permeability are mostly in the same range of classification of normal and good cover concrete excepted for the 17 month value of RAC-E at SEDE. This value can be explained by a rough surface of the cylindrical specimen, what can lead to a falsification of results. In general the air-permeability of NAC is slightly better than for RAC-E.

3.2 Carbonation Depth

The carbonation depth was determined by the indicator phenolphthalein sprayed on a split surface on the concrete according to NC 355: 2004 [9]. The results of the carbonation depths are listed in Table 4. With the determined carbonation depth values after 5 and 17 month, it can be observed a tendency of a slower carbonation

Table 4 Carbonation depth

Carbonation depth (mm)		Pt. Mat.	SEDE	UCLV
NAC	5 month	0.0	1.5	2.0
	17 month	0.9	3.0	1.9
RAC-E	5 month	0.0	1.1	0.7
	17 month	0.0	0.5	0.0

of RAC with the encapsulated fraction from 5 to 9 mm. In general, carbonation of concretes at SEDE and UCLV proceed faster because of more favorable ambience in terms of relative humidity, in comparison to concretes exposed at seashore, where carbonates are washed out because of tides. The highest value with 3 mm is reached in NAC exposed at SEDE, whereas the carbonation process seems to be stagnant at the UCLV.

3.3 *Chloride Profile*

The chloride ingress after five month was determined by the laboratory MPA of Stuttgart in Germany, using ion-chromatography. Values after 17 month were determined at the laboratory CEQA at the Central University "Marta Abreu" of las Villas in Cuba according to the standard ASTM C1218 [10]. Concrete cores of 50 mm diameter were pulverized in 5 mm steps until a depth of 20 mm. The chloride content of concretes exposed at SEDE and Punta Matamoros are presented in Figs. 3 and 4. Values are expressed in percentage of mass.

The values obtained for concretes exposed at Punta Matamoros (Fig. 4) are much higher than the values of concretes at SEDE (Fig. 4) due to the influence of the seawater. Due to the tides at the seashore and the wet-dry cycles, chlorides increases much faster in the concrete pore structure. The trends of chloride profile are similar

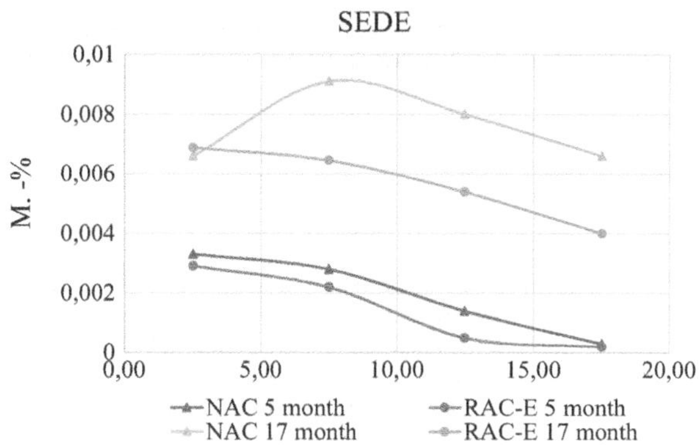

Fig. 3 Chloride profile after 5 and 17 month for concretes exposed at *SEDE*

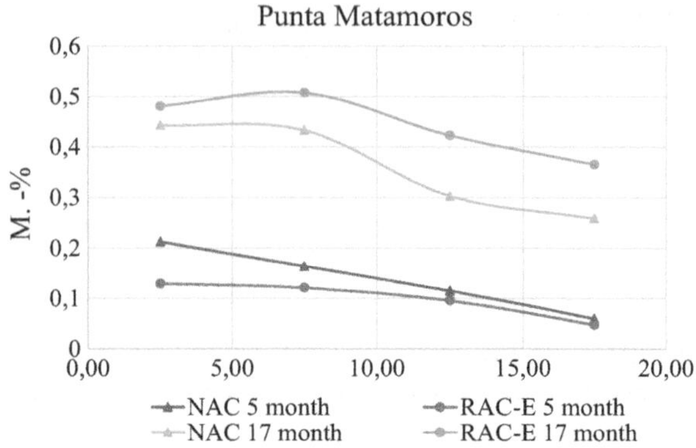

Fig. 4 Chloride profile after 5 and 17 month for concretes exposed at *Punta Matamoros*

for both concretes. Beside the chloride content after 17 month at SEDE, values in NAC are slightly higher.

4 Conclusions

1. The implementation of encapsulation of the 5–9 mm fraction of RCA impacts on the pore structure of concrete matrix. The values of air permeability are slightly higher than those of NAC; both show moderate permeability.
2. With the second durability series it is showed an acceptable behavior of RAC-E with respect to the time compared to NAC.
3. Results of air-permeability, carbonation and chloride ingress shows the tendency of a durable concrete respected to the ingress of CO_2 and Chlorides. In comparison to NAC, the achieved results for RAC-E are slightly better. Nevertheless, further results should be awaited for a more precise declaration relating to the future.
4. Concrete made entirely with RCA with the fraction 5–9 encapsulated fulfils durability requirements for elements situated in medium and low aggressiveness class according to Cuban standard NC 120.2014. More than half of concrete cast in Cuba is used in areas under these exposure conditions.

References

1. Behera, M., Bhattacharyya, S.K., Minocha, A.K., Deoliya, R., Maiti, S.: Recycled aggregate from C&D waste & its use in concrete—a breakthrough towards sustainability in construction sector: a review. Constr. Build. Mater. **68**, 501–516 (2014)
2. Xiao, J., Li, W., Fan, Y., Huang, X.: An overview of study on recycled aggregate concrete in China (1996–2011). Constr. Build. Mater. **31**, 364–383 (2012)

3. Rao, A., Jha, K.N., Misra, S.: Use of aggregates from recycled construction and demolition waste in concrete. **50**, 71–81 (2007)
4. Radonjanin, V.: Recycled concrete as aggregate for structural concrete production. Sustainability, pp. 1204–1225 (2010)
5. Castano-Cardoza, T., Linsel, S., Alujas-Diaz, A., Orozco-Morales, R., Martirena-Hernández, J.F.: Influence of very fine fraction of mixed recycled durability of mortars and concretes. Rev. Fac. Ing. Univ. Antioquia **80**, 4–8 (2016)
6. Lichtblau, S., Hernández, J.F.M., Linsel, S.: Improvement of the coarse fraction of recycled concrete aggregate to produce a durable concrete made entirely with recycled. In: Ibausil 2018 (2018)
7. NC.: NC 120: 2014—Hydraulic concrete—specifications. Cuban National Bureau of Standards, pp. 1–74 (2014)
8. Proceq: Operating Instructions—permeability Tester TORRENT (1997)
9. NC: NC 355: 2004—Determination of carbonation depth in hardened concrete. Cuban National Bureau of Standards, pp. 1–7 (2004)
10. ASTM: C 1218/C 1218 M—standard test method for water-soluble chloride in mortar and concrete. American Society for Testing and Materials, pp. 1–3 (1999)

Potential Use of a Quarry Waste for the Production of Self-compacting Concrete

Anahí López, Claudio J. Zega, Gabriela S. Coelho Dos Santos, Leandro E. Carrizo, Juan M. Etcheverry and Yury A. Villagrán Zaccardi

Abstract The main feature of Self-Compacting Concrete (SCC) is its compactability under the action of its own weight, filling formworks and covering reinforcement without the need of applying external energy. Limestone filler is the most used supplementary cementitious materials for the production of SCC in the Buenos Aires Metropolitan Area, Argentina. Alternative sources for fillers can be found in residues from the exploitation of rocks. A good example is the quartzite mud resulting as a residue from the washing of manufactured sand in the Buenos Aires Province. The use of this waste would benefit the environment in reducing the demand for non-renewable natural resources and in adding value to the residue. In this paper, the performance in the fresh and hardened states of SCC made with 25, 50, 75, and 100% of the quartzite residue as a partial substitute of limestone filler is evaluated. Tests include spread, passing time, compressive strength, and capillary water absorption. Results show that the superplasticizer demand increases considerably with the content of the residue. Nevertheless, the properties of concrete in the fresh and hardened states show the feasibility of using the material to produce SCC.

Keywords Quarry waste · Environmental impact · Sustainability · Quartzite filler · Self-compacting concrete

A. López (✉) · C. J. Zega · G. S. C. Dos Santos · L. E. Carrizo · J. M. Etcheverry · Y. A. Villagrán Zaccardi
LEMIT, CICPBA, 52 Entre 121 y 122 s/n, La Plata, Argentina
e-mail: hormigones@lemit.gov.com

A. López
UTN-FRLP/LEMaC, 60 y 124, Berisso, Argentina

C. J. Zega · Y. A. Villagrán Zaccardi
CONICET, La Plata, Argentina

© RILEM 2020

J. F. Martirena-Hernandez et al. (eds.), *Proceedings of the International Conference of Sustainable Production and Use of Cement and Concrete*, RILEM Bookseries 22,
https://doi.org/10.1007/978-3-030-22034-1_46

1 Introduction

The current need for structural members of different sizes and complex forms that maintain mechanical properties and adequate durability performance makes the use of self-compacting concrete (SCC) more necessary.

Several kinds of supplementary cementitious materials (SCMs) can be used as constituents of SCC to achieve the necessary workability [1–3], and limestone filler or ground blast-furnace slag are among the most used. In Argentina, some local ready-mix concrete manufacturers use only limestone filler to achieve a moderate viscosity level due to the convenient presence of quarries nearby. However, these SCMs are commercially unavailable to all the producers in the Argentine market. This situation is forcing some manufacturers to use only cement to comply with the required content of fines in SCC, with cement contents above 500 kg/m^3. Such high levels of cement content result in inconveniences such as low dimensional stability, high costs, and high environmental impact. Therefore, alternative sources of filler materials are essential for more convenient practices.

A candidate for application as filler is the quartzite mud resulting as a residue from the washing of manufactured sand produced in the Buenos Aires Province, Argentina. The annual production of the mud from the supplier quarry in the period May 2016-May 2017 was 10,800 tons, which represents approximately 10% of the weight of the manufactured sand being washed. The quartzite filler used in this study has different characteristics from those of other rock powders originated from the washing of aggregates. The main distinction lies in the content of clay components [4, 5]. Clays are well known for increasing water demand and viscosity in the fresh state. Therefore, applying this type of filler in the production of SCC might turn convenient the use of a smaller amount of viscosity-modifying agent.

Moreover, the use of quartzite filler in partial substitution of limestone filler (or Portland cement in the best cases) in the manufacturing of SCC may allow a reduction of costs, in relation with the transport of limestone filler. At the same time, it may also help in the valorization of this waste and the reduction of the environmental impact of SCC because of less generation of greenhouse gases. The quantification of these benefits will be accomplished in future work. This paper presents a proof of concept with the analysis of different fresh and hardened state properties of self-compacting concretes produced with variable contents of quartzite filler.

2 Methodology

2.1 Materials and Mixtures

Four concretes, C25, C50, C75, and C100, were made with varying quartzite filler contents: 25, 50, 75 and 100% v/v of total filler content, respectively. The substitution by volume of limestone filler (L) by quartzite filler (Q) resulted in small decreases in

Table 1 PSDs by laser diffraction of Q and L

Filler	Dv10 (μm) $\overline{X}$ (S)	Dv50 (μm) $\overline{X}$ (S)	Dv90 (μm) $\overline{X}$ (S)
Q	3.0 (0.3)	14.0 (0.9)	61.3 (3.2)
L	1.4 (0.1)	7.7 (0.6)	72.9 (10.4)

$\overline{X}$ average value. (*S*): standard deviation. dv10, dv50, and dv90: particle sizes (assumed spherical) corresponding to accumulated passing material of 10, 50, and 90% w/w, respectively

Table 2 Chemical compositions [%] by FRX of Q and L. nd: not determined

Filler	CaO	SiO_2	Al_2O_3	Fe_2O_3	TiO_2	SO_3	MgO	K_2O	Na_2O	P_2O_3	LOI]
Q	0.34	74.18	15.95	0.76	0.42	0.10	1.46	1.62	2.10	0.70	3.08
L	48.85	8.15	1.28	0.88	nd	0.05	1.14	0.28	1.25	nd	37.29

the filler content w/w, due to the slight difference in their densities (2.50 g/cm^3 for Q vs. 2.60 g/cm^3 for L). Limestone Portland Cement was used for concrete production (LPC, density 3.1 g/cm^3, equivalent to CEM II/A-L 42.5 N according to EN 197-1). Table 1 shows particle size distributions (PSDs) for Q and L [6]. Q was slightly coarser than L. It is worth mentioning that the PSD of Q was not optimized, as the only applied mechanical processing was intended for disaggregation only (milled in laboratory for a period of 30 min). The aim of the present study was to verify its suitability as received, whereas additional processing will be analyzed in future work. Table 2 shows the chemical compositions of Q and L.

Three natural aggregates commonly used in the Buenos Aires Province were combined for obtaining a continuous (well-graded) skeleton according to limits indicated in the CIRSOC 201 guidelines [7]. The fine aggregates were a siliceous sand (FA1, fineness modulus = 1.67, density = 2.65 g/cm^3, water absorption = 0.39% w/w), and a 0–6 mm manufactured granite sand (FA2, fineness modulus = 3.59, density = 2.72 g/cm^3, water absorption = 0.6% w/w). Coarse aggregate was crushed granite (CA, density 2.70 g/cm^3; maximum size of 12 mm). The (coarse aggregate volume)/(total aggregate volume) was 0.50. An ultra-high range water-reducing admixture was used. The dosage of superplasticizer (Sp) was nearby the recommended dosage by the manufacturer (0.4–1.5% w/w of cement).

Table 3 shows the mixture proportions for C25, C50, C75, and C100. Water/cement ratio was slightly varied among concretes, between 0.486 and 0.528, with the aim of improving the fresh properties of each mix. In correspondence, water/fines ratio varied between 0.298 and 0.320. Additional optimization of proportioning seems possible, and it will be conducted in future work. The content of superplasticizer was increased with the increase in Q content.

Table 3 Mixture proportions of concretes (kg/m^3)

Concrete	Water	LPC	Q	L	Sp	FA1	FA2	CA
C25	184	378	66	167	4.4	527	228	773
C50	184	349	116	111	4.5	527	230	774
C75	187	354	175	55	6.3	532	230	779
C100	176	356	237	0	6.4	538	234	790

2.2 *Experimental*

SCC properties in the fresh state were evaluated by the slump flow test (with parameters: T_{50} = flow time up to a 50 cm diameter, and D_f = final spread diameter), the V-funnel test (parameter: T_V = flow time for the complete discharge of the funnel) [8, 9], the unit weight (UW), and the air content.

In the hardened state, compressive strength at 7, 28, and 90 days, and capillary water absorption at 28 days, were evaluated. All samples were cured in a conditioned room (23 °C and RH > 95%) up to the corresponding testing age. All the reported compressive strength results are the average values of three 10-cm cylinders. For capillary water absorption, five 5-cm thick samples were cut from the middle part of 10-cm probes and tested according to IRAM 1871 standard [10].

3 Results and Discussion

The fresh properties of concrete mixtures are given in Table 4. T_{50} values were similar for C25, C50, and C75. C100 did not show enough filling ability for T_{50} to be able to be measured. Conversely, D_f values were very different among all concrete mixtures. This contrast would show that, whereas the viscosity does not change as the quartzite filler content increases, the filling ability of concrete is progressively reduced. Moreover, a higher dose of superplasticizer (6.3 kg/m^3) was necessary for C75 in comparison with C25 and C50 for achieving a similar T_{50} value. Therefore, Q increased the superplasticizer demand; meaning Q increased the viscosity of the mixes and a compensation by the superplasticizer was necessary. Further studies are necessary in this regard, but it seems clear that the nature of Q requires the use of a different type of superplasticizer, as the one used in this study was unable to counterweigh the affectation of viscosity as required for appropriate filling ability. This effect should not be necessarily considered a disadvantage of Q over L, as the former apparently allows obtaining more stable mixtures, making it possible to design very fluid mixtures without significant risks of segregation. These capabilities will be explored in future research.

Final spread diameters obtained for C25, C50, and C75 permit to classify them as self-compacting [8, 9, 11]. All mixtures showed good stability and deformability

Table 4 Test results

Concrete			C25	C50	C75	C100
Fresh concrete properties						
Slump flow	T_{50}	(s)	3.70	3.44	3.74	–
	D_f	(mm)	700	550	620	–
V-Funnel	T_V	(s)	13.59	15.34	14.07	nd
Unit weight	UW	(kg/m^3)	2358	2358	2358	2330
Air content		(%)	3.00	4.00	3.00	3.00
Hard concrete properties						
Compressive strength	7d	(MPa)	26.8 (0.8)	21.8 (0.6)	20.7 (0.6)	18.2 (0.4)
	28d	(MPa)	33.5 (3.0)	27.0 (2.3)	27.0 (1.3)	26.0 (0.5)
	90d	(MPa)	38.2 (0.3)	32.2 (0.8)	31.0 (0.9)	28.5 (1.3)
Capillary absorption	CAC	(g/m^2)	3540	4460	4237	4930
	CAR	(g/m^2/s$^{0.5}$)	4.40	6.53	5.91	7.68
		(R^2)	0.971	0.981	0.979	0.975

(even C100), with homogeneous surface, uniformity in coarse aggregate distribution, and well-defined edges without segregation. The UW and air content values were in the range of values for conventional concrete.

The results of compressive strength at 7, 28 and 90 days for C25, C50, C75, and C100 suggest a more significant dilution effect of Q than of L. This is deduced from the decrease in compressive strength as the content of Q filler increases. The same effect is observed at the three evaluated ages, with a significant impact for C50, C75, and C100. In relative terms, the rate of increase in compressive strength is similar for all concrete mixes, which is indicative of a similar filler effect (promotion of nucleation and hydration of cement). The impact of Q on microstructure development is an aspect that requires further investigations, especially in relation with the effect of the clay contained in the filler. Nevertheless, the concretes designed in the present study are suitable for structural applications, as compressive strength was higher than 25 MPa at 28 days in all cases.

Table 4 also shows the results of capillary absorption capacity (CAC), capillary absorption rate (CAR), and the coefficients of determination (R^2) for the fitting to the square-root-of-time model. In agreement with the compressive strength, Q increases both capillary absorption parameters, suggesting an increased porosity. No increase in bleeding with increasing Q content was observed, so the effect on water absorption is solely attributed to the impact on the development of microstructure. Microstructural analysis for confirming this outcome will follow in the future. The values for R^2 are similar to those normally obtained for conventional concrete. Results from water absorption test indicate that the designed mixtures do not comply with the requirements for concrete exposed to aggressive environments. These values are however quite expected, as durability performance was certainly not among the objectives of this investigation, and this is why a water/cement ratio near 0.50 was used. Future work will focus on the effect of the relative paste volume of SCC on transport properties.

4 Conclusions

The properties in the fresh and hardened states of self-compacting concrete (SCC) made with 25, 50, 75, and 100% v/v of quartzite filler (Q) produced from quarry waste, as partial substitute of limestone filler (L), are evaluated.

The use of quartzite filler up to 75% replacement ratio in SCC allows convenient workability, whereas unsuitable results were obtained for 100% replacement ratio. The viscosity of mixtures increased with the Q content, in association with an increased superplasticizer demand. Studies with other water reducing admixtures that do not modify the viscosity of the mix are necessary. The effect of Q is very likely linked with the colloidal activity of clay contained in it, and further investigation in this regard will follow. In practical terms, the potential additional sorption capacity of Q due to its clay content can also offer the advantage of increased stability of the mix. The experimental quantification of this capability also seems convenient.

Using Q reduced the compressive strength and increased the capillary absorption of SCC, demonstrating an increase in porosity. The dilution effect seems to be stronger for Q than for L. The distinction can be due to the particle size distribution of each filler, but some effect of the respective compositions is also anticipated. Further studies for improving the performance of SCC with Q, especially in the hardened state, are necessary. A possible upgrading could be achieved by mechanical optimization of the particle size distribution of Q for making it more compatible with the cement.

Acknowledgements The authors acknowledge Natalia Alderete (Researcher at Ghent University), Alejandro Pittori (Associate Technician at CICPBA-LEMIT) and María Eva Sosa (Researcher at UTN-FRLP) for their collaboration during the experimental program. This work was partially funded by the Commission of Scientific Research of the Province of Buenos Aires (CICPBA) through the IPI 2019/2020 Program.

References

1. Desnerck, P., Craeye, C., Boel, V., Van Itterbeeck, P.: Mixture compositions and fresh properties of self-compacting concrete: Analysis of 25 years of Research. In: Proceeding 8th International RILEM Symposium on SCC, pp. 65–75. Washington, D.C (2016)
2. Schankoski, R.A., Pilar, R., Prudêncio, L.R., Ferron, R.D.: Evaluation of quarry powders as viscosity-modifying material in cement mixtures. In: Proceeding of 8th International RILEM

Symposium on SCC, pp. 187–196. Washington, D.C (2016)
3. Türkmenoğlu, Z.F., Türkmenoğlu, M., Yavuz, D.: Using Waste Marbles in Self Compacting Lightweight Concrete. World Acad. Sci., Eng. Technol. Int. J. Civ. Environ. Eng. **10**(12), 1636–1639 (2016)
4. Forte, L., Etcheverry, R., Garrido, L., del Río, J.L., Hurtado, M., Couyoupetrou, L., Rolny, C.: Evaluación de las propiedades cerámicas de sub-productos sólidos de explotaciones de rocas de aplicación. In: 2º Congreso Argentino de Áridos. Los áridos son mucho más que piedras, vol. II, 535–545, Cámara de La Piedra de la Provincia de Buenos Aires (2014)
5. Coelho dos Santos, G.S., Benito, D., Marfil, S., Maiza, P., del Río, J.L.: Caracterización mineralógica de los materiales arcillosos procedentes de las canteras de piedra Mar del Plata. Potencial utilización como adición mineral. In: XX Congreso Geológico Argentino, pp. 20–25, San Miguel de Tucumán (2017)
6. Alderete, N., Villagrán Zaccardi, Y., De Belie, N.: Effect of supplementary cementitious material on capillary sorption: Relation with drying rate and testing time, Durability of Building Materials and Components (XIV DBMC). RILEM PRO **107**, 1–10 (2017)
7. CIRSOC 201 Reglamento Argentino de Estructuras de Hormigón, INTI, Buenos Aires, Argentina (2005)
8. EFNARC: http://www.efnarc.org/pdf/SandGforSCC.PDF (2002). Last accessed 20 Dec 2006
9. EFNARC : https://www.theconcreteinitiative.eu/images/ECP_Documents/EuropeanGuidelinesSelfCompactingConcrete.pdf (2005). Last accessed 20 Dec 2006
10. IRAM 1871: Hormigón. Método de ensayo para determinar la capacidad y la velocidad de succión capilar de agua del hormigón endurecido. IRAM, Buenos Aires, Argentina (2004)
11. ACI 237 Self-Consolidating Concrete, American Concrete Institute, USA (2007)

Successively Recycled Concretes Exposed to Sulfate Soil During 11 Years

Claudio J. Zega, Alejandro Pittori and Ángel A. Di Maio

Abstract Recycled concrete is still a material with some questionings about its true durable performance. Studies about recycled concretes exposed to external sulfate attack are very scarce. Generally, accelerated tests focused on the chemical attack are performed. However, studies on conventional concretes point out that structures deterioration by external sulfate attack is mainly due to the physical attack, and not merely by the chemical attack. Then, it is necessary to have results based on real exposure conditions, where transport mechanism and kinetics attack may be very different to that in accelerated test. In this paper, the performance of recycled concretes subjected to four recycling cycles and exposed to sulfate soil during eleven years is evaluated. Concretes were periodically assessed by visual inspection, weight variation and determination of dynamic modulus of elasticity. The results show that recycled concretes presented a satisfactory performance after eleven years of exposure, even better than pattern conventional concrete.

Keywords Recycled concrete · Recycling cycles · Sulfate soil · Durability

1 Introduction

Although the compressive strength level of recycled concrete (RC) made with coarse recycled aggregate (CRA) is equal to that of concrete made with natural aggregates [1–4], its use in structures exposed to aggressive environments is restricted [5–7].

One of the major durability problems of concrete structures is associated with external sulfate attack, mainly due to salts crystallization [8–12]. For RC concretes, studies regarding this topic are very scarce. In previous studies, RC with different matrix quality (different porosity), and made with variable content of CRA, per-

C. J. Zega (✉)
LEMIT, CONICET, 52 entre 121 y 122 s/n, La Plata, Argentina
e-mail: cj.zega@conicet.gov.ar

A. Pittori · Á. A. Di Maio
LEMIT-CIC, 52 entre 121 y 122 s/n, La Plata, Argentina

© RILEM 2020

J. F. Martirena-Hernandez et al. (eds.), *Proceedings of the International Conference of Sustainable Production and Use of Cement and Concrete*, RILEM Bookseries 22,
https://doi.org/10.1007/978-3-030-22034-1_47

formed in a similar way than reference concrete made only with natural aggregates [13].

The need to recycle concrete structures and elements, made with recycled concrete, leads to obtain recycled aggregates of second generation. This fact could cause the characteristics and properties of these recycled aggregates to differ even more from those of natural aggregates, with the consequent modification in the properties of concretes with them produced. The objective of this work is to evaluate the properties of CRA obtained after four cycles of recycling, and their influence in the performance of RCs exposed to sulfate soil during eleven years.

2 Experiences

2.1 Methodology

From a conventional concrete (CC), which was kept in fog room during 28 days (T: 23 ± 2 °C; RH > 95%), a first coarse recycled aggregate (CRA-1) was obtained after crushing and sieving it through a sieve with 4.75 mm mesh opening. This CRA-1 was used as replacement of coarse natural aggregate in the production of a first recycled concrete (RC-1), complying with a first recycling cycle. From RC-1, a second recycling cycle was started, obtaining a second coarse recycled aggregate (CRA-2) and a second recycled concrete (RC-2) was produced with it. The process was repeated up to produce four recycling cycles. In all concretes, the mixtures proportions were kept constant, using in each recycled concrete 75% (by volume) of the CRA obtained from crushing the previous concrete. The replacement percentage was adopted based in conclusions obtained from previous studies [14, 15].

2.2 Materials and Mixtures

In concrete production, blended Portland cement (equivalent to CEM II/B-M (L-S) 42.5 N according to EN 197-1) and river siliceous sand as fine aggregate were used. The coarse natural aggregate (CNA) used in all concretes was crushed granite stone, while each CRA was obtained from crushing and sieving the previous recycled concrete. The properties evaluated to each CRA and CNA are presented in Table 1.

As expected, CRA-1 presented lower specific gravity, higher water absorption and Los Angeles weight loss than CNA used, due to the cement mortar attached to the recycled aggregate particles. As recycling cycle progress, the evaluated properties for CRAs differed significantly from those of CNA, although from the third recycling cycle (CRA-3) the differences were slightly lower. This fact must be attributed to the high content of CRA (75%) used in the production of each new recycled con-

Table 1 Properties of coarse natural and recycled aggregates

Property	CNA	CRA-1	CRA-2	CRA-3	CRA-4
Fineness modulus (FM)	6.73	6.78	6.79	6.92	6.79
Specific gravity (SGssd)	2.62	2.46	2.41	2.37	2.34
Absorption (%)	0.4	4.4	5.9	6.9	7.7
"Los Angeles" weight loss (%)	28	36	45	48	47
Particles lower than 75 μm (%)	0.55	0.10	0.13	0.11	0.93

Table 2 Mixture proportions (kg/m^3), fresh and hardened state properties

Materials	Concretes				
	CC	RC-1	RC-2	RC-3	RC-4
Water	155				
Cement	311				
Fine aggregate	885				
Coarse natural aggregate	1032	258			
Coarse recycled aggregate	–	708	694	683	674
Properties	CC	RC-1	RC-2	RC-3	RC-4
Slump (mm)	95	90	85	70	70
Unit weight (kg/m^3)	2423	2323	2295	2286	2281
Air content (%)	2.0	2.5	2.2	2.5	2.2
28-days compressive strength (MPa)	25.6	25.4	26.4	27.3	25.5
Water absorption (%)	5.1	6.6	7.0	7.2	7.4

crete, which is mainly constituted by mortar, as consequence of the recycling cycles. Although Los Angeles weight loss increased with each new cycle, the values obtained in all CRA were lower to the limit (50%) specified in the Argentina Regulation [16].

Mixture proportions and fresh and hardened state properties are presented in Table 2.

To avoid modifications in the fresh state of mixtures, as consequence of higher absorption of CRAs, all coarse aggregates (recycled and natural) were used in saturated and surface dry (ssd) condition. After 28-days curing period, 75 × 100 × 430 mm prismatic specimens were weighed and the initial dynamic modulus of elasticity were determined. Subsequently, specimens were exposed in a saturated silty soil containing 1% of sodium sulfate [13], equivalent to a strong aggression environment according to the Argentina Regulation [16].

3 Results and Discussion

From the fresh state properties, slump values were similar for all concretes, due to the use of CRA in ssd condition. Air content was slightly higher for recycled concretes, respect to conventional one, without differences among RC in the successive cycles. Unit weight decreased as the recycling cycles progressed, because of the diminution in the specific gravity of CRA in the successive cycles of recycling.

Similar compressive strength level was obtained for all recycled concretes and conventional one. Water absorption increased for RC with each recycling cycle, in higher proportion for RC-1 compared to CC, and to a lesser extent for following cycles (Table 2). This fact is consequence of mortar attached to recycled aggregate particles and higher content of CRA used as replacement for natural aggregate.

3.1 Visual Inspection

Deterioration of concrete due to physical sulfate attack occurs in the air exposed zone of specimens, as this zone is affected by the wick action that cause salt crystallization in pores and micro-cracking near the surface of concrete [17]. Spalling of cover concrete first, and cracking near the edges and corners later, are typical characteristics of this type of sulfate attack. The modification in the visual aspect of concretes samples occurred as consequence of the exposure can be observed in Fig. 1.

In conventional concrete, the spalling of cover concrete allows coarse natural aggregates to be exposed. With increasing exposure time, cracking progressed and coarse aggregates were directly exposed to sulfate attack, causing their detachment when covering mortar was deteriorated. During the production of recycled aggregates, and recycling cycles progressed, natural rock in recycled concrete decreased, and mortar content increased, due to the mortar attached to recycled aggregate particles. Therefore, detaching of natural aggregate in recycled concretes is more difficult, and the progress of deterioration will depend on the deterioration of the mortar.

3.2 Dynamic Modulus of Elasticity

Figure 2 shows the variation of the dynamic modulus of elasticity for conventional and recycled concretes. Each informed value corresponds to the average of three samples, except in RC-4 for which two samples were assessed. The dynamic modulus at earlier ages for conventional concrete was different to those for recycled ones. RC concretes showed a decrease in dynamic modulus with each recycling cycle. This difference is consequence of the diminution in the specific gravity of CRA compared to natural one (Table 1). As the exposure period progressed, the different concretes performed in a different manner. Until 2 years, the dynamic modulus remained constant for CC

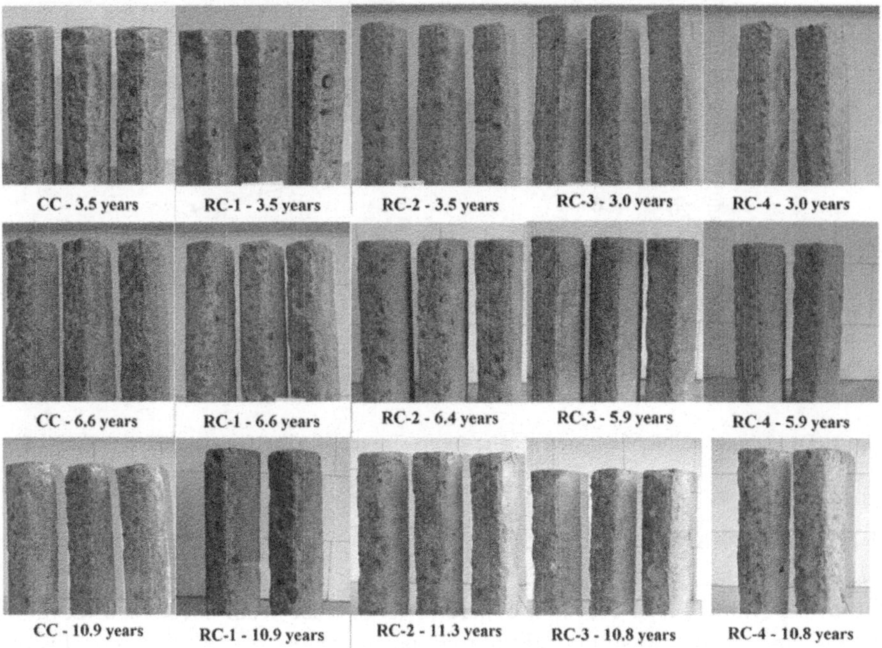

Fig. 1 Visual aspect of concretes under successive recycled cycles exposed to sulfate soil

and RC-1 concretes. Afterward, the value for dynamic modulus gradually decreased for both concretes, due to the degradation of the specimens by the external sulfate attack. For RC-2 concrete, dynamic modulus remained constant until one year, then, it started to decrease more gradually than CC and RC-1 concretes. RC-3 and RC-4 concretes presented a similar performance among them, where dynamic modulus remained practically constant until eight years, and them it started to decline slightly up to the last age of evaluation.

From literature, ITZ for CRA is denser than for natural one [18]. Therefore, for CC concrete, sulfate should penetrate easily through the ITZ of CNA than through the mortar matrix, producing CNA detaching. For RC, in spite of the higher content of mortar, sulfate ions penetration should depend on matrix porosity.

The variation in dynamic modulus was consistent with the visual inspection of the samples, were CC and RC-1 concretes specimens were curved along its longitudinal axis, with cracking around of CNA. From RC-2 up to RC-4, this aspect was not revealed for any of the concrete specimens, and the observed hurt was consisting with the degradation of the mortar.

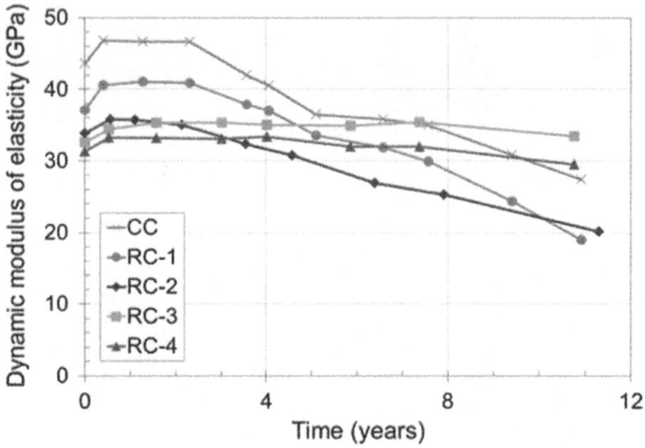

Fig. 2 Dynamic modulus of concretes under successive recycled cycles exposed to sulfate soil

4 Conclusions

From the evaluation of recycled concretes (RC) arise from successive recycling cycles, and exposed to sulfate soil during eleven years, the following conclusions emerge:

- Properties of coarse recycled aggregate (CRA) modify with recycling cycles, but the difference diminishes as the cycles progress, being similar to that of mortar matrix. This fact must be attributed to a high content (75%) of CRA used as replacement of the coarse natural aggregate in recycled concretes.
- All RCs obtained after each recycling cycle present identical compressive strength, which is equal to original concrete from which the first CRA was obtained.
- The changes in the visual aspect of RC are less significant as the cycles of recycling progress, due to the mortar in CRA particles and the higher content of it.
- Decrease in dynamic modulus is more pronounced for conventional concrete and less significant for successive recycled concretes. This behavior is consequence of a lower content of coarse natural aggregate in recycled concretes.
- Besides RCs were not design to satisfy durable requirements, they show very good performance after eleven years exposed in sulfate soil. RC performed similar or better than pattern conventional concrete, which encourage to continue the studies on durability of recycled concrete in order to increase its employment.

Acknowledgements This work was partially funded by FONCyT through PICT 2015-3339 Prest BID.

References

1. Hansen, T.C., Narud, H.: Strength of recycled concrete made from crushed concrete coarse aggregate. Concr. Int. **5**(1), 79–83 (1983)
2. Ajdukiewicz, A., Kliszczewicz, A.: Influence of recycled aggregates on mechanical properties of HS/HPC. Cem. Concr. Comp. **24**(2), 269–279 (2002)
3. Di Maio, A.A., Zega, C.J., Traversa, L.P.: Estimation of compressive strength of recycled concrete with the ultrasonic method. J. ASTM Int. **2**(5) (2005). www.astm.org
4. Tabsh, S.W., Abdelfatah, A.S.: Influence of recycled concrete aggregates on strength properties of concrete. Constr. Build. Mater. **23**(2), 1163–1167 (2009)
5. RILEM TC 121-DRG: Specifications for concrete with recycled aggregates. Mater. Struct. **27**, 557–559 (1994)
6. Grübl, P., Rühl, M.: German Committee for reinforced concrete (DafStb) code: concrete with recycled aggregate. In: Dhir, R.K., et al. (eds.) International Symposium Sustainable Construction: Use of Recycled Concrete Aggregates, pp. 409–418. Thomas Telford Publishing, London (1998). Available in www.b-i-m.de
7. EHE: Instrucción de Hormigón Estructural. Ministerio de Fomento, España (2008). Available in www.fomento.es
8. Eriksen, K.: Thaumasite attack on concrete at Marbjerg Waterworks. Cem. Concr. Compos. **25**(8), 1147–1150 (2003)
9. Ma, B., Gao, X., Byars, E.A., Zhou, Q.: Thaumasite formation in a tunnel of Bapanxia Dam in Western China. Cem. Concr. Res. **36**(4), 716–722 (2006)
10. Drimalas, T., Clement, J.C., Folliard, K.J., Dhole, R., Thomas, M.D.A.: Laboratory and field evaluations of external sulfate attack in concrete. Technical Report, FHWA/TX-11/0-4889-1, Center for Transportation Research, The University of Texas at Austin, 1616 Guadalupe, Ste. 4.202, Austin, TX 78701, 172p (2011)
11. Ma, K., Long, G., Xie, Y.: Railway tunnel concrete lining damaged by formation of gypsum, thaumasite and sulfate crystallization products in Southwest of China. J. Cent. South Univ. **19**(8), 2340–2347 (2012)
12. Lei, M., Peng, L., Shi, C., Wang, S.: Experimental study on the damage mechanism of tunnel structure suffering from sulfate attack. Tunn. Undergr. Sp. Technol. **36**, 5–13 (2013)
13. Zega, C.J., Coelho Dos Santos, G.S., Villagrán Zaccardi, Y.A., Di Maio, A.A.: Performance of recycled concretes exposed to sulphate soil for 10 years. Const. Build. Mater. **102**(1), 714–721 (2016)
14. Di Maio, A., Giaccio, G., Zerbino, R.: Hormigones con agregados reciclados. Ciencia y Tecnología del Hormigón **9**, 5–10 (2002). (In Spanish)
15. Zega, C.J., Di Maio, A.A.: Influencia de las características de los agregados reciclados en la elaboración de hormigones. In: Proceedings 15° Reunión Técnica AATH, Santa Fe, Argentina (2003). (Edited in CD) (In Spanish)
16. CIRSOC 201: Reglamento Argentino de Estructuras de Hormigón. Instituto Nacional de Tecnología Industrial, Argentina (2005)
17. Neville, A.: The confused word of sulfate attack on concrete. Cem. Concr. Res. **34**(8), 1275–1296 (2004)
18. Poon, C.S., Shui, Z.H., Lam, L.: Effect of microstructure of ITZ on compressive strength of concrete prepared with recycled aggregates. Const. Build. Mater. **18**(6), 461–468 (2004)

Sand Replacement by Fine Recycled Concrete Aggregates as an Approach for Sustainable Cementitious Materials

Iveta Nováková and Boy-Arne Buyle

Abstract Waste management and conservation of natural resources are highly discussed topics today. If these subjects are investigated closer, using tools such as Environmental Product Declaration or Live Cycle Assessment, exact numbers considering the CO_2 production or amount of produced construction and demolition waste (CDW) can be gained. Despite the existence of those techniques and the promotion of recycling, people are skeptical, or scared, to use recycled building materials. The research focuses on recycled concrete aggregates (F-RCA), which can replace natural aggregates (NA) in concrete production. An urgent need for solutions is approaching fast and many countries, especially islands, lacking capacity in CDW landfill and start to have a shortage of NA resources. Another favourable fact for higher utilization of F-RCA is an approaching lack of sand world-wide. Experimental work is focused on the evaluation of properties of various RCA fraction 0–4 mm, the design of two different series of mixes with replacement ratio 0, 20, 40 and 100% of NA by F-RCA. The rheological properties of fresh mortar and the properties of hardened concrete are examined. Our assumption of higher water absorption and a greater fines content of F-RCA was validated in some cases. Despite this fact, rheological properties of all mixes varied only slightly, and compressive strength differed by 7%. In the second series was recorded significant compressive strength drop of the mix with 100% replacement ratio. Overall it can be concluded that replacement ratio up to 40% has no negative influence on fresh nor hardened properties.

Keywords Fine recycled concrete aggregates (F-RCA) · Construction and demolition waste (CDW) · Rheological properties · Strength properties · Replacement ratio

---

I. Nováková (✉)
Innovation Centre Iceland, Árleynir 2-8, 112 Reykjavík, Iceland
e-mail: iveta.n@nmi.is

B.-A. Buyle
Arctic University of Norway, Lodve Langesgata 2, 8514 Narvik, Norway

© RILEM 2020

J. F. Martirena-Hernandez et al. (eds.), *Proceedings of the International Conference of Sustainable Production and Use of Cement and Concrete*, RILEM Bookseries 22,
https://doi.org/10.1007/978-3-030-22034-1_48

1 Introduction

The construction industry is experiencing quick growth in recent years, and integration of secondary raw materials as an ordinarily used building materials for newly built structures is highly recommended to secure sustainable development of cities and road infrastructure. New residential and commercial buildings are frequently built in densely build-up location on rather limited building space, or, as replacement of a demolished structure. An integral part of the city development is higher production of construction and demolition waste (CDW) which arise from the demolition and refurbishment of existing buildings that do not meet modern urban planning requirements but also waste material on-site during the building phase.

Particle size distribution of RCA produced by crushing of concrete CDW depends on crusher settings, but in general the composition of produced RCA is: 20 to 50% of fraction 0–4 mm, up to 65% of fraction 0–8 mm and the bigger fraction. As the EU standard [1] does not allow usage of fine recycled concrete aggregates (F-RCA), those 20 to 50% have no further application. Nevertheless, thanks to the national annexes and guidelines of individual countries, fraction 0–4 mm can be used in a certain replacement ratio, mostly for non-structural concretes.

Documents about RCA are commonly divided into three evaluation levels, starting with an evaluation of input material which is CDW from demolition sites, defected elements directly from production or residual concrete from concrete trucks. If any information is available about the origin of CDW (concrete debris), further RCA characteristics might be predicted according to quality evaluation system proposed by author Nováková et al. [2] in the earlier published study. Following evaluation level is connected to the properties of RCA produced by crushing and sieving into required fractures. The properties of RCA are tested in the same manner as NA with additional tests developed exclusively for RCA and assorted into classes based on their properties [3, 4]. The final level of evaluation is specifying allowed replacement ratio, strength and exposure class in which RCA could substitute NA in new concrete. Allowed replacement of coarse NA by coarse RCA vary from 20 up to 100%, but in case of F-RCA the allowed replacement ratio is more strict, and only several countries such as Brazil, Denmark, Japan, Russia, Switzerland and others permit its utilization.

Those national annexes and guidelines gave an opportunity for some outstanding examples such as the building of Trimli hospital in Zurich and Richi-haus in Switzerland, embankment wall in the Netherlands [5] or 6 km long concrete pavement highway section in the Czech Republic [6] (Fig. 1).

The most significant characteristic and the difference between fine NA and F-RCA is water absorption, fines content and shape of the particles. Rheological properties of fresh concrete were investigated by the team around de Brito [7, 8] and concluded that higher w/c ratio is needed to obtain same slump value as reference mix with only NA. Other studies concluded that SCC with various F-RCA replacement ratios has faster workability loss of fresh and could be modified by adjustment of free w/c ratio or addition of plasticizers [9]. In general, the tendency of decreasing compressive

Fig. 1 From left—Trimli hospital Zurich in Switzerland, middle—embankment wall in the Netherlands; right—concrete pavement of highway in the Czech Republic [5, 6]

strength based on increasing replacement ratio is recorded throughout all reviewed articles and explained by lower strength of F-RCA itself. The replacement up to 50% was in most cases acceptable, and therefore in our experimental testing replacement ration of 20, 40 and 100% was proposed and tested.

2 Materials and Methods

The experimental program consists of a description of input materials, the design of mortar and concrete for evaluation of fresh and hardened properties. Two different series of mixes with same replacement ratios 0, 20, 40 and 100% were designed; one with maximum particle size 4 mm and second with 8 mm. Rheological properties of fresh mortar with max. particle size 4 mm were measured by Viscometer 6 from ConTec. Properties such as density, compressive and flexure strength of hardened concrete were tested on prisms with dimensions 40 × 40 × 160 mm in case of both series.

Testing of aggregates properties was conducted on two types of natural sand and three types of fine recycled concrete aggregates listed and documented (Fig. 2).

- NANO: Natural non-washed minded quartz sand origin from Norway.
- NAIS: Natural washed crushed basalt sand origin from Iceland.
- RCA R1: CDW composed of more than 90% of concrete debris from concrete panel house, concrete outer walls with cinder aggregates. RCA obtained in the Czech Republic.
- RCA R2: CDW composed of more than 90% of concrete debris from the industrial concrete floor, high strength class concrete obtained from Norway.
- RCA R3: CDW composed of more than 90% of concrete debris from concrete bridge pavement. Concrete contained dispersed steel fiber and exposed to chlorides from seawater as the bridge is located at the coast of Iceland.

As further mortar and concrete testing was performed at Innovation Center Iceland, both used aggregate types NAIS and RCA R3 origin from Iceland. To have a clear idea of RCA influence on cementitious based materials, only cement, water and aggregates (NA and/or RCA) were used. Various admixtures and additions were

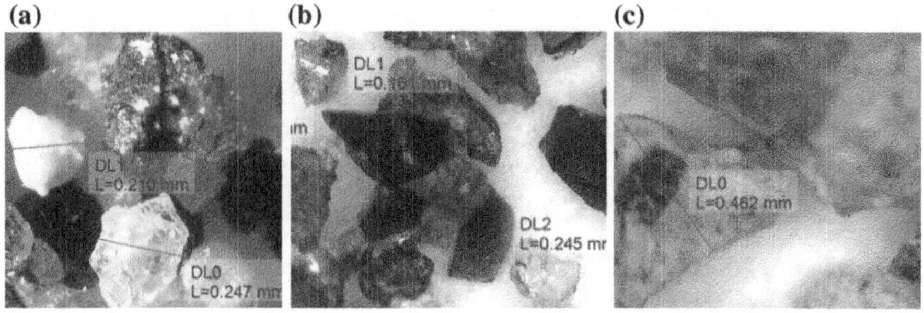

Fig. 2 Microscopic images of 150× magnification of **a** NANO; **b** NAIS; **c** RCA R3 (listed from left)

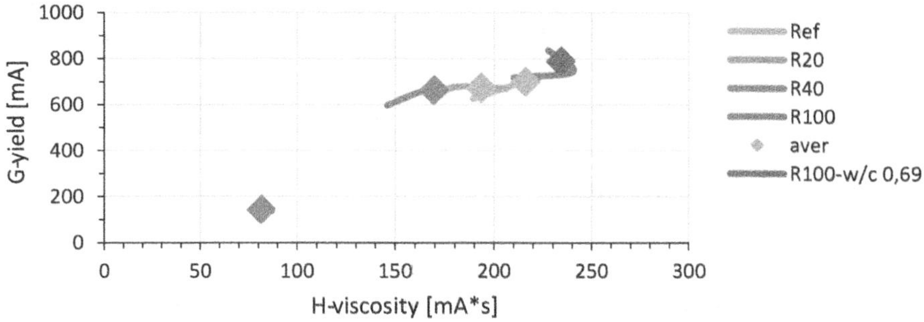

Fig. 3 Rheological properties of cement mortar with a various replacement ratio of F-RCA

not used to omit potential side reaction between them and RCA or NA. Mortar for rheological properties was designed according to the reference composition for Viscometer 6 used at Innovation Center Iceland: cement content 600 kg per 1 m^3 and w/c ration 0.52. The water absorption of aggregates was not taken into account, after initial testing with consideration of the water absorption of RCA R3 (higher by 8.7% than NAIS), the obtained mix was extremely flowable with slide segregation (Fig. 3: R100-w/c 0.69). Concrete for hardened properties testing contained fractures 0–4 mm and 4–8 mm in the ratio 50:50 and cement content 350 kg per 1 m^3. The w/c ratio was kept on 0.6 for all mixes besides the mix R100 where w/c ratio was increased to 0.7.

As it was mentioned earlier, mortar with maximum particle size 4 mm was prepared for evaluation of fresh state properties such as density and rheological properties: viscosity (H-viscosity) and yield value (G-yield). The H-viscosity and G-yield are rheological parameters obtained by ConTec Viscometer 6 and are equal to plastic viscosity and yield stress [10]. Samples with dimension 40 × 40 × 160 mm were cast after measuring density and recording the rheological data of the fresh mortar at 8, 15 and 18 min from initiation of the mixing. Same testing samples were prepared in case of the second series and aged for 7 and 28 days in water storage.

3 Results and Discussion

3.1 Properties Testing of Natural Sand and Fine Recycled Concrete Aggregates

Analysis of the two NA and three F-RCA showed that high water absorption is not necessarily connected to high fines content for both, NA and RCA. The dry specific density of RCA is on average lower by 750 kg/m^3 and water absorption approx. 8 time higher which is usually sign for workability reduction of fresh concrete. Fines content of RCA R3 is very low and might be caused by high compressive strength of parent concrete and influence of carbonation of concrete due to long-term exposure to seawater (Table 1).

3.2 Fresh Mortar Testing

For evaluation of rheological properties were selected RCA R3 and NAIS as both are crushed and have a low content of fines which is beneficial for evaluation of fresh mortar as the shape and fines content are usually extensively influencing factors. Water absorption of NAIS was measured 1.2%, but this sand has ability to retain a high amount of water on its surface and up to 18% of moisture content in normal conditions can be measured. That fact most probably contributed to small spread of measured date from rheological testing.

Furthermore, replacement ratio 20 and 40% of natural sand by F-RCA contributed to better workability by adjusting the particle size distribution of natural sand. As cement content of tested mixes was high (600 kg per 1 m^3) angular particles had enough space to manipulate and did not cause blocking during the measurements.

Table 1 Properties of natural and recycled concrete aggregates

	NANO	NAIS	RCA R1	RCA R2	RCA R3
Dry specific density (kg/m^3)	2679	2881	2021	2021	2054
Surface dry spec. density (kg/m^3)	2692	2920	2207	2248	2257
Water absorption (%)	0.5	1.2	9.2	11.2	9.9
Fines content (%)	10.0 (f10)	3.0 (f3)	16.1 (f16)	7.1 (f10)	3.0 (f3)

Table 2 Hardened mortar and concrete properties

Mortar	Ref	R20	R40	R100
Density (kg/m^3)	2496	2483	2445	2317
Compressive strength 28 days (N/mm^2)	69.7	73.6	74.5	72.0
Compressive strength 28 days (%)	100	106	107	103
Concrete	Ref	R20	R40	R100
Density saturated surface dry (kg/m^3)	2318	2193	2061	1858
Flexure strength 28 days (N/mm^2)	6.0	6.6	5.6	4.0
Compressive strength 28 days (N/mm^2)	44.8	47.6	42.1	23.4
Compressive strength 28 days (%)	100	106	94	52

3.3 Hardened Mortar and Concrete Testing

Results obtained from hardened mortar testing showed that F-RCA has no negative influence on compressive strength, on the contrary, it is increased. The density of tested mixes was decreasing along with increasing replacement ratio, particularly by 179 kg/m^3 in case of mortar R100, and 256 kg/m^3 in case of concrete R40 in comparison to Ref mixes. Despite density reduction, compressive strength increased by 3% respectively dropped by 6%, see Table 2. The only significant reduction of compressive strength was measured in case of concrete R100 where the samples were very porous and visibly lacking binding, also demonstrated by low density.

4 Conclusions

Properties of fine recycled concrete aggregates and their influence on fresh mortar, and hardened mortar and concrete were investigated. Aggregate testing showed that F-RCA not necessarily have to contain a high amount of fines and have a direct link between fines content and water absorption, as demonstrated by RCA R1 and RCA R3. Favorable particle size distribution of FRCA might contribute to better workability and adjust overall particle size distribution of aggregates in the mortar/concrete mix, and lead towards better strength properties. Rheological testing showed that replacement ratio 20, 40 and 100% had no critical influence on water demand of fresh mortar even though the water absorption of RCA R3 was significantly higher than natural sand NAIS.

All mortar mixes with F-RCA gained slightly higher compressive strength than Ref mix which may indicate that strength of fine aggregates is not decisive. In the other hand, when fraction 4–8 mm was added compressive strength was acceptable up to 40% replacement ratio. Reduction of the density of mortar/concrete by F-RCA utilization was proven and can beneficially serve for non-structural concrete.

Further research focused on replacement ratio between 40 and 100% is necessary to verify if the maximum acceptable replacement ratio is 40% or could be higher.

References

1. ČSN EN 206+A1: 2018. Concrete-Specification, Performance, Production and Conformity. Czech Standard Institute, Czech Republic (2018)
2. Nováková, I., Hájková, I.: Quality evaluation system for description of recycled concrete aggregates and verification of its functionality. In: Materials Science Forum, vol. 902, pp. 14–19, ISSN 0255-5476 (2017)
3. Gonçalves, P., de Brito, J.: Recycled aggregate concrete (RCA)—comparative analysis of existing specifications. In: Magazine of Concrete Research, vol. 62, pp. 339–346, ICE Virtual Library (2010)
4. McNeil, K., Kang, H.-K.T.: Recycled concrete aggregates: a review. Int. J. Concr. Struct. Mater. **7**, 61–69. ISSN 1976-0485 (2013)
5. Sika Products & Solutions Homepage. https://www.sika.com/. Last accessed 2019/03/31
6. Birnbaumová, M.: Použití recyklovaného kameniva do cementobetonových krytů, Silnice + mosty, 4nd 2018, Sdružení pro výstavbu silnic, Praha (2018)
7. Cartuxo, F., de Brito, J., Evangelista, L., Jiménez, J.R., Ledesma, E.F.: Rheological behaviour of concrete made with fine recycled concrete aggregates—influence of the superplasticizer. In: Construction and building materials, vol. 89, pp. 36–47, Elsevier (2015)
8. Carro-Lopéz, D., González-Fonteboa, B., de Brito, J., Martínez-Abella, F., González-Taboada, I., Silva, P.: Study of the rheology of self-compacting concrete with fine recycled concrete aggregates. In: Construction and Building Materials, vol. 96, pp. 491–501, Elsevier (2015)
9. Kou, S.C., Chi-Sun, P.: Properties of self-compacting concrete prepared with coarse and fine recycled concrete aggregates. In: Cement and Concrete Composites, vol. 31, pp. 622–627, Scientific Research (2009)
10. Kubens, S.: Interaction of cement and admixtures and its effect on rheological properties. Dissertation thesis, Weimar University (2010)

Encapsulation of Recycled Aggregates Using Mixed Construction and Demolition Wastes

Isbel Chávez Díaz and Lesday Martínez Fernández

Abstract Construction and demolition waste (RCD) represents an important contribution to socio-economic development with respect to the depletion of non-renewable resources, where aggregates are approximately 60–75% of the concrete mix and can be reused, therefore the present investigation Its purpose is to study the behavior of hydraulic concretes by encapsulating the intermediate fraction of recycled aggregates. For this purpose, the materials to be used were characterized, being governed, in the case of aggregates, by NC 251 "Aggregates for hydraulic concretes. Requirements." The fraction of 5–9 mm was encapsulated with the objective that this thin layer covers the pores of the aggregate that cause a high absorption of water, affecting the properties of the concrete. For this we tested with LC3 cement and two proportions of these materials (60 and 40 kg/m^3) deciding among them the best mixture for the encapsulation. From this concrete was made, demonstrating that said encapsulation, increases the compressive strength of it.

Keywords Recycled aggregate · Durability · Permeability · Resistivity · Carbonation · Absorption · Chlorides

1 Introduction

Construction is one of the most important spheres of society. In this area, concrete represents one of the most used products worldwide, it is known that about 12 billion cubic meters of it are produced annually. This is composed of at least three basic elements, cement, water and aggregates, the latter provide a high mechanical resistance to the mixture and decrease the amount of cement of the same, thus contributing economically. That is why they constitute the bulk of the volume of concrete.

I. C. Díaz (✉)
Empresa de Materiales de Construcción de Villa Clara, Villa Clara, Cuba
e-mail: isbel@matvcons.co.cu

L. M. Fernández
Universidad Marta Abreu de las Villas, CIDEM, Santa Clara, Cuba
e-mail: lesday@uclv.cu

© RILEM 2020

J. F. Martirena-Hernandez et al. (eds.), *Proceedings of the International Conference of Sustainable Production and Use of Cement and Concrete*, RILEM Bookseries 22,
https://doi.org/10.1007/978-3-030-22034-1_49

Due to the great exploitation of aggregates for construction purposes, the quarries have fallen into decay, so a large number of these are exhausted. On the other hand, buildings that have fulfilled their useful life are in a state of demolition. There are also construction waste that can still fulfill various functions, these are called Construction Waste and Demolition (RCD). The debris that remains after the passage of meteorological phenomena, leaving great destructions, is also very useful. The recycled aggregates are one of the alternatives to continue building to pass the exhaustion of natural aggregates.

2 Materials and Methods

2.1 Materials

Two types of cement were used to carry out this research: Cement Portland 35 (P-35.), From the cement factory Carlos Marx, of Cienfuegos and the low carbon cement LC3 formed from the fusion to parts equal of the LC2 produced with the clay from the Yaguajay field and with cement P-35, as an additive the chemical additive Dynamon SX 32 was used. While the selection of debris that make up the recycled aggregates were collected in the Prefabricated Company "Luis Ramírez López "from Santa Clara, Villa Clara. The data from the chemical analysis of the Yaguajay clay are shown below (Tables 1 and 2).

2.1.1 Additive

The additive to be used is the Dynamon SX 32 since, due to its properties, it is a superplasticizer for high quality concretes with a reduced loss of workability. It is an additive based on non-sulfonated acrylic polymers, free of formaldehyde. It allows the concrete to significantly reduce the mixing water.

Recycled aggregates

After the material was collected, it was transferred to a block factory where the milling of the material would be carried out, thus obtaining the three fractions of aggregates required. For the crushing it was necessary to fragment the concrete blocks into smaller pieces for grinding in a automated jaw mill which is capable of crushing the aggregate in the three corresponding fractions. Then a second manual sieving process is carried out in order to extract the fractions exceeding 19 mm.

Table 1 Chemical analysis of LC2 cement

Clay	SiO_2	Al_2O_3	Fe_2O_3	CaO	MgO	SO_3	Na_2O	K_2O	TiO_2	P_2O_5	Mn_2O_3	Others	PPI
Yaguajay	46.58	20.06	14.41	2.94	0.74	0.04	0.11	0.06	1.12	0.13	0.73	0.14	12.74

Table 2. Cement P-35. Test results, cement, Carlos Marx factory, Cienfuegos

<table>
<tr><th colspan="7">Mechanical flows</th><th>Observations</th></tr>
<tr><th>Test</th><th colspan="3">Results</th><th colspan="3">According to NC 96: 2011</th><th></th></tr>
<tr><td>Compression resistance (MPa)</td><td>3 days</td><td>7 days</td><td>28 days</td><td>3 days</td><td>7 days</td><td>28 days</td><td rowspan="2">The product complies with the specifications of the standard with respect to mechanical tests</td></tr>
<tr><td>Compression resistance (MPa)</td><td>23.5</td><td>31.6</td><td>46.2</td><td>17</td><td>25</td><td>35</td></tr>
<tr><th colspan="7">Physical trials</th><th>Observations</th></tr>
<tr><th colspan="3">Test</th><th colspan="2">Results</th><th colspan="2">According to NC 96: 2011</th><th></th></tr>
<tr><td colspan="3">Specific Surface (cm²/g)</td><td colspan="2">3671</td><td colspan="2">2800</td><td rowspan="3">The product meets the specifications of the noma with respect to physical tests</td></tr>
<tr><td colspan="3">Initial Setting Time (min)</td><td colspan="2">3.67</td><td colspan="2">45</td></tr>
<tr><td colspan="3">Final Setting Time (h)</td><td colspan="2">5.79</td><td colspan="2">10</td></tr>
<tr><th colspan="7">Chemical tests</th><th>Observations</th></tr>
<tr><th colspan="3">Test</th><th colspan="2">Results</th><th colspan="2">According to NC 96: 2011</th><th></th></tr>
<tr><td colspan="3">Loss on ignition (max.)%</td><td colspan="2">3.67</td><td colspan="2">4.0</td><td rowspan="3">The product meets the specifications of the noma with respect to chemical tests</td></tr>
<tr><td colspan="3">Insoluble residue (max.)%</td><td colspan="2">5.79</td><td colspan="2">4.0</td></tr>
<tr><td colspan="3">Sulfur trioxide (max.)%</td><td colspan="2">0.72</td><td colspan="2">3.5</td></tr>
</table>

2.2 *Methods*

In order to achieve a correct characterization of the recycled aggregates to be used in the manufacture of concrete, the corresponding laboratory tests must be carried out, based on the Cuban standard. These tests are as follows: Grain size Analysis (NC 178: 2002) [6] Volumetric Weight (NC 181: 2002) [7] Specific Weight and Water Absorption (NC 186 and 187: 2002) [9, 10] and Sieve 200 (NC 182: 2002) [8].

2.2.1 Design of the Experimental Plan

In the present investigation it is proposed to make a concrete with 100% recycled aggregates. The experimental design was obtained from the Excel Toufar sheet, it is designed for a minimum content of holes.

The encapsulation consists of adding a thin layer of hydrated cement to the recycled aggregate of the 5–9 mm fraction in order to increase the compressive strength of the concrete to be produced with this type of aggregate. The main objective is to reach the optimum mixture resulting from two types of cement (LC3 and PP-35) for two quantities of it (60 and 40) kg/m^3 of concrete, it was made with a palletizer located in the facilities of the Department of Mechanical Engineering at the UCLV

The process consists in placing the material in the rotating drum with a 30° inclination that is responsible for mixing the aggregate, this is saturated with the objective that it does not absorb the water of the mixture, then proceeds to spray the cement with the help of a sieve# 600 at the same time that the cement is hydrated until it has been completely exhausted. Then the resulting material is placed in nylon bags and covered with wet blankets that are able to absorb and filter water to comply with the process of curing the material that is performed every 24 h with a duration of 7 days.

From the design in the Toufar and the optimal mixture resulting from the encapsulation, the necessary samples will be made to perform the compression resistance tests at 7 and then 28 days. For this, cubic molds of 100 mm × 100 mm × 100 mm are used.

2.2.2 Calculation of Encapsulation

An initial analysis is carried out where it is decided to carry out the encapsulation with LC3 cement, each with an amount of 60 and 40 kg/m^3 of concrete, to obtain the optimum mixture from these variants.

The calculation of the encapsulation was made based on the amount of cement and the amount of aggregate obtained from the Toufar for each m^3 of concrete (Table 3).

Determination of the optimal mixture

After these encapsulations and the curing process of the material were carried out for seven days, the 100 mm × 100 mm × 100 mm cube test pieces were made to perform the resistance tests. Table 4 shows the compressive strength of concrete with encapsulated aggregates after seven days.

Strength results is higher and also the water/cement ratio is 0.36 which is favorable because at a lower ratio a/c there is a smaller number of pores in the mixture that makes up the layer that covers the intermediate fraction of the aggregate.

2.2.3 Concrete

The concretes were made in the materials laboratories of the CIDEM in the faculty of Constructions of the Central University "Marta Abreu" of Las Villas and for it a horizontal axis mixer of 350 dm^3 of capacity was used.

The procedure for making the concrete was as follows.

Table 3. Encapsulation of the material with LC3

		Material 5-9 mm	cemento [g]	a/c calc.	agua calc.	agua calc. Antes [g]	agua calc. Después	agua usado	pomo sat. ante [g]	pomo sat. ante [g]	agua usado add. Saturar	agua total [g]	a/c real
LC3	60 kg/m^3	2750	388	0.5	194	194	128.9	65.1	218.3	142.4	75.9	141	0.36
	40 kg/m^3	2750	259	0.5	129.5	129.5	97.3	32.2	210	114.2	95.8	128	0.49

Table 4. Compression resistance results

Series	$f_{c,\ cube(100\ mm)}$ 7d
LC3-60	**10**
	9
	10
LC3-40	**9**
	9
	9

First, the aggregates are poured into the mixer starting with the coarse, then the intermediate and finally the fine aggregate, then saturated with a process of homogenization for about 2 min, let stand for 10 min keeping them covered so they do not lose moisture, this process is carried out because the recycled aggregate has a high absorption of water and if they are saturated before and they keep absorbing this water then they would not absorb the water of the mixture and thus the a/c ratio does not vary. After this period of time, add the cement and water with the additive, allowing to mix for approximately 2 min. It proceeds to measure the temperature and to make the cone of Abrams according to the NC 174-2002 Fresh concrete. Determination of settlement by the cone where three settlements are measured at different heights.

3 Results and Discussion

The results obtained from the characterization of the aggregates, analyzing their behavior and are compared with the standard established for natural aggregates NC 251 "Aggregates for hydraulic concretes. Requirements." In addition, the physico-mechanical properties of the concretes made with these recycled materials were determined from the values of resistance to compression at 7 and 28 days.

Properties of recycled aggregates.

Grain size distribution

The granulometries of the recycled aggregates are presented in the following graphs.

Testing of hardened concrete

In Fig. 10 we do not take into account the series 1 for encapsulated aggregates, which shows an increase in resistance by 15% for 7 days and by 20% for 28 days when the intermediate fraction of aggregates is encapsulated.

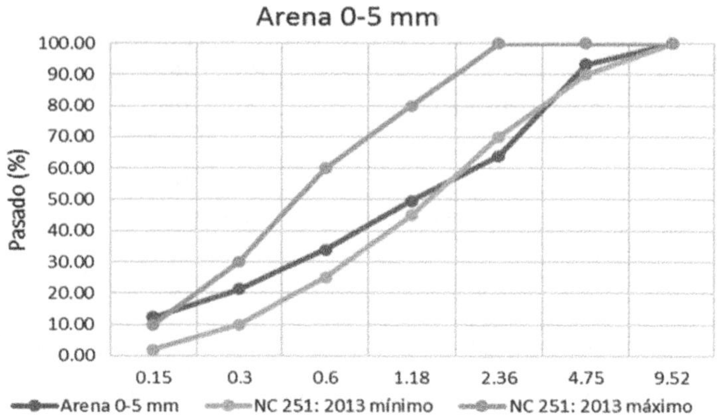

Graph 1 Results of the analysis granulometric fraction: 0-5 mm

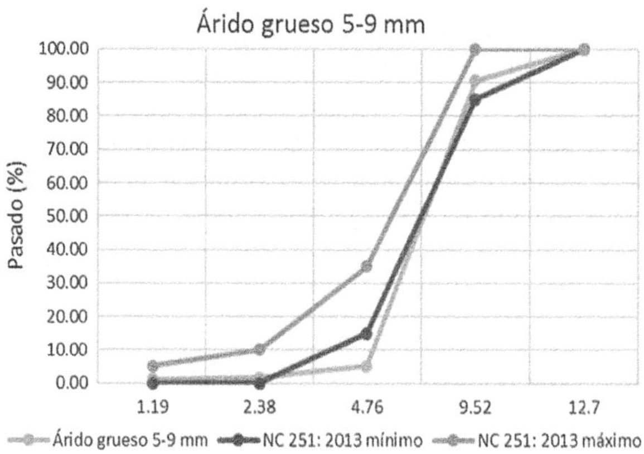

Graph 2 Results of the granulometric fraction: 5-9 mm

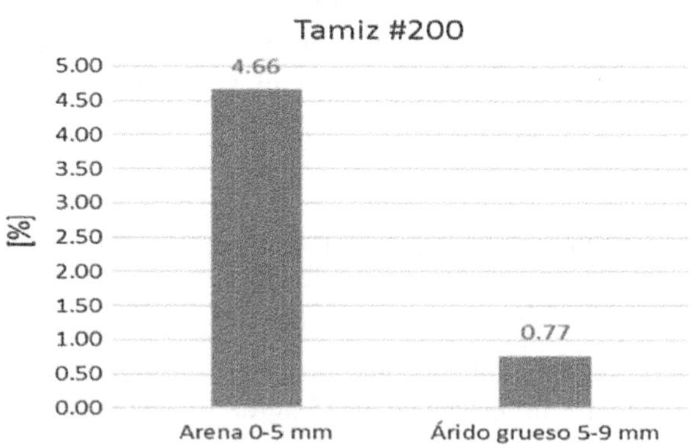

Graph 3 Results of the 200 sieve test. Volumetric Weight

Graph 4 Test results weight specific water absorption (10 min)

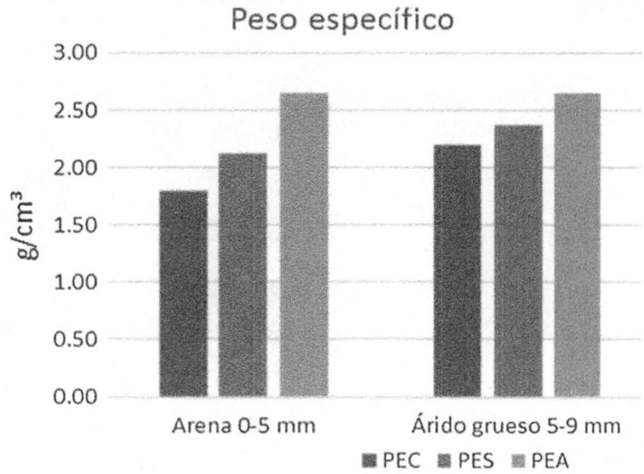

Graph 5 Test results volumetric weight

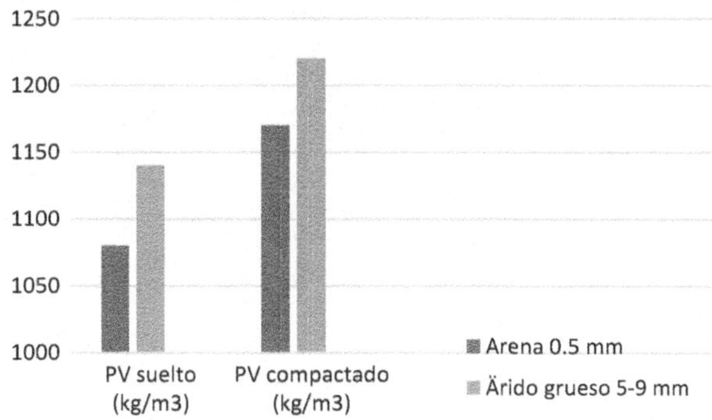

Fig. 6 Test results absorption of water (10 min)

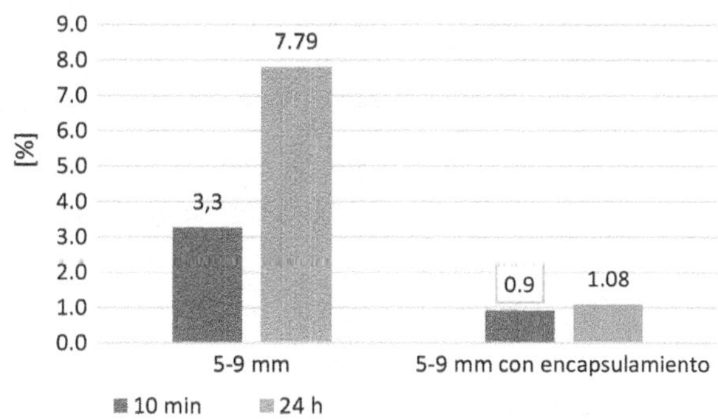

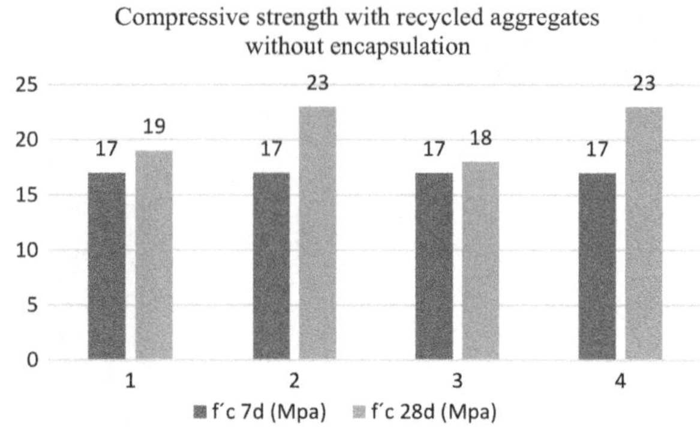

Graph 7 Compression resistance

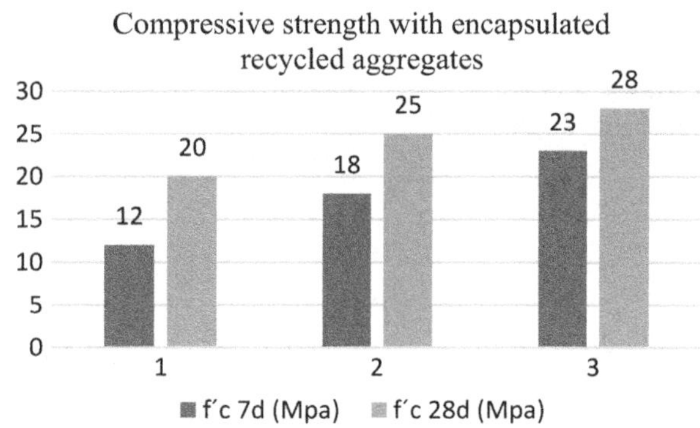

Graph 8 Compression resistance

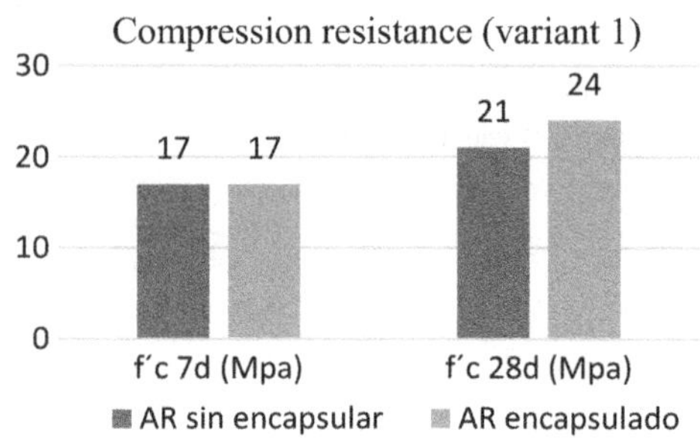

Graph 9 Results compression resistance with encapsulated recycled aggregates (variant 1)

4 Conclusions

Based on the results obtained in this research, the following conclusions can be reached:

1. The constant use of natural aggregates in construction is generating the depletion of them, so that recycling in this sector would largely solve environmental and economic problems.

Graph 10 Results compression resistance with encapsulated recycled aggregates (variant 2)

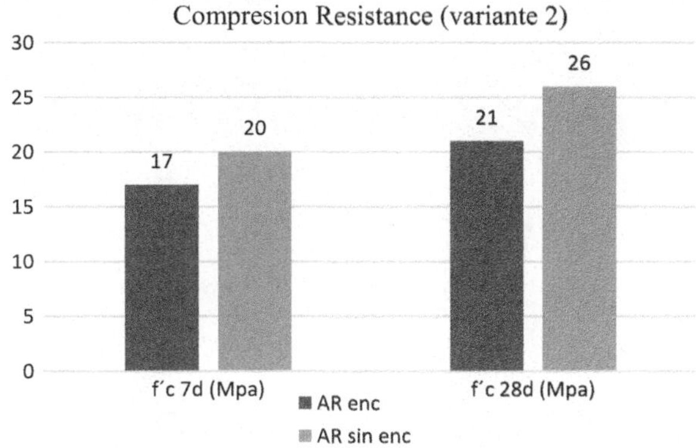

2. The mortar adhered to the natural aggregate and the crushing process have a negative effect on the properties of this aggregate, contributing to the granulometry and its density. Its porous shape and texture elevates the percentage of water absorption, decreasing the physico-mechanical properties of hydraulic concrete
3. The tests show that the encapsulation of the material represents a viable procedure to increase the resistance to compression reaching increases of up to 20% approximately in a concrete composed of recycled aggregates, forming a layer that considerably prevents the passage of water into the pores of the gravel, decreasing by 70% the absorption of water from it.

References

1. Anon.: Construction and Demolition Waste. s.l.:s.n (2006)
2. Baéz, A.d.G.: Study of the Fundamental Properties of Non-Structural Prefabricated Concrete Elements, with the Incorporation of Recycled Aggregate in its Coarse and Fine Fraction. Madrid: s.n (2010)
3. Left, I. S.: s.f. Use of RCDs as Recycled Aggregates and Their Use in Structural Concretes. S.l.:s.n
4. Juan, M.S.d.: Study on the Use of Recycled Aggregate for the Manufacture of Structural Concrete. s.l.:s.n (2004)
5. Martinera, F.: Influence of Aggregates on Mortar and Concrete Properties. s.l.:s.n (2015)
6. NC 178.: Aggregates. Granulometric analysis.. s.l.:s.n (2002)
7. NC 181.: Aggregates. Determination of the Volumetric Weight. Test Method.. s.l.:s.n (2002)
8. NC 182.: Aggregates. Determination of the Material Finer than the 0.074 mm Sieve (No. 200). Testing Method. s.l.:s.n (2002)
9. NC 186.: Arena. Specific Weight and Water Absorption. Test Method.. s.l.:s.n (2002)
10. NC 187.: Arid Thickness. Specific Weight and Water Absorption. Test Method. s.l.:s.n (2002)
11. NC 244.: Hardened Concrete. Determination of the Compressive Strength in Cylindrical Specimens. s.l.:s.n (2004)
12. NC 251.: Aggregates for Hydraulic Concretes. Requirements. s.l.:s.n (2013)

Combining Reactivity Test, Isothermal Calorimetry, and Compressive Strength Measurements to Study Conventional and Alternative Supplementary Cementitious Materials

Sivakumar Ramanathan, Prannoy Suraneni, Ying Wang, Hongyou Shan, Amir Hajibabaee and Jason Weiss

Abstract In this study, we show the results of reactivity testing for a large variety of supplementary cementitious materials (SCMs) and fillers. The reactivity test is performed by measuring the heat release (using isothermal calorimetry) and calcium hydroxide consumption (using thermogravimetric analysis) of SCMs mixed with calcium hydroxide (3:1 ratio of calcium hydroxide and SCM) at 50 °C in a 0.5 M KOH environment. Based on the response in the test, SCMs may be classified into inert, pozzolanic, and latent hydraulic; the pozzolanic and latent hydraulic materials may be further classified into less reactive and more reactive materials. Reactivity test results are compared with isothermal calorimetry and compressive strength measurements on cementitious pastes made with a subset of SCMs. Good correlations are obtained between isothermal calorimetry and compressive strength. Clear correlations between reactivity test results and compressive strength were not seen.

Keywords Supplementary cementitious materials · Reactivity testing · Isothermal calorimetry

S. Ramanathan · P. Suraneni (✉) · Y. Wang · H. Shan
Department of Civil, Architectural, and Environmental Engineering, University of Miami, Coral Gables, FL 33146, USA
e-mail: suranenip@miami.edu

A. Hajibabaee
Ozinga Ready Mix Concrete Inc, Chicago, IL 60616, USA

J. Weiss
School of Civil and Construction Engineering, Oregon State University, Corvallis, OR 97331, USA

© RILEM 2020

J. F. Martirena-Hernandez et al. (eds.), *Proceedings of the International Conference of Sustainable Production and Use of Cement and Concrete*, RILEM Bookseries 22,
https://doi.org/10.1007/978-3-030-22034-1_50

1 Introduction

Supplementary cementitious materials (SCMs) offer various benefits when used in concrete, including reductions in cost, increase in sustainability, and increased concrete durability [1–5]. The supply of conventional SCMs such as fly ash, ground granulated blast furnace slag, and silica fume are limited and increasing shortfalls in the availability of these conventional SCMs have been noted [1, 3, 4]. Various alterative SCMs have been tested and reported in literature. These include, but are not limited to calcined clays, natural pozzolans, and ground lightweight aggregates [2, 5–7]. Before these materials can be used as SCMs, it is necessary to a) determine whether these materials are reactive or inert and b) determine the reactivity of these materials. Two types of SCMs may be distinguished. Pozzolanic SCMs react with the calcium hydroxide produced from the hydration of cement, whereas latent hydraulic SCMs react with water, once activated [5]. Reactivity of the SCMs is complex, and depends on chemical composition, amorphous content, fineness, replacement levels, among other factors [5].

Various tests to classify the reactivity of these SCMs exist but each of these methods have their own drawbacks. Tests such as the strength activity index test may give false positives as they do not adequately consider fineness [8]. A recently developed reactivity test involves the measurement of two important parameters-namely, the heat release and calcium hydroxide consumption due to SCM reaction [4, 5]. Calcium hydroxide and the SCM are mixed together in a ratio of 3:1 in a 0.5M potassium hydroxide solution and the aforementioned parameters are measured at 50 °C—heat release for 10 days and calcium hydroxide consumption at 10 days. This test is a complement to the "R^3 test" [2, 3, 9]; however, no additional sulfates or carbonates are added in this variant, which is also run at a higher temperature (50 °C as compared to 40 °C). Further differences between the tests and the justification for these differences are discussed elsewhere [4, 5].

In the current study, almost 50 SCMs are classified into various categories using the reactivity test. For a smaller subset of these materials, reactivity test, isothermal calorimetry (on cementitious pastes), and compressive strength (on cementitious pastes) are compared. The data that is presented here has in part been published in other journal publications [4, 5, 7, 10].

2 Materials and Methods

Twenty one fly ashes (FA including both Class C and Class F fly ashes), nine ground granulated blast furnace slags (SL), eleven silica fumes (SF), three calcined clays (CC), three ground lightweight aggregates (GLWA), one ground limestone (LS), one basic oxygen furnace slag (BOFS), one municipal solid waste incineration fly ash (MSWIFA), and one ground pumice (P), were studied using the reactivity test [4, 5, 7, 10]. The CaO-SiO_2-Al_2O_3 ternary phase diagram of the SCMs using bulk oxide con-

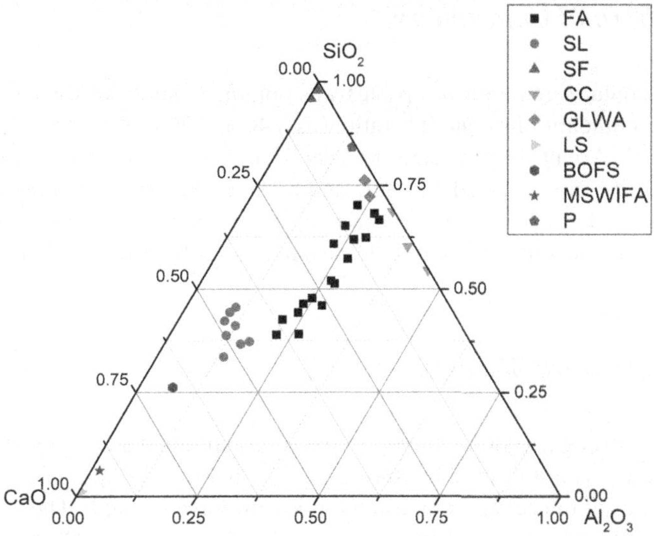

Fig. 1 CaO-SiO_2-Al_2O_3 ternary phase diagram using bulk oxide content for the SCMs studied

tent is shown in Fig. 1 [4, 5]. Further details about these SCMs can be found elsewhere [4, 5, 7, 10]. Out of these, eight materials were chosen for further study—isothermal calorimetry and compressive strength evolution over 91 days. These tests were done using a Type I/II cement, similar in composition to that presented elsewhere [10]. The testing is still in progress and preliminary reports of correlations are presented here.

2.1 *Reactivity Test*

The reactivity test was performed by mixing a 3:1 mass ratio of reagent grade calcium hydroxide and SCM with solutions of 0.5M reagent grade potassium hydroxide, maintaining a liquid-to-solid ratio of 0.9 [4, 5, 7, 10]. Approximately 40 g of the materials were hand mixed for 4 min using a spatula in a plastic container [4, 5, 7, 10]. Approximately 6–7 g of material was placed in a glass ampoule and lowered into an isothermal calorimeter which had been conditioned at 50 ± 0.05 °C [4, 5, 7, 10]. After 10 days, the samples from the sealed ampoules were removed and thermogravimetric analysis was performed by heating approximately 40 mg of material at 10 °C/min to 600 °C in a nitrogen purged atmosphere [4, 5, 7, 10]. The calcium hydroxide consumption was calculated by obtaining the initial and final calcium hydroxide contents of the mixture. Further details are presented elsewhere [4, 5, 7, 10].

2.2 *Isothermal Calorimetry*

Heat release data of cementitious pastes was obtained using isothermal calorimetry. The water to cementitious materials ratio was 0.40 and 20% of the OPC was replaced by SCM [10]. About 40 g of samples were hand mixed for 4 min and 6–7 g of sample was placed in a sealed ampoule and lowered into the calorimeter which had been conditioned at 23 ± 0.05 °C. The first 45 min of data was disregarded due to temperature differentials. The heat release data was collected for 168 h [10].

2.3 *Compressive Strength*

Compressive strength evolution over 35 days was studied here. Cementitious paste cubes (50 mm) with the same mixture design as before were prepared and at least three replicates of the cubes were tested for compressive strength. The samples were demolded after 24 h and cured in a moist curing chamber at a RH of 95% until the testing duration. The test was performed according to standard ASTM procedures.

3 Results

3.1 *Reactivity Test*

Figure 2 [4, 5] shows the plot of cumulative heat release versus the calcium hydroxide consumption of the SCMs. The heat release of FA ranges from 192–351 J/g SCM. The heat release of SL has a range of 377-519 J/g SCM, and SF and CC have a much broader range of 324–611 J/g SCM and 366–711 J/g SCM respectively [4, 5]. The calcium hydroxide consumption of SCMs (per 100 g of SCM) were 43–81 g (FA), 21–47 g (SL), 61–176 g (SF), 80–131 g (CC) [4, 5]. The classification into zones was carried out based on the heat release and calcium hydroxide ratio (the slope), and the distance from the origin [4, 5]. As a large number of SCMs were tested in the study, the classification can be suggested with reasonable confidence [4, 5].

The SCMs were classified into inert; pozzolanic, less reactive; pozzolanic, more reactive; latent hydraulic, less reactive; and latent hydraulic, more reactive [4, 5]. Materials that lie close to the origin, namely, limestone, and municipal solid waste fly ash are considered "inert" [4, 5]. Slags are classified as "latent hydraulic, more reactive", with the BOFS, being classified as "latent hydraulic, less reactive" [4, 5, 10]. This is likely due to the high Fe_2O_3 content and low SiO_2 and Al_2O_3 content in the BOFS [4, 5, 10]. Ten SF and two CC are classified as "Pozzolanic, more reactive" [4, 5]. FA and other SCMs such as P, GLWA are classified as "Pozzolanic, less reactive" [4, 5]. The difference between the two types of pozzolanic materials is the distance from the origin, however, the exact boundaries may be somewhat more

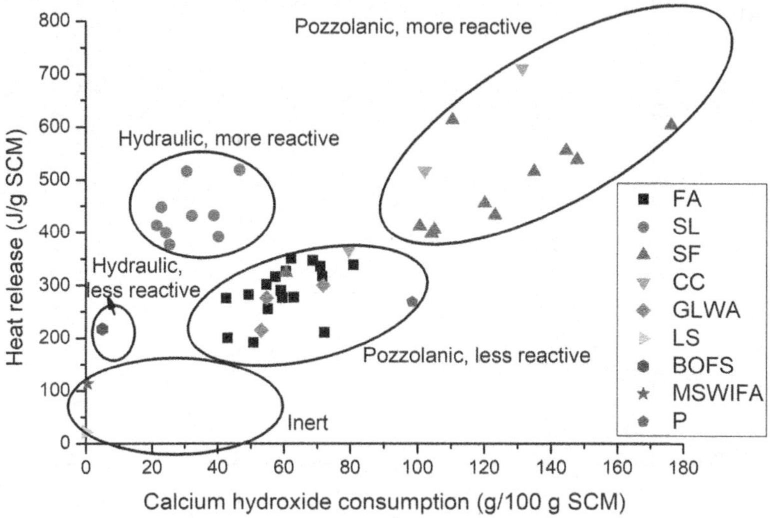

Fig. 2 Plot of heat release versus calcium hydroxide consumption for the reactivity test

subjective [4, 5]. The classification, and its dependence on chemical composition is discussed in detail elsewhere [4, 5].

3.2 *Tests on Cementitious Pastes*

Isothermal calorimetry

Figure 3a, b show the heat flow and (cumulative) heat release of the cementitious pastes for a subset of these SCMs tested. The silicate and sulfate depletion peaks of FA, MK, and P are lower in magnitude compared to that of the control, while MK has higher magnitudes compared to that of the control. The time to reach the silicate peak is accelerated for all SCMs except the FA. Similar analyses can be performed for the sulfate depletion peak also. The heat release at 168 h has a range of 257 (FA)–318 (MK) (J/g cementitious materials). MK has a heat release value that is 9% higher than that of control, whereas the heat release for the other SCMs is less than the control mixture by 12% (FA), 12% (P), and 3% (SL). This behavior from FA, P, and SL is expected when the effect of dilution of OPC and low early-age SCM degree of reaction is taken into consideration [11]. The effect of SCMs on isothermal calorimetry heat flow and heat release is discussed in detail elsewhere [11]. It is not trivial to identify SCMs from inert materials using isothermal calorimetry as carried out here because of the filler effect and low SCM degree of reaction at early ages [7].

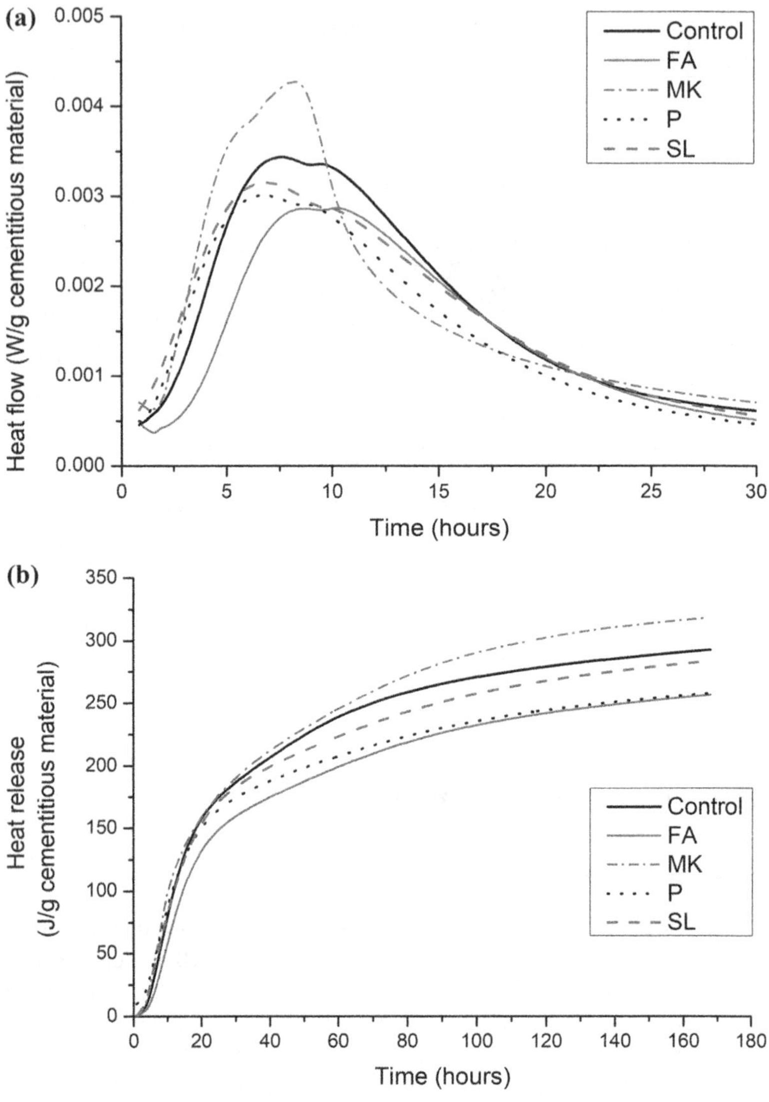

Fig. 3 **a** Heat flow for cementitious pastes; **b** Cumulative heat release of cementitious pastes

Compressive strength

The compressive strength evolution for a period of 35 days of a subset of the SCMs tested is shown in Fig. 4. Coefficient of variability values were around 5% and are not shown here for visual clarity. At one day, the replacement of OPC with FA, P,

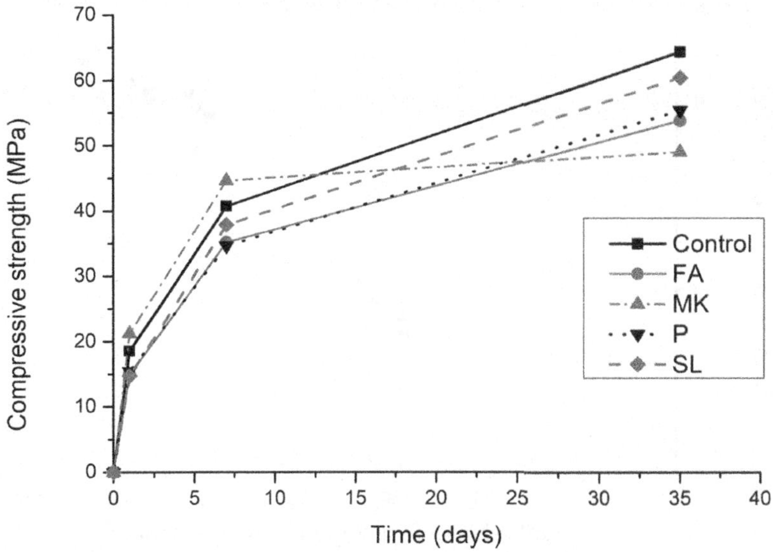

Fig. 4 Evolution of compressive strength for cementitious pastes

and SL causes a decrease in compressive strength by 19, 17, and 21%, respectively, which are similar to amounts expected from pure dilution. It is presumed that at 1 day, the reactivity of these SCMs is low, although SCM reactivity in cement pastes is complex. The addition of MK increases the compressive strength by 14%, which is attributed to its fineness and reactivity.

At 7 days, MK has the highest strength (9% higher than control). P, FA, and SL compressive strengths 15, 14, and 7% lesser than control.

At 35 days, the strength of SL is lower than the control mixture by 6%. P and FA mixtures have strengths 14 and 16% lower than control respectively. The compressive strength of MK is lower than control by 24%. This could be due to lack of space for pozzolanic reaction to continue. For other SCMs, as time increases, the strength difference compared to the control mixture generally reduces. At 35 days, the increasing order of compressive strength is MK < FA < P < SL < control. This order is somewhat unexpected and maybe due to a complex interplay of factors such as dilution, filler effect, SCM reactivity, and experimental variability. While strength is commonly used to differentiate SCMs from inert materials, the lack of fineness considerations in standard strength testing can cause inaccuracies in such differentiation [8].

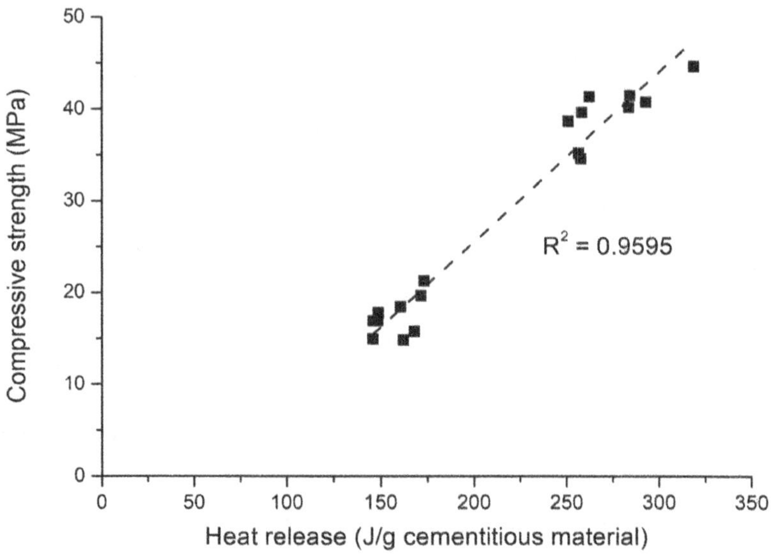

Fig. 5 Correlation between heat release and compressive strength of cementitious pastes

4 Discussion

Figure 5 shows the correlation between heat release of cementitious paste and corresponding compressive strength at 1 and 7 days for seven SCMs and one control mixture. There is good correlation ($R^2 = 0.94$) between the heat release and the compressive strength, suggesting that the former can be used to predict the latter [10], although this must be (clearly) done with care for mixtures with varying water contents. This is a somewhat expected trend as the heat release is an indicator of the degree of hydration and the degree of reaction. It is pointed out that when one considering one age (say 7 days), there is some data scatter, therefore, caution must be exercised when making strength predictions.

Figure 6 shows the correlation between the heat release from the reactivity test at 1 day and 3 days and the relative compressive strength of the cementitious pastes at 7 days (calculated as the difference from the control cement paste sample). The correlation is not obvious nor particularly strong. Various other correlations were explored, however, most of these were not strong. We were unable to predict the compressive strength of cement pastes using the reactivity test response, which is in contrast to other research [9]. The following reasons for this discrepancy are proposed:

1. The differences could be related to differences in SCM replacement level—20% used here and 30% elsewhere [5, 9]. We are currently in the process of studying SCM replacement level and combining SCM replacement with reactivity test response.

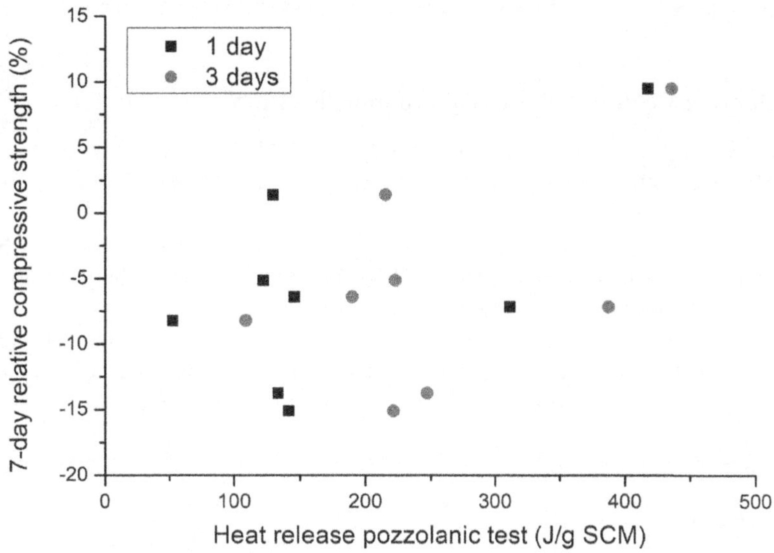

Fig. 6 Relative compressive strength at 7 days versus heat release from reactivity test at 1 and 3 days

2. The differences could be related to differences in the way the reactivity test is performed—higher temperatures are used here; additional sulfate and carbonate are used elsewhere [5, 9].
3. The differences could be related to the way strength testing is done—paste was tested here, mortar elsewhere [5, 9].
4. The differences could be related to the choice of reference material for strength—quartz may be a better reference (inert) as compared to Portland cement [5, 9].

Nevertheless, it appears from these results that the reactivity test cannot be used to predict strength gain in cementitious pastes (when the tests are performed in the manner outlined here). While this is somewhat unfortunate, it should be pointed out that strength gain does not depend only on reactivity. False positives with tests which rely on strength gain and chemical composition to determine whether materials could be used as SCMs have been reported [8]. Such findings do make sense—very fine filler materials may pass a strength gain test; however, they will be classified as inert in the reactivity test. Therefore, it may make more sense to correlate reactivity test response with concrete durability, instead of strength, although this needs to be conclusively demonstrated.

5 Conclusions

The following are the conclusions drawn from this study:

- The reactivity test response of a wide range of SCMs have been studied and these SCMs have been classified into different categories based on their heat release and calcium hydroxide consumption.
- The heat release and the compressive strength evolution of a subset of these SCMs in cementitious pastes was studied. There is good correlation between the heat release and compressive strength of cementitious pastes.
- No strong correlations between the reactivity test heat release and the compressive strength of the cementitious pastes could be found, somewhat contrary to the findings of others. Reasons for this discrepancy have been proposed.

References

1. Snellings, R.: Assessing, understanding and unlocking supplementary cementitious materials. RILEM Tech. Lett. **1**, 50–55 (2016)
2. Avet, F., Snellings, R., Diaz, A.A., Haha, M.B., Scrivener, K.: Development of a new rapid, relevant and reliable (R^3) test method to evaluate the pozzolanic reactivity of calcined kaolinitic clays. Cem. Concr. Res. **85**, 1–11 (2016)
3. Snellings, R., Scrivener, K.L.: Rapid screening tests for supplementary cementitious materials: past and future. Mater. Struct. **49**, 3265–3279 (2016)
4. Suraneni, P., Weiss, W.J.: Examining the pozzolanicity of supplementary cementitious materials using isothermal calorimetry and thermogravimetric analysis. Cem. Concr. Compos. **83**, 273–278 (2017)
5. Suraneni, P., Hajibabaee, A., Ramanathan, S., Wang, Y., Weiss, J.: New insights from reactivity testing of supplementary cementitious materials. Cem. Concr. Compos. **103**, 331–338 (2019)
6. Burris, L.E., Juenger, M.C.G.: Milling as a pretreatment method for increasing the reactivity of natural zeolites for use as supplementary cementitious materials. Cem. Concr. Compos. **65**, 163–170 (2016)
7. Suraneni, P., Fu, T., Azad, V., Isgor, B., Weiss, J.: Pozzolanicity of finely ground lightweight aggregates. Cem. Concr. Compos. **88**, 115–120 (2018)
8. Kalina, R.D., Al- Shmaisani, S. Ferron, R.D., Juenger, M.C.G.: False positives in ASTM C618 specifications for natural pozzolans. ACI Mat. J. **116**, 165–172 (2019)
9. Li, X., et al.: Reactivity tests for supplementary cementitious materials: RILEM TC 267-TRM phase 1. Mater. Struct. **51**, 151–164 (2018)
10. Wang, Y., Suraneni, P.: Experimental methods to determine the feasibility of steel slags as supplementary cementitious materials. Constr. Build. Mater. **204**, 458–467 (2019)
11. Ramanathan, S., Moon, H., Croly, M., Chung, C.-W., Suraneni, P.: Predicting the degree of reaction of supplementary cementitious materials in cementitious pastes using a pozzolanic test. Constr. Build. Mater. **204**, 621–630 (2019)

The manufacturer's authorised representative in the EU is Springer Nature Customer Service Centre GmbH, Europaplatz 3, 69115 Heidelberg, Germany. If you have any concerns regarding our products, please contact ProductSafety@springernature.com

Printed and bound by CPI Group (UK) Ltd, Croydon, CR0 4YY
30/09/2026
02225494-0019